THE BRONTËS AND THE IDEA OF THE HUMAN

What does it mean to be human? The Brontë novels and poetry are fascinated by what lies at the core – and limits – of the human. *The Brontës and the Idea of the Human* presents a significant re-evaluation of how Charlotte, Emily, and Anne Brontë each responded to scientific, legal, political, theological, literary, and cultural concerns in ways that redraw the boundaries of the human for the nineteenth century. Proposing innovative modes of approach for the twenty-first century, leading scholars shed light on the relationship between the role of the imagination and new definitions of the human subject. This important interdisciplinary study scrutinises the notion of the embodied human and moves beyond it to explore the force and potential of the mental and imaginative powers for constructions of selfhood, community, spirituality, degradation, cruelty, and ethical behaviour in the nineteenth century and its fictional worlds.

ALEXANDRA LEWIS is Senior Lecturer in English Literature, and Director of the Centre for the Novel, at the University of Aberdeen. She is editor of the Norton Critical Edition of *Wuthering Heights*, and has published extensively on the Brontës, memory and trauma, and nineteenth-century literature and psychology.

CAMBRIDGE STUDIES IN NINETEENTH-CENTURY
LITERATURE AND CULTURE

General editor
Gillian Beer, *University of Cambridge*

Editorial board
Isobel Armstrong, Birkbeck, *University of London*
Kate Flint, *University of Southern California*
Catherine Gallagher, *University of California, Berkeley*
D. A. Miller, *University of California, Berkeley*
J. Hillis Miller, *University of California, Irvine*
Daniel Pick, *Birkbeck, University of London*
Mary Poovey, *New York University*
Sally Shuttleworth, *University of Oxford*
Herbert Tucker, *University of Virginia*

Nineteenth-century British literature and culture have been rich fields for interdisciplinary studies. Since the turn of the twentieth century, scholars and critics have tracked the intersections and tensions between Victorian literature and the visual arts, politics, social organization, economic life, technical innovations, scientific thought – in short, culture in its broadest sense. In recent years, theoretical challenges and historiographical shifts have unsettled the assumptions of previous scholarly synthesis and called into question the terms of older debates. Whereas the tendency in much past literary critical interpretation was to use the metaphor of culture as 'background', feminist, Foucauldian, and other analyses have employed more dynamic models that raise questions of power and of circulation. Such developments have reanimated the field. This series aims to accommodate and promote the most interesting work being undertaken on the frontiers of the field of nineteenth-century literary studies: work which intersects fruitfully with other fields of study such as history, or literary theory, or the history of science. Comparative as well as interdisciplinary approaches are welcomed.

A complete list of titles published will be found at the end of the book.

THE BRONTËS AND THE IDEA OF THE HUMAN

Science, Ethics, and the Victorian Imagination

EDITED BY

ALEXANDRA LEWIS

University of Aberdeen

CAMBRIDGE
UNIVERSITY PRESS

University Printing House, Cambridge CB2 8BS, United Kingdom

One Liberty Plaza, 20th Floor, New York, NY 10006, USA

477 Williamstown Road, Port Melbourne, VIC 3207, Australia

314-321, 3rd Floor, Plot 3, Splendor Forum, Jasola District Centre, New Delhi - 110025, India

79 Anson Road, #06-04/06, Singapore 079906

Cambridge University Press is part of the University of Cambridge.

It furthers the University's mission by disseminating knowledge in the pursuit of education, learning and research at the highest international levels of excellence.

www.cambridge.org
Information on this title: www.cambridge.org/9781316608371
DOI: 10.1017/9781316651063

© Cambridge University Press 2019

This publication is in copyright. Subject to statutory exception and to the provisions of relevant collective licensing agreements, no reproduction of any part may take place without the written permission of Cambridge University Press.

First published 2019
First paperback edition 2021

A catalogue record for this publication is available from the British Library

ISBN 978-1-107-15481-0 Hardback
ISBN 978-1-316-60837-1 Paperback

Cambridge University Press has no responsibility for the persistence or accuracy of URLs for external or third-party internet websites referred to in this publication, and does not guarantee that any content on such websites is, or will remain, accurate or appropriate.

Contents

List of Figures	page vii
List of Contributors	ix
List of Abbreviations	xiii

    Introduction. Human Subjects: Reimagining the
    Brontës for Twenty-First-Century Scholarship   1
    Alexandra Lewis

1 Hanging, Crushing, and Shooting: Animals,
   Violence, and Child-Rearing in Brontë Fiction   27
   Sally Shuttleworth

2 Learning to Imagine: The Brontës and Nineteenth-Century
   Educational Ideals   48
   Dinah Birch

3 Charlotte Brontë and the Science of the Imagination   67
   Janis McLarren Caldwell

4 Being Human: De-Gendering Mental Anxiety; or
   Hysteria, Hypochondriasis, and Traumatic Memory in
   Charlotte Brontë's *Villette*   84
   Alexandra Lewis

5 Charlotte Brontë and the Listening Reader   107
   Helen Groth

6 Burning Art and Political Resistance: Anne Brontë's
   Radical Imaginary of Wives, Enslaved People, and Animals
   in *The Tenant of Wildfell Hall*   125
   Deborah Denenholz Morse

Contents

7 Degraded Nature: *Wuthering Heights* and the Last Poems
   of Emily Brontë   147
   Helen Small

8 'Angels ... Recognize Our Innocence': On Theology and
   'Human Rights' in the Fiction of the Brontës   167
   Jan-Melissa Schramm

9 'A Strange Change Approaching': Ontology,
   Reconciliation, and Eschatology in *Wuthering Heights*   189
   Simon Marsden

10 'Surely Some Oracle Has Been with Me': Women's
   Prophecy and Ethical Rebuke in Poems by Charlotte,
   Emily, and Anne Brontë   207
   Rebecca Styler

11 *Jane Eyre*, A Teaching Experiment   226
   Isobel Armstrong

12 Fiction as Critique: Postscripts to *Jane Eyre* and *Villette* 251
   Barbara Hardy

13 *We Are Three Sisters*: The Lives of the Brontës as
   a Chekhovian Play   260
   Blake Morrison

Bibliography   269
Index   285

Figures

1.1	William Hogarth, 'First Stage of Cruelty', *The Four Stages of Cruelty*, 1751	*page* 29
1.2	[Mary Elliot] 'The History of a Goldfinch', reproduced in Andrew W. Tuer, *Stories from Old-Fashioned Children's Books*	37
11.1	The first tableau: Mrs. Fairfax, Jane and Adèle	230
11.2	The second tableau: Rochester and Blanche	231
11.3	Collage: Chapter 25 (Bertha and Jane)	242

Contributors

Isobel Armstrong is Emeritus Professor of English Literature, Birkbeck, University of London and a Fellow of the British Academy. Her publications include *Victorian Poetry: Poetry, Politics and Poetics* (1993), *The Radical Aesthetic* (2000), *Victorian Glassworlds: Glass Culture and the Imagination* (2008) and *Novel Politics: Democratic Imaginations in Nineteenth-Century Fiction* (2016).

Dinah Birch is Pro-Vice-Chancellor for Cultural Engagement at the University of Liverpool. She is the general editor of the *Oxford Companion to English Literature* (2009). Her publications include *Our Victorian Education* (2007) and editions of *John Ruskin: Selected Writings* (2004), *Elizabeth Gaskell's Cranford* (2011), Anthony Trollope's *Can You Forgive Her?* (2012) and *The Small House at Allington* (2015). She co-edited, with Mark Llewellyn, *Conflict and Difference in Nineteenth-Century Literature* (2010) and, with Francis O'Gorman, *Ruskin and Gender* (2002), and contributed the chapter on Emily Brontë to the *Cambridge Companion to English Poets* (Cambridge University Press, 2011).

Janis McLarren Caldwell is Associate Professor of English at the University of California, Santa Barbara, United States. She is the author of *Literature and Medicine in Nineteenth-Century Britain: from Mary Shelley to George Eliot* (Cambridge University Press, 2004). She has published widely on science, sympathy and disease in nineteenth-century literature.

Helen Groth is Professor of English at the University of New South Wales. Her publications include *Victorian Photography and Literary Nostalgia* (2003), *Dreams and Modernity: A Cultural History*, co-authored with Natalya Lusty (2013), *Moving Images: Nineteenth Century Reading and Screen Practices* (2013) and, as co-editor with Chris Danta, *Mindful Aesthetics: Literature and the Science of Mind* (2014).

The late **Barbara Hardy** was Emeritus Professor of English Literature, Birkbeck, University of London, a Senior Fellow of the British Academy and a Fellow of the Royal Society of Literature. Her publications include *Forms of Feeling in Victorian Fiction* (1985), *Thomas Hardy: Imagining Imagination in Hardy's Poetry and Fiction* (2000), *Dickens and Creativity* (2008) and *Dorothea's Daughter and Other Nineteenth-Century Postscripts* (2011).

Alexandra Lewis is Senior Lecturer in English Literature and Director of the Centre for the Novel at the University of Aberdeen. She is editor of the 5th *Norton Critical Edition of Wuthering Heights*. Her publications on the Brontës include chapters in *Acts of Memory: The Victorians and Beyond* (2010), *The Brontës in Context* (Cambridge University Press, 2012), *Picturing Women's Health* (2014), *Feminist Moments* (2016) and *Charlotte Brontë: Legacies and Afterlives* (2017). She contributed the chapter on 'Psychology and Psychiatry' for the *Blackwell Encyclopedia of Victorian Literature* (2015) and serves on the executive committees of the British Association for Victorian Studies and the Australasian Victorian Studies Association.

Simon Marsden is Senior Lecturer in English at the University of Liverpool, and the author of *Emily Brontë and the Religious Imagination* (2014). His writing on aspects of theological and scientific discourse in nineteenth-century literature includes articles on Charlotte Brontë and creation theology, and imagination, materiality and Emily Brontë's diary papers.

Blake Morrison is Professor of Creative and Life Writing at Goldsmiths, University of London, and a Fellow of the Royal Society of Literature. Blake is an award-winning poet, novelist and journalist, best-known for two family memoirs and a study of the Bulger case, *As If*. He has also translated and adapted plays, written libretti and critical studies, and edited anthologies of contemporary writing.

Deborah Denenholz Morse is Sarah E. Nance Professor of English and Emerita Inaugural Fellow, Center for the Liberal Arts at The College of William and Mary. She is the author of *Women in Trollope's Palliser Novels* (1987), and *Reforming Trollope: Race, Gender, and Englishness in the Novels of Anthony Trollope* (2013). Deborah is the co-editor of several collections including *Victorian Animal Dreams: Representations of Animals in Victorian Literature and Culture* with Martin Danahay (2007), *A Companion to the Brontës*, with Diane Long Hoeveler (2016) and *Time, Space, and Place in Charlotte Brontë*, with Diane Long Hoeveler (2016).

She contributed the chapters on Emily Brontë and Charlotte Brontë for the *Blackwell Encyclopedia of Victorian Literature* (2015).

Jan-Melissa Schramm worked as a lawyer before undertaking a PhD on the changing idea of evidence in the long nineteenth century. She is now University Reader in Literature and Law, and Deputy Director of the Centre for Research in the Arts, Social Sciences, and Humanities (CRASSH), at the University of Cambridge. She is the author of *Testimony and Advocacy in Victorian Law, Literature, and Theology* (Cambridge University Press, 2000), *Atonement and Self-Sacrifice in Nineteenth-Century Narrative* (Cambridge University Press, 2012), and *Censorship and the Representation of the Sacred in Nineteenth-Century England* (forthcoming, 2018). She has also co-edited two volumes of essays: *Fictions of Knowledge: Fact, Evidence, Doubt* (2011), and *Sacrifice and the Modern Literature of War* (2018).

Sally Shuttleworth is Professorial Fellow in English, St Anne's College, University of Oxford, and a Fellow of the British Academy. Her publications include *George Eliot and Nineteenth-Century Science: The Make-Believe of a Beginning* (Cambridge University Press, 1984), *Charlotte Brontë and Victorian Psychology* (Cambridge University Press, 1996) and *The Mind of the Child: Child Development in Literature, Science and Medicine, 1840–1900* (2010). She co-edited, with Jenny Bourne Taylor, *Embodied Selves: An Anthology of Psychological Texts, 1830–1890*.

Helen Small is Merton Professor of English Language and Literature at the University of Oxford, and a Fellow of Pembroke College. Her publications include *Love's Madness: Medicine, the Novel, and Female Insanity, 1800–1865* (1996), *The Long Life* (2007), *The Value of the Humanities* (2013) and as co-editor, with Trudi Tate, *Literature, Science, Psychoanalysis, 1830–1970: Essays in Honour of Gillian Beer* (2003).

Rebecca Styler is Principal Lecturer in English at the University of Lincoln. She is the author of *Literary Theology by Women Writers of the Nineteenth Century* (2010) and has published articles on nineteenth-century women's religious literature including spiritual auto/biography, the fiction of Elizabeth Gaskell and the poetry of Anne Brontë.

Abbreviations

Unless otherwise stated in the notes to a chapter, references to the Brontës' novels are to the following Oxford World's Classics editions and the below abbreviations are used in parentheses where required for the sake of clarity.

Anne Brontë

AG   *Agnes Grey*, Robert Inglesfield and Hilda Marsden, eds., with an introduction and additional notes by Sally Shuttleworth (2010)

T   *The Tenant of Wildfell Hall*, Herbert Rosengarten, ed., with an introduction and additional notes by Josephine McDonagh (2008)

Charlotte Brontë

JE   *Jane Eyre*, Margaret Smith, ed., with an introduction and revised notes by Sally Shuttleworth (2008)

P   *The Professor*, Margaret Smith and Herbert Rosengarten, eds., with an introduction by Margaret Smith (2008)

S   *Shirley*, Margaret Smith and Herbert Rosengarten, eds., with an introduction and additional notes by Janet Gezari (2007)

V   *Villette*, Margaret Smith and Herbert Rosengarten, eds., with an introduction and notes by Tim Dolin (2008)

Emily Brontë

WH   *Wuthering Heights*, Ian Jack, ed., with an introduction and additional notes by Helen Small (2009)

INTRODUCTION

Human Subjects
Reimagining the Brontës for Twenty-First-Century Scholarship

Alexandra Lewis

> What it was, whether beast or human being, one could not, at first sight, tell.
> – Charlotte Brontë, *Jane Eyre*

> [T]hese human nerves would thrill.
> – Anne Brontë, 'Severed and Gone'

> The world within I doubly prize.
> – Emily Brontë, 'To Imagination'

> When imagination once runs riot where do we stop?
> – Charlotte Brontë, *Villette*

What does it mean to be human? What does it mean to be something other than human? Where and how are we to (indeed, ought we at all to) draw the lines? The Brontë novels and poetry are fascinated by what lies at the core – and limits – of the human. Ranging across explorations of the 'human fallibility' that separates men and women from Gods (and from each other) – including 'the whole burden of human egotism' (*Villette*, p. 197, p. 361); the depths of human suffering and the private solitude of the 'human heart' (*Villette*, p. 366); what it means to be degraded to the status of an 'embruted partner' or to affirm one's own 'free' human spirit and 'independent will', acting above and beyond 'mere human law' (*Jane Eyre*, p. 292, p. 253, p. 317); and even what it might mean to exist as an 'automaton', 'a machine in the shape of a human being' (*Shirley*, p. 334) – Charlotte, Emily and Anne Brontë each respond to wider scientific, legal, political, theological, literary and cultural concerns in ways that redraw the boundaries of the human for the nineteenth century.

The lines often tenuously dividing humankind and the animal kingdom, mortal and divine, child and adult, and self and other are mapped by the Brontës through explorations of violence, education, spirituality, and the science and ethics of imagining what it means to be human. The Brontë novels and poetry move between considerations of deficit, cruelty and the social outcast (class, slavery, illegitimacy, young impoverished women) to ideas about human goodness, mercy, morality and community, catching up complex notions about human thought and behaviour, about conscience and consciousness. Throughout this volume's account of the Brontës' creative interrogation of what it means to be human there emerges a strong sense of the importance for twenty-first-century scholarship of a re-evaluation of the Brontës on these terms and at this cultural moment. This collection brings together timely and exciting viewpoints from eminent Brontë scholars and creative practitioners that foreground, for the first time, the link between the role of the imagination and new definitions of the human subject in the nineteenth century. The importance of interpretation and of the reader's active engagement with the text, illuminated most directly by the reflective analysis of Isobel Armstrong, Blake Morrison and the late Barbara Hardy, is pointed up by other chapters' shared concern with the various methodologies and theoretical or historicist perspectives that have shaped critical decoding of references to the human – or less than human – in the Brontës' works since the time of their publication. Readerly complicity (in depictions of violent or 'inhumane' acts) and the valuing of inconsistency emerge as powerful interpretive forces and imaginative tools regarding the deployment of which the authors would have been well aware, actively drawing nineteenth-century and twenty-first-century readers into the vital and ongoing debate about what it means to be human.

Writing in the context of twenty-first-century fiction, Peter Boxall describes the 'dismantling' of the human: 'we are now living through a historical period in which the meaning of the human is radically uncertain – as uncertain, perhaps, as it has ever been'.[1] For Boxall, mid- to late-twentieth-century global political transformations including decolonisation and feminism have led to 'a rejection of enlightenment humanism' – that 'racist humanism', in Jean-Paul Sartre's phrase, and a sexist one – heralding the 'expiry of the western conception of the human'.[2] Developments in biotechnology and information technology have further disrupted 'the way that we conceive of the spatial, temporal and physical limits of our being'.[3] Are we, then, post-human? Jeff Wallace, following Bart Simon in distinguishing between popular (often dystopian futurist) and critical post-humanism, defines the latter – 'a way of *thinking* the human' – as 'a critique, both of an essentializing conception of human nature, and of human *exceptionalism*'.[4] So far, so good. There is, Wallace asserts, 'a sense that posthumanism might be read back into history as a more tenacious and meaningful idea of the human itself'; on this, he cites Katherine Hayles's claim that 'we have always been posthuman'.[5] Wallace in one sentence affirms that 'the best current work in critical posthumanism offers itself neither as a transcendence nor as a rejection of humanism, but as part of an ongoing critique of "what it means to be human"', yet in the next figures the 'significance' of humanism as 'residual'.[6] Andy Mousley, meanwhile, calls precisely for 'a new literary humanism', stating that 'literature's human significance has remained under-theorized' and calling for a reinvigorated recognition that 'literature is one of the places where humanity's grounding values and characteristics are explored, put to the test, "lost-and-re-found" and sometimes lost again. This is not a triumphal humanism, but a humanism nonetheless',[7] and it is certainly a claim for something more than residual.

If we are to entertain the thought experiment that we have always been post-human, we ought also to acknowledge traffic in the opposite direction: that the Victorians remain 'close to us', a claim made in 1989 by Charles Taylor in *Sources of the Self: The Making of Modern Identity* and much reiterated in more recent Victorian Studies scholarship. Like the Victorians, writes Taylor, 'we still instinctively reach for the old vocabularies, the ones we owe' – for better or worse – 'to Enlightenment and Romanticism'; even as, again like the Victorians, we seek to assert our own advancement and 'superiority'.[8] Rick Rylance's study of *Victorian Psychology and British Culture, 1850–1880* reinforces the notion that we are still, for all our differences, in very real and active conversation with Victorian thinkers and writers at least in the realm of human psychology,[9] a major arena, as several essays in this volume show, for fruitful discussion of what it means to be human. Post-humanism is a sound framework for interpreting and, through action, shaping our world in so far as it challenges centrist views, exposes the narrowly Western and male (rights of 'man') version of liberal-humanist individualism, and calls into clearer view the subjugation of externalised 'others'. Post-human perspectives, particularly in work influenced by Michel Foucault, Donna Haraway, and Jacques Derrida, have been instrumental in dissolving the distinction between animal and human – and, as Cary Wolfe puts it, 'racist and sexist hierarchies have always been tacitly grounded in the deepest – and often most invisible – hierarchy of all: the ontological divide between human and animal life'.[10] However, as Deborah Denenholz Morse and Martin A. Danahay astutely observe, even the seemingly radical Derridean meditation on 'animal autobiography' and voice extends upon late-nineteenth-century literary and scientific efforts to refigure understandings of the human/animal continuum (such as Anna Sewell's *Black Beauty* (1877) and Charles Darwin's *The Expression of the Emotion in Man and Animals* (1872)). Morse and Danahay reiterate the warning sounded by Harriet Ritvo in her ground-breaking *The Animal Estate: The English and Other Creatures in the Victorian Age* (1987): 'a term like "post-human" carries with it its own assumption and could exemplify "the same kind of wishful thinking that the term 'late capitalism' does" if it simply recycles the same old metaphors and clichés'.[11]

Taken in the wider view, these ('post-') terminologies – aiming to assist – can risk making treacherous terrain even slipperier: there is already before us the problem that 'we need to acknowledge the unknowability of all beings, including human ones'.[12] For historian Joanna Bourke, author of the magisterial *What it Means to be Human: Reflections from 1791 to the Present* (2011), 'in the end, all we know for certain is that we don't'.[13] Bourke concludes her study with a discussion of radical alterity and the need to challenge 'tyrannical dichotomies such as biology/culture, animal/human, colonizer/ed, and fe/male', but in such a way that we don't deny through abstraction the reality of experience or collapse all differences to the point of 'a radical flattening out of the contours of our world'. For Bourke, avoiding the tendency to 'invent other creatures (human or animal) in our own image or to use them simply as pawns in our own ideological or material battles' will point us towards 'a politics that is as committed to uniqueness of all life forms as much as to the creative, exhilarating desire and struggle for community and communion, authenticity and certainty'. There is, she writes, 'more – much more – always to come'.[14] Yuval Noah Harari endows this (post-) human forecast with a darker twist: the near-future transformations in human technology, organisation, consciousness and identity will be so profound that the curtain is 'about to drop on Sapiens history': the very natures of the theoretical, political and philosophical debates we are engaged in today 'will in all likelihood disappear along with *Homo Sapiens*'.[15] In Harari's closing view, 'despite the astonishing things that humans are capable of doing, we remain unsure of our goals and we seem to be as discontented as ever'. Pause for thought: 'Is there anything more dangerous than dissatisfied and irresponsible gods who don't know what they want?'[16]

Stepping back from – but keeping in our view – the possibility of bioethics run amok, and of the political and ecological disasters of human making that threaten us daily on so many levels, what can we learn from the writing of the Brontës – other than finding an affirmation (elsewhere in their writing complicated and contradicted) that (in line with Harari's 'Afterword') 'human nature is perverse' (*The Professor*, p. 41) or that 'human natur', taking it i' th' lump, is naught but selfishness' (*Shirley*, p. 275)? This Introduction identifies three key interlinked areas – science, psychology and education; human rights, ethics and religion; and creativity – where the imagination and the idea of the human intersect in the Brontës' works. It outlines why a re-evaluation of the Brontës' works along these lines is important for twenty-first-century scholarship in Victorian studies and literary theory and criticism more broadly.

In *Charlotte Brontë: The Imagination in History*, Heather Glen makes a strong argument for locating apparently universal themes in the literature of Charlotte Brontë (love, loss, pain, joy) within the deeply specific historical framework of the writer's imagination.[17] This volume extends Glen's important work, acknowledging at the outset that recourse to the concept of 'the human' could, if superficially applied, counterproductively mask important structural, cultural and aesthetic differences between the Brontës' actual and fictional worlds and our own. Rather, the volume's chapters explore, vitally, the ways in which complex questions raised by the problematic idea of the human (and what it meant to be human at a particular time and place in the nineteenth century) were central to the Brontës' fictional enterprise. The chapters in this volume address, from a range of perspectives and with innovative contextualised close readings, the ways the Brontës' writing (novels and poetry, juvenilia and essays) influenced and was influenced by wider developing conceptions of the human subject in science, philosophy, political economy, religious thought and other works of literature. Together they provide a thought-provoking and integrated treatment of the role of the imagination in Charlotte, Emily and Anne Brontës' explorations of what it means to be human, and open the way for further re-evaluation of nineteenth-century understandings of the human.

The idea of the human motivated many writers of realist fiction in the nineteenth century and received distinctive and complex treatment within the Brontës' particular ethical and aesthetic visions. As George Levine eloquently reminds us, Victorian literature is marked by a moral energy, 'a drive to find a way to move beyond the narrow limits of individual consciousness into a sympathetic and empathetic relation to others, to the not-self'.[18] For Levine, this concerted campaign to 'thrust the reader into an intimacy with' a broad variety of possibilities of selfhood arises in part in response to troubling questions swirling in the space between scientific endeavour, theology and secularism: (how) is it 'possible to sustain the moral life without religion'?[19] Andrew H. Miller's study of nineteenth-century moral perfectionism affirms that 'the period's literature was inescapably ethical in orientation: ethical in its form, its motivation, its aims, its tonality, its diction, its very style, ethical in ways that remain to be adequately assessed'.[20] Neither Levine's *Realism, Ethics and Secularism* nor Miller's *The Burdens of Perfection* deal at length with any of the Brontës' novels. Levine glances over the 'wildly non-human' characters in *Wuthering Heights*, claiming for Heathcliff and 'partially' for Catherine an 'utter incompatibility with realist narrative subjects', rooted in incomprehensibility.[21] Yet the Brontës' works evince passionate and detailed fascination with human purpose and the ethics of being; human vulnerability; human excess; and existential singularity. Though Levine asserts that 'there's little space in [Victorian realism] for dangerously unknowable forms of life' (such as Heathcliff),[22] I am proposing here that the Brontës' capacious and energetic explorations within the realist genre (and including elements of the Gothic) led them to pull back the veil and acknowledge precisely the 'dangerous unknowability' of what it is to be human, even as their imaginative dealings in extremes of violence, affect and degraded or diminished personhood (be it fleeting or life-long degradation) attempted to reach in the dark towards the heart, the source, of the danger (that is, the point of epistemological crisis). As John Bowen rightly observes in his article on the transformations of Romanticism, the Brontës' fiction is so distinctive partly because of the way it engages with the marginal, the excluded, the dispossessed, 'at a time when that great admirer of the Brontës' work, Karl Marx, saw bourgeois and proletarian males as the representative figures of nineteenth-century social life'.[23] Bowen finds close parallels between the Brontës and Danish philosopher Søren Kierkegaard, their contemporary. These range from a central interest in love and desire; a Protestant faith that, sincere, is 'independent-minded to the point of heresy'; an interest in individual conduct (which might include freedom and responsibility); and 'the overwhelming sense of human life being studded with moral choices of world historical import'.[24]

This volume addresses a gap in existing scholarship by arguing for the centrality of the idea of the human within the works of the Brontës. As this volume hopes to show, the Brontës' investigation and redefinition of the nature, agency and complex limitations of the human subject was made possible by their fascination with (and exercise of) one central human power: that of the imagination. The vital new understanding of the Brontës, the imagination and the idea of the human offered by this volume opens out to two wider considerations, taken up by a number of the chapters. Firstly, building upon the different ways of reading (theory, history, textual analysis) that make possible original critical appraisals, what are the possibilities for the role of creative imagination in our re-evaluation of the Brontës' works in, and for, the twenty-first century? And secondly, how might a fresh appreciation of the Brontës' textual perspectives on such subjects as violence, childhood, psychology, belief, multiple ways of reading, and the valuing of inconsistency inform our present conceptualisations of what it means to be human?

Across each of the chapters, a timely preoccupation with delineating new ways of evaluating the Brontës' works, and their nineteenth-century literary and cultural contexts, emerges. In drawing together path-breaking research in the field and giving new emphasis to understandings of the human in the early- to mid-nineteenth century, this volume provides a thorough and structured review of the science, ethics, and aesthetics of what it means to be human in the context of the Brontës. *The Brontës and the Idea of the Human: Science, Ethics, and the Victorian Imagination* encompasses close readings and historical and theoretical analyses of the novels, poetry and juvenilia, investigating human nature, human rights, human behaviour, and, most importantly, and central to each of these areas, human imagination.

Current and developing interest in the idea of the human (as understood at different historical junctures and at the present moment) is reflected in such overarching titles as *Humanity 2.0: What it Means to be Human Past, Present and Future* (Steve Fuller, Palgrave Macmillan, 2011), *Beyond Human: From Animality to Transhumanism* (edited by Charlie Blake, Claire Molloy and Steven Shakespeare, Continuum, 2012), and, as already discussed, *Literature and the Human: Theory, Criticism, Practice* (Andy Mousley, Routledge, 2013), and *What it Means to be Human: Reflections from 1791 to the Present* (Joanna Bourke, Virago, 2011), as well as works ranging from Hannah Arendt's *The Human Condition* (University of Chicago Press, 1958) to Kate Soper's *What is Nature? Culture, Politics and the Non-Human* (Blackwell, 1995) and Giorgio Agamben's *Homo Sacer: Sovereign Power and Bare Life* (1995; Daniel Heller-Roazen, trans., Stanford University Press, 1998). What makes this volume distinctive is its specific historical focus on the nineteenth century and its foregrounding of the concept of imagination. The focus on the works and cultural context of the Brontës allows for an original and detailed examination of how three writers of fiction and their milieu were confronting issues about what it meant to be human before Darwinian theory took firm hold.

Ideas about post-Brontëan biological identity are taken up in such works as Cannon Schmitt's *Darwin and the Memory of the Human: Evolution, Savages, and South America*. For Schmitt, late-nineteenth-century engagement with Darwinian theory resulted in 'the invention of a new and enduring human subject': the modern 'human-as-animal'. Memory, a human technology, 'enabled the invention of the human as natural'.[25] This idea meshes in interesting ways with Gillian Beer's reflections on the belief of geologist Charles Lyell – friend of Darwin and contributor to his thinking on evolution – that 'the power to re-imagine the remote past was the characterizing property of human reason': a narrative of the past not only physical, of volcanic action and deep time, but cultural, with classical reference for Lyell forming 'a bridge back towards the earliest reaches of human civilisation'.[26] Ideas about civilised sociality and savage natures – bound up in destructive ideologies of race and colonial plunder – fuelled fears about the capacity for human degeneration (on individual and broader scales) at the Victorian fin de siècle in ways that have been well documented.[27] But how was human memory, and human creatureliness, figured by the Brontës? As Laura Brown's *Homeless Dogs and Melancholy Apes: Humans and Other Animals in the Modern Literary Imagination* attests, 'literary animals have inspired modes of thought that question conventional hierarchies' since the eighteenth century, and – moving beyond their use as metaphors for human needs and aspirations – have given rise to animal rights discussions, the founding of the Society for the Prevention of Cruelty to Animals in 1824, and the anti-vivisection movement.[28] In addition to Morse and Danahay's *Victorian Animal Dreams* (2007), a second collection of essays on *Animals in Victorian Literature and Culture* (Laurence W. Mazzeno and Ronald D. Morrison, eds., 2017) traces some of these shifting boundaries and viewpoints, with earlier work by Barbara Munson Goff, Ivan Kreilkamp and others contributing to discussions of the Brontës' writing within the interdisciplinary field of animal studies.[29]

In the Brontës' writing, boundaries between the human animal, the non-human animal and even the shapes through which humans conceive of the spiritual and supernatural realms are often blurred and often also connected with powers of memory and story-telling, gendered expectations, and the language of rights. Jane Eyre is at pains to move beyond a gendered discussion of the needs and feelings of 'men' and 'women' towards the true equality she sees as residing within the more capacious category of the 'human' in her famous rumination on factors constraining the lives of women in the nineteenth century: 'human beings…must have action' (*JE*, p. 109).[30] No more is she willing to acquiesce to being positioned in Rochester's (presciently envisaged) 'Harem' (p. 269) – a slavery of sorts – than does she wish to be ensnared, as she knows so many women to be, by the trappings of domesticity, a caged 'bird' (p. 253). When Jane flees the bounds of human society, the maternal moon assumes a 'white human form' (p. 319); Mrs Reed's deathbed reminiscence reveals her struggle to comprehend the materially powerless child Jane's disturbing inner strength, transmuting her to a male-speaking non-human entity: 'I felt fear, as if an animal that I had struck or pushed had looked up at me with human eyes and cursed me in a man's voice' (p. 239). Rochester asks Jane whether she is 'altogether a human being' when they are reunited (p. 436), calling her a 'malicious elf', 'changeling', 'fairy-born and human-bred' (p. 274, p. 438) when she does not submit entirely to his assumed mastery. Despite Nelly's protestations in *Wuthering Heights* – 'Hush! Hush! He's a human being' (*WH*, p. 152) – Heathcliff is seen variously as brutal monster, fiend and ghoul, 'preter-human' (p. 157) in a vengeance that through its very reasoned and grimly calculating control seems to push beyond the usual extremities of human passion and mourning. Depictions of humans as non-human shock the reader into considerations of stark social realities concerning disability and mental illness. We witness the horrifying falling-short of both physical care and emotional compassion for the vulnerable, and in these failures of recognition and adequate provision we are (particularly in the first-person fictional autobiographical form) implicated, complicit. For Lucy Snowe in *Villette*, caring for the 'crétin' was 'more like being prisoned with some strange tameless animal, than associating with a human being' (*V*, p. 157); in *Jane Eyre* the suffering first wife has been so grossly mistreated by Rochester and perhaps also the narrative that she has been stripped of voice, memory and identity. Bertha Mason, a growling, bellowing 'it', seems to the stunned Jane no longer a woman but a 'clothed hyena' – caged in a room without a window – who can only rise up, scratch, bite, gaze 'wildly' at her visitors and stand 'tall on its hind feet' (*JE*, p. 293).

The Brontës' interest in the idea of the human extends beyond illustrating or troubling the divisions between human, animal and divine – whether based on speech and language (Aristotle), thought (René Descartes), rationality (Immanuel Kant), fallibility – to a deep-set interest in understanding the functioning of language, thought, fallibility: the workings of body and mind, text and self. For all the slippage, and uncertainty, about soul, mind, bodily frame, and human ability and limitation that we find in their works, there is an emphasis on what might constitute basic human needs: being 'a human being' and having 'a human being's wants': food, water, shelter, affective community. There is, as I've suggested, an abiding interest in, to borrow Jane Eyre's phrase, 'what human beings do instinctively when they are driven to utter extremity' (for the governess, as for Helen Burns, that is to turn to God and prayer; in other of the novels and poems, different instincts emerge). There is also an emphasis on the human causes of – and failures to alleviate – human suffering. In the sketch Lucy Snowe prepares for those who seek to judge her, but not to provide assistance:

> 'Human Justice' rushed before me in novel guise, a red, random beldame with arms akimbo. I saw her in her house, the den of confusion: servants called to her for orders or help which she did not give; beggars stood at her door waiting and starving unnoticed; a swarm of children, sick and quarrelsome, crawled round her feet, and yelled in her ears appeals for notice, sympathy, cure, redress. The honest woman cared for none of these things. She had a warm seat of her own by the fire, she had her own solace in a short black pipe, and a bottle of Mrs. Sweeny's soothing syrup; she smoked and she sipped, and she enjoyed her paradise; and whenever a cry of the suffering souls about her pierced her ears too keenly – my jolly dame seized the poker or the hearth-brush: if the offender was weak, wronged, and sickly, she effectually settled him: if he was strong, lively, and violent, she only menaced, then plunged her hand in her deep pouch, and flung a liberal shower of sugar-plums. (*V*, p. 402)

Crucially, as Lucy's provocative and sardonic thought-piece on injustice demonstrates, the Brontës' exploration of what it meant to be human was largely conducted in terms of the powers of the imagination. This volume offers a re-evaluation of those understandings of the mind, memory, affect, and imagination that were emerging in the early- to mid-nineteenth century and shaping cultural conceptions of the human subject and the ethical self. In *Embodied: Victorian Literature and the Senses* (2008), William Cohen suggests that Victorian writers answered the question 'what does it mean to be human?' in terms of material existence: that what is human is nothing more or less than the human body itself.[31] This states the case for the body too strongly, smoothing over the rich complexities of Victorian positions on the mind, definitions of the soul, and the crisis of faith. The writers of *The Brontës and the Idea of the*

Human scrutinise this notion of the embodied human (with attention to pain and the senses) and move beyond it to explore the force and potential of the mental and imaginative powers for constructions of selfhood, community, spirituality and ethical behaviour in the nineteenth century and its fictional (or representative) worlds.

Seeking neither to settle readers of the Brontës with an easily brandished hearth-brush, nor to sprinkle them liberally with sugarplums, *The Brontës and the Idea of the Human* engages with a range of voices, emanating both from inside and outside the worlds of the novels and poems. It presents a re-evaluation of significant critical responses to the Brontës' novels, poetry, essays, and juvenilia, as well as suggesting new modes of approach for the twenty-first century based in critical analysis of the human subject and the imagination.

Science, Psychology, and Education

Chapters 1 to 5 cover the terrain of science, childhood, violence, education, and psychology. Ranging across novels by Anne, Emily and Charlotte Brontë, Sally Shuttleworth's chapter establishes at the outset a key area of investigation relating to science, the human and the imagination: the animal/human (and animal/child) divide as envisaged through violence, complicity and humane necessity. The idea of the human was interwoven with that of the animal for the Brontës more so than for any other novelists in the nineteenth century, and Shuttleworth's chapter addresses, in detail, three key moments of animal cruelty: the hanging of Isabella's dog in *Wuthering Heights*, the crushing of birds in *Agnes Grey*, and the shooting of Victor's dog in *The Professor*. Shuttleworth sets her close reading of these scenes of torture in the context of contemporary medical and educational literature on child development, the vivisection debates, and the visual imagination (William Hogarth's prints depicting *The Four Stages of Cruelty*). As Shuttleworth observes, in the context of rabies in the Victorian cultural imaginary (breaking down bodily boundaries by introducing fluids and life forms from canine to human bodies), 'passion, which transgresses social bounds, is figured as the crossing of a species barrier' – and, in *The Professor*, passion in a child is, like rabies, 'an infection that must needs be exterminated'. The question of the boundary – or overlap – between animal and human, together with the Brontës' challenge to Victorian ideological projections of 'civilised' humanity, is further explored in subsequent essays (Chapters 6–8) in the different contexts of slavery and other forms of oppression, and theological and spiritual approaches to human rights. In this first section of the volume, however, Shuttleworth's focus on the interface between the child and the animal (and practices of child-rearing, particularly the training of boys, with kindness to animals taken up in domestic conduct manuals as a form of moral education) leads into a further chapter on learning and adolescence.

For Dinah Birch, debates surrounding the education of women in the first half of the nineteenth century highlight growing conflict between models of education as a process of social control and as a pathway to self-determination. Contradictory experiences of the schooling of creativity (into adherence to public values or intense individualism), shared by many women of the Brontës' generation and class, are fundamental to the Brontës' divergent fictional realisations of the concept of human imagination. Birch considers the influence of educational reformers such as Sir James Kay-Shuttleworth and the early feminist writer Mary Wollstonecraft, showing how the rapidity of industrialisation and urban development early in the century increased the urgency of questions about the instrumental utility of discipline set against imaginative autonomy. Ought schooling be designed to enable individual students to reach their 'fullest human potential'? As Birch notes, the Brontë juvenilia shows how the boundary between education and creativity was 'always porous' in the Brontë family: the narratives of the elaborate fictional worlds of Gondal and Angria reflected the children's reading in literature, current affairs, politics and history. The 'transformative power of education' was central to the Brontës' lives and ambitions. Tracing the experiences of the Brontës as pupils and teachers, Birch demonstrates their range of novelistic approaches to didacticism and creativity, and to the importance of home (rather than schoolroom) education of children into the values of honesty, sympathy and charity, with particular reference to Charlotte's *Jane Eyre*, Emily's *Wuthering Heights*, and Anne's *Agnes Grey* and *The Tenant of Wildfell Hall*. To an extent, Birch suggests, Anne's novels may be seen as acts of considered resistance to the educational ideals advocated by Charlotte and Emily. Anne is prepared more forcefully to 'attack' the cultural context that limits opportunities for women in an 'unprincipled world'; has 'no truck' with the seductions of 'quasi-Byronic heroes'; and holds the inner life of the imagination to be best informed by a foundation of evangelical religion.

Chapters 3–5 turn from the schooling of creativity, and medical and educational understandings of childhood cruelty, to the sciences of mind and imagination – and to the acoustic imaginary of the reader. Janis McLarren Caldwell compares Charlotte Brontë's writing about her experiences of 'imaginative transport' (in the Roe Head Journal, which sustained engagement with the Angria of her earlier juvenilia, and in *Jane Eyre*) with Victorian scientific writing about the imagination. The work of early psychologists suggests that the creative imagination was increasingly described not as Romantic transcendence, but as a function of the automatic mind. For Caldwell, the 'discontinuity' experienced by Charlotte Brontë between creative unconscious and conscious states – two dimensions of the experience of being human – may be related to her frequent mature pattern of presenting Romantic interludes only to ironise or deflate them. With close attention to the Roe Head manuscripts, and Charlotte Brontë's 'intensely sensate' and primarily visual accounts of imaginary travel to Angria while teaching at Roe Head, Caldwell shows how Brontë's early writings reserved 'her most exalted language' for the experiences of imagining. Furthermore, as Caldwell argues, 'though she says farewell to Angria, Brontë's mature writing doesn't smooth over her characters' tendencies to both imaginative rapture and frustration'. Drawing upon Elaine Scarry's notion of 'vivacity', Caldwell explores how Brontë's work is placed – within the history of imagination – in relation to pictorialist accounts of mental imaging. Charlotte Brontë, while influenced by materialist thought, rejected the wholesale erasure of the soul. While Brontë was attuned to ideas about embodiment and scientific accounts of the mind, she also employed a scriptural language of body and soul, and was preoccupied with moments of 'soaring imagination' clashing with a sinking return to 'being embodied'. As Caldwell's chapter shows, Charlotte Brontë's introspective accounts of creativity are valuable for our contemporary neurological understandings of verbal and visual cognition and for continued Western philosophical investigations of the 'mind/body problem'.

Alexandra Lewis takes the complex life of the mind – and resulting mental and bodily markers of happiness and distress – as central to Charlotte Brontë's conception of what it means to be human. Lewis demonstrates how, by negotiating between unconscious cerebration and ideologies of self-will in *Villette*, Brontë devises a narrative of traumatic memory which complicates gendered notions of anxiety (through the medical terms hysteria and hypochondriasis); considers the role of imagination and empathy in diagnosis; and tests the possibilities of that creative form devoted to the exploration of the individual human subject – fictional autobiography. As Lewis shows, where scientists and physicians such as Sir Henry Holland – a distant cousin of Charles Darwin – mobilise the discourse of self-control to establish the distinction between simple memory and willed recollection as one of the strongest demarcations between the human intellect and that of other animals, Brontë takes the problem of loss of control over deliberate recollection to penetrate to the heart of one of the most difficult – and least well understood – elements of human experience: the recurrence of traumatic images or flashbacks which are simultaneously unrecoverable (in any meaningful totality) and yet irrepressible. Brontë's final novel brings the language and ideas of contemporary physiological theory into a more potent and expansive metaphorical usage. In so doing, the novel challenges theories which reduce mind to body (claiming always a physical cause for psychic disturbance). For Charlotte Brontë, cultural understandings of the idea of the human are constrained by highly gendered medical categories which – through their delimited perspective – risk transforming sufferers into something less than human. As Lewis shows, Brontë's fiction provides a startlingly forceful original argument for understanding mental distress and mnemonic dysfunction in terms that take an appreciation of the 'human' beyond the limiting boundaries of entrenched gender stereotype.

Where Lewis's chapter discusses the limitations of language (existing medical and cultural vocabularies) and failures of communication in the face of trauma and mnemonic dysfunction, Helen Groth is concerned to trace sounds, voices and acoustic stresses in Charlotte Brontë's writing. Groth's chapter shows how scenes from *Jane Eyre* and *Villette*, as well as selected verse, exemplify Charlotte Brontë's use of literary soundscapes to train her readers to listen to and empathise with the unfamiliar or previously unheard – using sounds as prompts, then, for 'thinking through the limits of human experience and cognition'. Groth shows how the listening reader of both poetry and prose is required to eavesdrop on conversations that take place 'in the resonant silence of literary form', requiring a particular kind of attentiveness. From distracting noise (or what Brontë referred to as 'the jargon of Conventionality') to more authentic silences, Brontë privileges what *Jane Eyre* famously calls the 'inward ear' and the alignment of narrative with the involuntary flow of consciousness by relying on first-person narration in both *Jane Eyre* and *Villette*. Groth demonstrates how this interest in the dynamics (and acoustics) of an interior life aligns with ethological theories of character formulated in the 1840s and traces the link between mind and sound throughout the Victorian reception of *Jane Eyre*. In Brontë's narrative poems, published in 1846, Groth finds 'the processes by which the ear records and recalls the voices of the past' to be one of the recurring motifs. Brontë's monologue forms reveal complex mindscapes and states of reverie attuned to environmental stimuli, and Groth discusses how 'The Teacher's Monologue' develops into a 'lyrical meditation on the porous boundaries' between dream and conscious nostalgia, suggesting striking affinities between the speaker and Jane Eyre. In 'Pilate's Wife's Dream', Brontë constructs a monologue out of silenced experience and uses sound to isolate the speaker (while summoning associations that will resonate with the reader's anticipated biblical repertoire). Much has been written about the visual aspects of Brontë's work, and Groth makes a strong case for attention to the sonic dimensions of her writing. At climactic moments, Brontë places particular emphasis on listening to non-verbal cues as a way of both responding ethically to other minds and spaces and fostering readerly awareness of the fragile and often elusive nature of human communication – another kind of training or education of the imagination.

Human Rights, Ethics, and Religion

Chapters 6–11 investigate ethics, law, religion, destitution and degradation (moral, social and financial), and the potential for literature's recognition of our shared humanity. These chapters develop ideas about the human imagination in terms of conscience, whereas Chapters 1–5 are primarily concerned with consciousness.

Deborah Denenholz Morse's chapter on Anne Brontë's 'radical imaginary' of wives, enslaved people, and animals links art and political resistance both within the novels and in terms of their creation. Morse extends the recent animal studies interest in Emily Brontë's work to a close consideration of how the mistreatment of animals, particularly dogs, in Anne's novels embodies interrogations of other forms of confinement and abuse. With a focus on *The Tenant of Wildfell Hall* and reference to *Agnes Grey*, Morse shows how Anne Brontë's writing 'exposes the cultural and societal license' that connects acts of violence against women and animals and, in doing so, undermines masculine privilege and human mastery. *The Tenant of Wildfell Hall* insists upon the humanity of enslaved people and also, by extension, of wives (another form of human property) as well as arguing for the humane treatment of animals. With detailed reference to race discourses in Brontë studies (including the range of post-colonial interpretations of Charlotte's *Jane Eyre* as either complicit in or resistant to the horrific history of British West Indian slavery) and specific Yorkshire and West Riding links to slavery and abolitionism, Morse links Huntingdon's views of tormented wife Helen and the long-suffering butler Benson (described as a 'brute', or not fully human, because he is of the servant class) to the master's relegation of enslaved people to animal or 'brute' status – which is then taken as a justification for vicious treatment. Morse traces

Helen's movement from servitude to incendiary rebellion through the changes in her art: from imitative painting to authentic self-expression; and in a diary that becomes a 'chronicle' at once personal and social. Both Helen's art and Anne Brontë's novel are powerful indictments of 'oppressive masculine prerogative' and the English law, which come together in the legally brutal control of gentlemen over enslaved female bodies. In evoking that related yet displaced narrative of anti-slavery, and equating the treatment of female and animal (particularly canine) bodies, Anne Brontë's novel provides a study in, and call for, various rights to freedom.

As Morse notes, Helen's diary in Anne Brontë's *Tenant of Wildfell Hall* painstakingly records not only her own experience of marital enslavement and abuse but also the process of her husband's degradation. For Helen Small, Emily Brontë's imaginative engagement with the concept of degradation is equally compelling – and no less powerful for being 'far from systematic'. Small traces Brontë's wider thinking about what is entailed in the experience of human degradation and whether or not it is possible for an animal to be degraded. Emily Brontë's novel abounds with physical and moral humiliations perpetrated along hierarchical lines: the powerful oppress the relatively disempowered. The language of abuse in *Wuthering Heights* is very much a language of 'animalistic debasement' and Small links this with Emily Brontë's treatment of verbal force as proximate to physical violence: a means of challenging politeness, sentiment, moralism, and conventional representation of the period. Making reference to pervasive nineteenth-century language of the animal 'kingdom' (with attendant sense that the 'higher' animals are deemed worthier as closer in intellectual and social capability to humans), Small asks us to question whether Emily Brontë would have agreed with the idea of ranking species. Degradation, for Brontë, Small posits, has to do not with any external taxonomy of worth (where the closer an animal sits in relation to humans, or the more trainable it proves for human purpose, the higher the place of importance it is seen to inhabit) but more simply with its own being – 'the damage done to the worth of a creature...against the standard of what it should be in and of its own individual nature'. Drawing on the work of Hannah Arendt and Kate Soper on human nature and social agreement, a core problem of human exceptionalism is clearly delineated: as Soper has outlined, to speak of 'human nature' is to deny the claimed exclusively 'human' ability to step outside one's 'nature' through the exercise of self-will. For Small, ethical questions posed of Emily Brontë's writing are apt to sound 'weakly uncomprehending of the forces they are contending with': one reason the novel is resistant to conventional moral readings is that, in its extreme 'natural determinism', it is 'not obvious what we owe to each other, and indeed to ourselves, in the way of nature'. Cathy's 'metaphysical gesturalism' (in talk of the soul) is a co-option of theological language to her own ends; and in Cathy and Heathcliff's absolute identification, 'narcissism and sado-masochism become two psychological manifestations of the one ontological outlook'. Small also attends to Emily's devoir piece on intemperance and denaturing ('La Palais de la Mort') and the last two poems in Brontë's 'Gondal Notebook', suggesting a progression from conservative to cynical thought on civil war, morality, and humanity's lust for power.

It is a shame, as Small acknowledges, that there is so slender a textual base on which to build speculative accounts of Emily's literary voice. Strange, too, as Small states, that so few accounts of *Wuthering Heights* seek to make comparative analysis with the poetry (beyond 'No Coward Soul') — a challenge taken up in both Small's and Simon Marsden's essays for this volume. In her thrilling reading of Poem 127, Small explores its 'dark meditation on the self-degrading consequences of all humanity in the pursuit of power', reaching that point of existential and epistemological crisis described earlier in this Introduction: that is, a recognition of 'the incoherent condition of all humanity'. For Small, Poems 126 and 127 put — as does *Wuthering Heights* — 'normative ethics under considerable strain' and may even point towards a move away from the human, of sorts: read in isolation, Emily's novel is 'incorrigibly double' (revenge/degradation narrative, and lyric expression of love that is at once the perfection of and escape from identity); but framed by the two poems it may seem a stage in a longer development away from character to a 'fully impersonal lyricism, quite abstracted from scenes of human action'.

We move then from 'human property', abolition, women's rights, animal treatment, degradation and ethics to three chapters explicitly engaging the religious aspects of each of the Brontës' ideas of the human and of imagination, across both poetry and the novels. Jan-Melissa Schramm considers the extent to which Anne, Charlotte and Emily Brontës' ideas of the human depend upon, and differ from, legal and theological ideas of 'rights' before the law and 'creatureliness' before God. For the Brontës, the idea of the human cannot be understood solely in legal terms. While, on the one hand, this can lead to a rhetoric of emotional extremity that appears almost to valorise cruelty, on the other, it allows the three writers to claim for the 'human' a richness beyond the rational, reasonable conception of personhood on which the law depends. For Schramm, the Brontës' 'recognisable rhetorical style' and emphasis on the human passions emerges from the conceptual battle between intellectual liberty (where self-development is 'a necessary precondition to full enfranchisement') and submissiveness to God (in line with 'inherited legacies of orthodox Christian thought'). Providing a detailed examination of the submissive self in Victorian Evangelical theology; Romantic autobiography and the language of experience; and nascent formulations of human rights frameworks, Schramm moves to a close reading of *Jane Eyre* and 'the extension of the franchise of the human', including an analysis of religious (William Ellery Channing) as opposed to later secular (Samuel Smiles) self-culture in Jane's 'ethical education'. Schramm shows how, in line with the Brontës' rejection of Calvinist predestination, both Helen Burns (in *Jane Eyre*) and Helen Huntingdon (in *The Tenant of Wildfell Hall*) 'believe in the doctrine of universal salvation', and argues that 'the Brontë protagonists are not governed wholly by sympathy to the detriment of doctrinal conviction'. Anne and Charlotte Brontës' inclusion of Biblical quotation both worked to underpin trajectories of female empowerment and helped to create a more liberal sense of Biblical meaning. As Schramm attests, the sense of equality of all before God had to be established in the public sphere before legal recognition could follow. The Brontës' complex ideas of the human – combining reason, the heart and Christian humility – serve for Schramm 'as a case study of the extent to which modern ideas of autonomy and self-development might be successfully accommodated alongside – or grafted onto – Christian ideas of humility and self-abnegation before God': and *Jane Eyre, Shirley, Agnes Grey, Wuthering Heights, Villette* and *The Tenant of Wildfell Hall* all play a part in this process.

With a focus on Emily Brontë's novel in the context of her poetry, Simon Marsden examines ontological identification, reconciliation and notions of community. Marsden suggests that the paradigm of romance is often silently privileged in readings of *Wuthering Heights* and that this has significantly shaped critical analysis of the novel's engagement with religion – resulting in neglect of Christian theology's concern with 'the ontological status of the human person, the nature of human flourishing and the relationship between language and meaning'. Marsden's account, with reference to recent theological accounts of human ontology and semiotics (Catherine Pickstock and John Milbank), examines both the refusal of the stranger and patterns of repetition and difference as ways into understanding representations of social fragmentation and the redemptive returns of the second generation. For Marsden, the failure of gift-exchange in *Wuthering Heights* parallels wider social fragmentation, and the novel, with its final imagined renewal of a model of reciprocity, brings into view discourses of human selfhood that are rooted in theological accounts of the human person; agapeic love (distinct from erotic love or simple affection); forgiveness and the refusal of vengeance; and eschatology. Marsden finds in Emily's final poem 'Why ask to know the date – the clime?' a critique of the hypocritical violation of Christian ethics, and discusses also how 'Shed no tears o'er that tomb' suggests a specific theological basis for Branderham's sermon in *Wuthering Heights*: that the refusal of mercy for others 'is simultaneously a refusal of God's mercy for oneself'. Marsden also connects his analysis to Brontë's essay 'Filial Love', where Brontë discusses the parent–child bond in terms that gesture to a broadly inclusive understanding of creation, bridging the animal/human divide (an impulse highlighted in other essays in this volume) using theological imagery: the protective instinct, Brontë wrote in 1842, 'is a particle of the divine spirit we share with every animal that exists'.

Rebecca Styler turns to the 1846 collection, *Poems by Currer, Ellis, and Acton Bell*, to ask: 'Is the imagination a reliable source for ethics?' By exploring the Brontës' conversation, through poetry, about the relationship between subjective dreaming (particularly the prophetic or ecstatic imagination) and social responsibility, Styler shows how the sisters offer distinct and nuanced perspectives on the idea of the human as both moral and creative being, and on the purpose of the imagination in light of their conception of the human as standing in relation to a divine reality that exists both within and outside the self. Writing within an inherited female prophetic literary tradition that 'eschewed escapist visionary flight', taking rather as its focus 'the world of human affairs and the advancement of humanity', the Brontës rework Romantic ideals and the Gothic symbol of the avenging ghost in combination with 'biblical and dissenting models of the prophet'. Styler closely examines a selection of poems in which the spiritual authority of a female outsider is imposed upon the 'masculine' cultural values of a patriarchal establishment, including Charlotte's 'Pilate's Wife's Dream' and 'Gilbert', Emily's 'The Prisoner' and 'The night was dark yet winter breathed', and Anne's 'A Word to the Calvinists'. For Styler, while Charlotte Brontë's poetry prophesies in the realm of gender politics by denouncing sinful male privilege and domination (a form of feminist critique or revenge), Emily and Anne evoke the female prophetic voice – even, in Emily's case, to the point of the 'ecofeminine' – to propound a model of fellowship, rather than hierarchy, and thus to work against the separation of humans from one another and from their fellow – animal – beings.

Creativity

The three final chapters, from Isobel Armstrong, the late Barbara Hardy, and Blake Morrison, draw together the prevailing themes of the imagination and the human explicitly to consider the role of creativity in criticism, through reflective analysis of three contemporary responses to the Brontëan vision of the human: Armstrong on teaching *Jane Eyre* through the students' creation of artworks, Hardy on fiction (*Dorothea's Daughter and Other Nineteenth-Century Postscripts*, 2011), and Morrison on theatre (*We Are Three Sisters*, 2011). Developing the attentiveness of the other chapters in the volume to the ways in which contemporary concerns have, at different stages, inflected critical approaches to the Brontës' fiction, Armstrong's, Hardy's and Morrison's chapters emphasise the burgeoning role of the creative imagination in our reading of the Brontës in the twenty-first century.

As the essays in this volume individually and together show, the category of the 'fully human' is questioned and its limits tested throughout the Brontë oeuvre. For Isobel Armstrong and the MA students she taught at the Middlebury Bread Loaf School of English in Vermont, Bertha Rochester's 'species being and her exclusion from the category of the fully human' are among the most pressing issues raised by *Jane Eyre* – itself 'almost always the first Brontë novel that people read'. How best, in the twenty-first-century seminar room, to foster the kind of intense engagement that allows readers to inhabit the spaces and questions of a 'multi-faceted and complex' text? In Armstrong's view, as far as *Jane Eyre* is concerned, 'the wild racism of the juvenilia and the colonial violence of the Angrian sagas that Charlotte and Branwell made together does not transfer unproblematically to the novel': so, detailed and reflective exploration will be necessary. As the novel itself is 'profoundly experimental', 'carefully shaped', and breathes 'through imagery that has the force of analysis', what better way to approach *Jane Eyre* than through a carefully planned pedagogical experiment where art, photography, sound, and movement produced by the students embody their critical analysis of Brontë's fictional evocation of the idea of the human? Armstrong's chapter reveals the readings that emerged of human need and Rochester's 'family'; Jane Eyre as non-subject (dropping out of personhood on the heath, a nineteenth-century King Lear); and Bertha's dehumanisation (so little recognised within the novel in comparison with that of Jane). Degradation, imagined interior worlds, the animal/human divide (including moments of 'total taxonomic confusion'), and speech and silence – explored in earlier chapters in the volume – explode into questions of racist and colonial contexts which could be 'endlessly debated'; as

Armstrong affirms, 'what is not in question is the text's constant attempt to calibrate Bertha and Jane, its attempt to compare the two women as species being and its understanding that human subjecthood can be arbitrarily taken away from both women'. Building upon her work on species being and the spatial politics of dispossession in *Novel Politics* (2016),[32] and drawing on (while troubling some of the fundamental implications of) Giorgio Agamben's discussion of 'bare life', Armstrong's chapter asserts that the 'primal human needs insistently reiterated' in *Jane Eyre* 'actually define what it is to be human': if food, clothing, and sleep are denied or distorted, whether through destitution or abuse, a person's 'species being is in question'. Armstrong and her students, through their creative responses, find in *Jane Eyre* an underlying egalitarianism: while Jane, as unaccommodated woman, a vagrant, is at risk of becoming a 'thing' or an animal, the language of the text picks up on a discourse of 'trans-historical and core human rights to which all have access without the need for justification' – Jane's status as deficit subject is not intrinsic to her, but a mantle cast upon her by the views of others. If the scandal of Bertha is, problematically, 'a kind of unconscious of Brontë's white woman's destitution', we may see that, much as lack challenges the fully human, textual silence troubles and challenges interpretation.

The late Barbara Hardy traces the ways in which the creative imagination allows access to the spaces between authorial design and academic interpretation, from whence new insights and ways of understanding the original text might emanate. Hardy provides authorial insight into her volume of nineteenth-century postscripts, which work precisely to tease out loose threads and weave material into 'hidden narrative secrets' and which act as 'an appropriately muted...recognition and celebration of Brontë's feminism'. For Hardy, in the process of writing *Dorothea's Daughter and Other Nineteenth-Century Postscripts*,[33] there emerged 'a purpose of making literary judgment and analysis through creative rather than critical discourse', and Hardy was well attuned to voice, lexicon, form, affect, and the sense of an ending in each work. Hardy's stories create space for, among other things, consideration of *Jane Eyre*'s Adèle Varens and *Villette*'s Paulina Bretton and were written not only as critique in the form of fiction but also in the spirit of homage. It is that spirit of homage that guides the inclusion of Hardy's chapter here, in loving memory of her sparkling company and sustained and brilliant contributions to nineteenth-century literary studies.

Blake Morrison considers the ways in which intertextual resonances can disperse the traditional 'gloom' of biographical interpretations of the Brontës and highlight instead qualities of resourceful authorship and human resilience. Questions of the interface between the human and the textual enlivened the Brontës' writing. The centrality of intertextuality in both the imaginative calling into being of text and its subsequent interpretation creates a powerful resonance between Armstrong's chapter (with its consideration of *King Lear* and *Jane Eyre*) and Morrison's chapter (which explores his reinterpretation of the Brontës' lives within the context of the theatre of Chekhov). Morrison's chapter outlines his imagining of the lives of the Brontës within the framework of a Chekhovian play, informed by themes and ideas that both preoccupied the Brontës' writing lives and feature in Chekhov's *Three Sisters* (1900). Using the Chekhov as a 'template, or launch pad' rather than something to be 'slavishly adhered to', Morrison found a way to negotiate the boundaries between biography and fiction(s): weaving lines from both the novels and the letters of the Brontës into the fabric of the play – and finding it important to listen to the words of the letters in particular. Neither did Morrison wish to fail to be 'true to the Brontës' lives' (by imposing too rigidly the Chekhovian framework), nor did he want to make the mistake of equating 'what fictional characters say with what their creators thought and believed'. What was (and remains) important, however, was to 'challenge the stereotypes of them as repressed, miserable and unworldly' and to show the Haworth of the Brontës' day as a place open to intellectual discussion and not the 'dead-end spot' of the Brontë myth.[34] Writing candidly of his preparation of *We Are Three Sisters* for production,[35] Morrison reveals the kinds of questions a new or unexpected vantage point on a seemingly well-known group of authors or texts can bring into the light.

An awareness of the impact of creative approaches and adaptations on the critical field can deepen and enhance ongoing scholarship. Morse's chapter, as it turns to the abolitionist context of Anne Brontë's *The Tenant of Wildfell Hall*, builds outwards from reference to Jean Rhys's influential post-colonial prequel to *Jane Eyre, Wide Sargasso Sea* (1966), and Andrea Arnold's 2011 screen adaptation of *Wuthering Heights* (which, as Morse notes, 'links Emily's novel with British Victorian slave history'). Groth's emphasis on the role of the listening and attentive reader of Charlotte Brontë's verse and prose soundscapes, and Birch's perspectives on creativity within (and perhaps in spite of) nineteenth-century models of education, provide further important points of exchange with Armstrong's, Hardy's and Morrison's perspectives on audience involvement, modes of reading, and creative interpretation. These possibilities are taken up in a plethora of responses to the Brontës' lives and works, from poetry (for example Anne Carson's 'The Glass Essay', in *Glass, Irony, and God* (1995)), to neo-Victorian novelistic reworkings by authors such as Emma Tennant, Jasper Fforde and Gail Jones,[36] to the arresting art installations and paper models of Su Blackwell, photographed by Simon Warner (one of which graces the cover of this volume).[37] The idea of the human has been translated and rewritten through different times and cultures, and the reader/audience (of the Brontës' works and of subsequent creative revisionings) is challenged to consider how our twenty-first-century perspectives on the human and imagination might be informed by looking back to lives and ideas of the nineteenth century – and to the interpretive history.

The novels and poetry of Charlotte, Emily and Anne Brontë redraw the boundaries of the human (and human rights), claiming for the 'human' a richness that expands beyond and troubles jurisprudential parameters, cultural modes, scientific understandings and certain Christian expectations, while also questioning the social efficacy and consequences of the very imaginative powers that make possible new understandings of human 'being' and potential. The Brontë novels, poetry, juvenilia, and essays on the one hand powerfully reassess nineteenth-century literature's role in critiquing conditions where some subjects are defined as not fully human, yet they also worry about – or on occasion even appear to fall short of – their own constantly evolving ideals (witness the different relations to degradation and to the upholding of inner worth of Jane and Bertha in *Jane Eyre*). Education, empire, ethics and ways of approaching the division between animal and human behaviour and treatment emerge as just some of the aspects which are at once central to the development of the Brontës' own creative voices, but which also incorporate inherent contradictions: causing each of the sisters to grapple in distinctive ways with how best to shape – and to communicate – their perspective on the idea of the human. Can an animal be degraded? When does a human become less than human? What are the boundaries of the human – and how are they policed, enforced or repositioned given temporal and cultural changes?

That very tension, productive and exciting, between different ideas of the human in Anne, Emily and Charlotte Brontës' work – from the juvenilia to mature writing, in the novels and the poetry, between the three sisters and within each individual oeuvre – and the ways that these ideas in fiction of what it means to be human are set against or build upon a range of definitions of personhood, human rights and human responsibilities in educational, medical, scientific, philosophical and theological texts, emerge as the underpinning strength, and source of much of the vitality, of this volume. Across the volume's chapters, close reading and textual awareness combine with historical, theoretical and creative contextualisations to provide path-breaking insights into this important topic. Science, ethics, and the Victorian imagination jostle for attention as the richness and complexity of ideas about, and attempted definitions of, 'the human' assert their dominance in the fiction of the Brontës.

Notes

1. Peter Boxall, 'The Limits of the Human', in *Twenty-First-Century Fiction: A Critical Introduction* (Cambridge: Cambridge University Press, 2013), pp. 84–122, p. 84.
2. Boxall, pp. 86–7, citing Jean-Paul Sartre's 'Preface' to Frantz Fanon's *The Wretched of the Earth* (New York: Grove Press, 1963), p. 26.
3. Boxall, p. 88.
4. Jeff Wallace, 'Literature and Posthumanism', *Literature Compass*, 7/8 (2010), 692–701, p. 692; Bart Simon, 'Introduction: Toward a Critique of Posthuman Futures', *Cultural Critique*, 53 (Winter 2003), 1–9, p. 8.
5. Wallace, p. 693, citing Katherine N. Hayles, *How We Became Posthuman: Virtual Bodies in Cybernetics, Literature, and Informatics* (Chicago: University of Chicago Press, 1999), p. 291.
6. Wallace, p. 697.
7. Andy Mousley, *Literature and the Human: Criticism, Theory, Practice* (London: Routledge, 2013), pp. 5, 9.
8. Charles Taylor, *Sources of the Self: The Making of the Modern Identity* (Cambridge, MA: Harvard University Press, 1989), pp. 393–4.
9. Rick Rylance, *Victorian Psychology and British Culture, 1850–1880* (Oxford: Oxford University Press, 2000), p. 2.
10. Cary Wolfe, 'Is Humanism Really Humane?', interview with Natasha Lennard, *The New York Times*, 9 January 2017. See further Cary Wolfe, *Animal Rites: American Culture, the Discourse of Species, and Posthumanist Theory* (Chicago: University of Chicago Press, 2003).
11. Deborah Denenholz Morse and Martin A. Dananhay, eds., *Victorian Animal Dreams: Representations of Animals in Victorian Culture and Literature* (Aldershot: Ashgate, 2007), p. 3, citing Harriet Ritvo, 'Afterword', in *The Animal Estate: The English and Other Creatures in the Victorian Age* (Cambridge, MA: Harvard University Press, 1987). See further Jacques Derrida, 'The Animal That Therefore I Am (More to Follow)', David Wills, trans., *Critical Inquiry*, 28:2 (2002), 369–418.
12. Joanna Bourke, *What it Means to be Human: Reflections from 1791 to the Present* (London: Virago, 2011), p. 378.
13. Ibid., at p. 380.
14. Ibid., at pp. 380, 385.

15 Yuval Noah Harari, *Sapiens: A Brief History of Humankind* (2011; London: Vintage, 2014), p. 463.
16 Ibid., at pp. 465–6.
17 Heather Glen, *Charlotte Brontë: The Imagination in History* (Oxford: Oxford University Press, 2002). For important historically grounded studies of the Brontës and religion, education, and psychology, see Marianne Thormählen's *The Brontës and Religion* (Cambridge: Cambridge University Press, 1999) and *The Brontës and Education* (Cambridge: Cambridge University Press, 2007) and Sally Shuttleworth's *Charlotte Brontë and Victorian Psychology* (Cambridge: Cambridge University Press, 1996). See also *The Brontës in Context* (Cambridge: Cambridge University Press, 2012), Marianne Thomählen, ed.
18 George Levine, *Realism, Ethics and Secularism: Essays on Victorian Literature and Science* (Cambridge: Cambridge University Press, 2008), p. viii.
19 Ibid.
20 Andrew H. Miller, *The Burdens of Perfection: On Ethics and Reading in Nineteenth-Century British Literature* (Ithaca: Cornell University Press, 2008), p. xi.
21 Levine, p. 258.
22 Ibid., at p. 259.
23 John Bowen, 'The Brontës and the Transformations of Romanticism', in John Kucich and Jenny Bourne Taylor, eds., *The Nineteenth-Century Novel, 1820–1880*, Volume 3 of *The Oxford History of the Novel in English* (Oxford: Oxford University Press, 2011), pp. 203–19, p. 205.
24 Ibid., at p. 207.
25 Cannon Schmitt, *Darwin and the Memory of the Human: Evolution, Savages, and South America* (Cambridge: Cambridge University Press, 2009), p. 3.
26 Gillian Beer, *Open Fields: Science in Cultural Encounter* (Oxford: Clarendon, 1996), p. 210.
27 See, for example, Stephen Arata, *Fictions of Loss in the Victorian Fin de Siècle* (Cambridge: Cambridge University Press, 1996), Christine Ferguson, *Language, Science and Popular Fiction in the Victorian Fin-de-Siècle: The Brutal Tongue* (Aldershot: Ashgate, 2006) and Patrick Brantlinger, *Taming Cannibals: Race and the Victorians* (Ithaca: Cornell University Press, 2011).
28 Laura Brown, *Homeless Dogs and Melancholy Apes: Humans and Other Animals in the Modern Literary Imagination* (Ithaca: Cornell University Press, 2010), p. 24.
29 Laurence W. Mazzeno and Ronald D. Morrison, eds., *Animals in Victorian Literature and Culture* (London: Palgrave Macmillan, 2017); see further Barbara Munson Goff, 'Between Natural Theology and Natural Selection: Breeding the Human Animal in *Wuthering Heights*', *Victorian Studies*, 27:4 (1984), 477–508 and Ivan Kreilkamp, 'Petted Things: *Wuthering Heights* and the Animal', *Yale Journal of Criticism*, 18:1 (2005), 87–110.

30 On the language of the human in this feminist moment see Alexandra Lewis, '"Supposed to be very calm generally": Anger, Narrative, and Unaccountable Sounds in Charlotte Brontë's *Jane Eyre*', in *Feminist Moments: Reading Feminist Texts*, Susan Bruce and Kathy Smits, eds. (London: Bloomsbury, 2016), pp. 67–74.
31 William Cohen, *Embodied: Victorian Literature and the Senses* (Minneapolis: University of Minnesota Press, 2008).
32 Isobel Armstrong, *Novel Politics: Democratic Imaginations in Nineteenth-Century Fiction* (Oxford: Oxford University Press, 2016).
33 Barbara Hardy, *Dorothea's Daughter and Other Nineteenth-Century Postscripts* (Brighton: Victorian Secrets Ltd., 2011).
34 On beliefs about the sisters and their town being cut off from intellectual discussion, see Lucasta Miller's *The Brontë Myth* (London: Vintage, 2002).
35 Blake Morrison, *We Are Three Sisters*, a Northern Broadsides production, directed by Barrie Rutter (Nick Hern Books, 2011).
36 On these neo-Victorian reworkings, see Alexandra Lewis, 'The Ethics of Appropriation; Or, the "mere spectre" of Jane Eyre: Emma Tennant's *Thornfield Hall*, Jasper Fforde's *The Eyre Affair* and Gail Jones's *Sixty Lights*', in *Charlotte Brontë: Legacies and Afterlives*, Amber Regis and Deborah Wynne, eds. (Manchester: Manchester University Press, 2017), pp. 197–220.
37 Su Blackwell's exhibition, 'Remnants', at the Brontë Parsonage Museum in 2010, featured site-specific installations that evoked the Brontës' imaginary worlds and drew connections between their material possessions and use of paper as a precious commodity. Su Blackwell's creative interventions ranged from moths fluttering out from Branwell's nightgown, to a reimagining of the juvenilia (toy soldiers at battle), to a beautifully lit *Wuthering Heights* tableau emerging from the pages of a book. I am very grateful to Su and to the photographer of her exhibition, Simon Warner, for graciously allowing one of these images to feature on the cover of this volume.

CHAPTER I

Hanging, Crushing, and Shooting
Animals, Violence, and Child-Rearing in Brontë Fiction

Sally Shuttleworth

For the Brontës, more than any other novelists in the nineteenth century, the idea of the human was bound up with that of the animal, not in any simple metaphorical mode, but through a fundamental, imaginative engagement which both draws upon, and cuts across, constructions of animal/human relations in the natural sciences, religion, and culture of the period. Animals abound in the novels. There are probably more dogs in *Wuthering Heights* than any other mainstream Victorian novel. When Lockwood first visits the Heights, dogs seem to emerge out of the very walls; in addition to the 'nursery' (p. 4) of the 'bitch pointer' and her squealing brood, 'other dogs haunted other recesses' (p. 3). Canine reproduction displaces that of the human, while their haunting anticipates the more uncanny human form to follow. The dogs' attack on Lockwood, re-enacted on his second visit, parallels that on Catherine as she and Heathcliff spy on Thrushcross Grange, and the threatened attack on Isabella when she first enters the Heights. *Wuthering Heights* is, famously, a novel preoccupied with barriers and thresholds, and dogs both demarcate and dissolve the boundaries of the human domain. In *Agnes Grey* and *Shirley*, dogs function more straightforwardly as touchstones of human character: individuals are repeatedly judged by the qualities of their interaction with animals. It is not, however, in the harmonious relations between human and animal that the interest of the novels lie, but rather in moments of violence between the two which challenge and fracture safe categorisations and distinctions. In this chapter, I address three very different moments of violence against animals in Brontë novels – the hanging of Isabella's dog by Heathcliff; the crushing of the birds in *Agnes Grey*, and the shooting of Victor's dog in *The Professor* – setting them in the context of nineteenth-century constructions of human/animal relations. In particular, the chapter explores the interface between the child and animal, with reference to medical and educational literature on child development.[1]

'Fanny, suspended to a handkerchief'

When Heathcliff absconds with Isabella, he leaves behind an elegantly perverse calling card. Nelly, in quest of a doctor for Catherine (who is at the height of her delirium), notices something unusual in the garden: 'at a place where a bridle hook is driven into the wall, I saw something white moved irregularly, evidently by another agent than the wind'. This is no ghost, however. On investigating, she discovers, 'by touch more than vision, Miss Isabella's springer, Fanny, suspended to a handkerchief, and nearly at its last gasp' (p. 114). It is an eerie moment, made even more striking by Nelly's matter of fact response. There are no loud exclamations about cruelty; she simply unties the dog, wonders briefly about how it got there, and 'what mischievous person had treated it so' (p. 114) and heads off to fetch the doctor (ignoring that other ominous animal sign, the drumming of horses' hooves). For the reader, however, Heathcliff is leaving a clear symbolic message, identifying himself with a stock figure in moral and educational literature, who starts off in youth by torturing and hanging animals, progresses to violence against people (particularly women), and ends up disgraced, and even hanged, as expressed most iconically in William Hogarth's series of prints, *The Four Stages of Cruelty* (1751).[2] The precise detail of the 'bridle hook', with its associative linguistic resonances, reinforces an ominous parallel: Isabella is well and truly hooked and bridled, her bridal veil the fluttering handkerchief which presages the cruelty of her marriage.

In Hogarth's first print (Figure 1.1) he depicts a crowded street scene, crammed with instances of animal cruelty: cats are hung from an improvised gibbet and flung from windows; birds have their eyes pierced with a hot needle, cocks and dogs are tormented. At the focal heart of the scene, however, is a more unusual detail: the sodomising of a dog with an arrow. Diagonal lines of shade track the arrow, guiding our eyes to the dog's anus; this disturbing detail highlights the elements of sexual sadism which underpin the whole. Nero, the central figure (who is, like Heathcliff, a 'charity' boy), progresses to abuse of horses, and then to that of women. He murders his pregnant lover, who, despite her conscience, had resolved, 'to venture body and soul to do as you would have me', and had robbed her mistress so that she could run away with him. With symbolic retributive justice, Nero is hung, and his body given to surgeons for public dissection – the ultimate denial of human status – while a dog feasts on his entrails. The prints had a lasting legacy in the culture and institutions of England, furnishing the structure for innumerable educational

Figure 1.1 William Hogarth, 'First Stage of Cruelty', *The Four Stages of Cruelty*, 1751. Reproduced with permission from the Trustees of the British Museum.

moral tales as well as paving the way for laws addressing animal cruelty, and the founding of the Society for the Prevention of Cruelty to Animals (SPCA) in 1824.[3] *Wuthering Heights* draws on this legacy, and many of the elements of the Hogarth tale – the charity child who goes to the bad, the animal/female parallel, and the slavish devotion of the deluded, absconding woman – but with a difference. Heathcliff self-consciously plays with, and undermines, the Hogarth script. Far from feeling moral shame, he taunts others with his mastery of the genre: a hanging dog signals his intentions with regard to the female in his power, although his forms of cruelty will eschew the simplistic form of murder for more sophisticated modes of psychological torture. As he informs Nelly, he is careful to ensure he keeps 'strictly within the limits of the law' (p. 133). Unlike Hogarth's Nero, he is not tracked down and punished by the law, but rather deliberately employs its framework of couverture, male ownership of the female within marriage, to further his ends.[4]

The scene of the dog hanging is not described directly – only that enigmatic signal of the white, wafting shape – but Heathcliff does subsequently explain himself to Nelly, in the presence, provocatively, of Isabella. Excusing himself from any deceit, he argues that he could not be accused of showing any tenderness towards her since, 'The first thing she saw me do, on coming out of the Grange, was to hang up her little dog' (p. 133) – an act then reinforced by his expressed wish that he could treat her family likewise. He voices no remorse, only contempt for his newly wedded wife: 'But no brutality disgusted her – I suppose she has an innate admiration of it, if only her precious person were secure from injury!' (p. 133). In an extraordinary move of sophistical displacement, perversion is fixed on Isabella, who is assigned a sadistic, and indeed masochistic love of brutality, as if she were the agent of the dog's hanging. The claim recalls our first sight of Isabella and Edgar (as mediated by Heathcliff, recounting the scene to Nelly), quarrelling over a little dog and virtually tearing it apart. Narrative placement and readerly association helps give subtle credence to his outrageous assertion. The suggestion is pushed even further as Isabella, in Heathcliff's self-vindicating account, becomes herself a dog – a brach (or bitch), a 'pitiful, slavish, mean-minded brach' (p. 133). Heathcliff orders Nelly to 'Tell your master' (a command that also aligns Nelly with a slavish dog), 'that I never, in all my life, met with such an abject thing as she is'. Such abjection seems to goad him on further, although he notes 'I've sometimes relented, from pure lack of invention, in my experiments on what she could endure, and still creep shamefully cringing back' (p. 133). In Heathcliff's twisted version of the master/slave dialectic, Isabella is at once cringing dog and human pervert, whose base nature thrives on Heathcliff's need to act as scientific experimenter, exposing the inner secrets of bourgeois life, where animal torture (or experimentation) is undertaken in the name of the pursuit of truth.

The specific resonance of 'experiments' is reinforced later when Heathcliff, watching Catherine and Linton together, observes, 'Had I been born where laws are less strict, and tastes less dainty, I should treat myself to a slow vivisection of those two, as an evening's amusement' (p. 238). The term vivisection, as I will discuss shortly, was unusual for the time, suggesting Brontë's explicit engagement with contemporary discussions of animal cruelty and physiological experimentation.[5] In aligning Heathcliff with the scientific experimenter, or vivisector, Brontë is inverting the structure of the moral tale outlined by Hogarth (and replicated in endless texts for children) on the dire consequences of torturing animals. Unlike the Lintons, or John Reed in *Jane Eyre*, who we are told 'twisted the necks of the pigeons, killed the little pea-chicks, and set the dogs at the sheep' (p. 15), Heathcliff does not appear to torture animals in childhood (although he does make traps for birds).[6] Rather, from a state of close union with nature as a child, social exclusion drives him in adulthood to adopt the practices of the supposedly civilised world. In a reversal of the Hogarth, he does not end up on the dissection table, but becomes himself a vivisector – an even more disturbing practice since he operates on live, rather than dead, bodies.

Heathcliff is at once callous and self-righteous – his stance partaking of that outlined by Charlotte Brontë in her preface to the second edition of *Jane Eyre* – one who does not wish to let 'white-washed walls vouch for clean shrines' and who wishes to 'penetrate the sepulchre, and reveal charnel relics' (p. 4). His self-proclaimed advancement to the position of scientific experimenter, is at once an assumption of the role of social exposer and scourge, and also a knowing embrace of his own debasement. He stands in judgment on his own child, Linton, and his ward, Hareton, even as he actively encourages their petty cruelties. Heathcliff speaks in utter scorn of his son Linton who 'can play the little tyrant well. He'll undertake to torture any number of cats if their teeth be drawn, and their claws pared' (p. 243). Cowardice, weakness and tyranny are rolled into one in this image of a child who can only torture animals if they have ceased to represent any form of danger. Hareton, for this part, is depicted as 'hanging a litter of puppies from a chair-back' (p. 161) and 'taking a pride in his brutishness', following Heathcliff's deliberate immersion of him in 'coarseness and ignorance' (p. 193). Barbara Munson Goff has argued that this act should not be seen as gratuitous cruelty, but rather part of a 'routine of culling', for people who treat animals as part of an economic system, rather than sentimentalising or anthropomorphising them.[7] This is to ignore, however, Heathcliff's exultation in Hareton's debasement, and the narrative reference back to his own, quite explicitly, gratuitous act. For Goff, '*Wuthering Heights* is about the colossal stupidity, arrogance, even impiety of anthropocentrism'. It is a tempting, even seductive hypothesis, but in assimilating the novel to a proto-Darwinian vision, it tends to flatten out the novel's more complex relationship to contemporary constructions of the animal/human relation.

Vivisection

In depicting Heathcliff's desire to 'treat myself to a slow vivisection' of Cathy and Linton, Brontë was using a term that was startling in its scientific specificity. It was to become commonplace in the 1870s, with the antivivisection campaigns led by Frances Power Cobbe, but in the 1830s and 1840s it was hardly ever used outside medical journals and textbooks, and even there, only rarely. Although the first citation of the term in the *OED* is from 1707, even medical discussions before the 1860s tended to use the formulation 'experiments on living animals', or other equivalent descriptive forms rather than the more specialised Latinate term.[8] Nor was animal experimentation a focus for widespread social concern at the time that Brontë was writing, although animal cruelty certainly was. In 1822 a law had been passed prohibiting cruelty to larger animals such as horses and cattle, and in 1824 the Society for the Prevention of Cruelty to Animals was founded (it was to become the 'Royal' Society in 1840 when Queen Victoria gave it her seal of approval). Partly as a result of the SPCA's campaigning, the 1822 act was extended to domestic animals, such as dogs and cats in 1835. As Harriet Ritvo has argued, however, the main focus of concern for the SPCA in its early years was the brutal treatment of horses and cattle by the lower orders. The critique of inhumane treatment of animals confounded two missions, she suggests: 'to rescue animal victims and to suppress dangerous elements of human society.'[9] The early decades of the Society were driven by a desire to educate, and morally reform, the cruelties of the lower classes. Vivisection, as an activity by the educated and specialised medical establishment, therefore drew little attention in these early years although, as Ritvo notes, the 1824 prospectus for the Society did raise the question of whether it was justifiable 'to conduct certain experiments of a painful nature', arguing that 'Providence cannot intend that the secrets of Nature should be discovered by means of cruelty'.[10] Newspaper accounts of the activities of the RSPCA at this period are overwhelmingly focused on the prosecutions they brought for the mistreatment of cab-horses (as in Hogarth's second 'Acts of Cruelty'), or other working animals.[11]

The issue of vivisection was raised, however, at the Annual General Meeting of the SPCA in 1837 in a speech by its president, the Earl of Carnarvon, and his attacks on the practice were reiterated in his speech to the organisation in 1843. The Society also sponsored an essay prize and the winning entry, the Revd. John Styles's *The Animal Creation: its Claims on our Humanity* (1839), took up the Earl's condemnation of vivisection, offering more graphic accounts of the torture inflicted on animals 'under the cloak of humanity and zeal for science by men who have received the highest education'.[12] In particular he highlights the Earl's example of an experiment in Edinburgh in which an iron bar was heated and thrust into the brain of a dog which 'with fiendish skill, was kept alive for sixteen days'.[13] In this, and other accounts, the emphasis is laid on the sheer gratuitousness of the act, and the seeming cold curiosity which prompts these scientific torturers to prolong their acts. Styles also quotes from John Elliotson's *Human Physiology* (1835), a work much used by Patrick Brontë, and probably one of the works available in the Parsonage, which appears to be one of the earliest medical textbooks to condemn the practice of experimentation on animals.[14] Clearly responding to demonstrations by Magendie, and others in the same school of French physiology, Elliotson argued that 'to torture animals unnecessarily is a most cowardly and cold-blooded act, and, in my opinion, one of the most utmost depravity and sin'. To reinforce his point, he offers an extraordinarily vivid account from a French physician, Dr. Brachet, of experimenting on a dog that started to rage against him. He therefore put out its eyes, destroyed its ear drums and filled them with wax, and having thus destroyed two of its senses, was pleased to note that the animal then seemed 'even sensible to my caresses' before he commenced further experiments.[15] The account is disturbing on many levels, not least the scientist's triumph in the fact that the mutilated creature would still respond gratefully to a caress from the very hand that had tortured it. The same psychological dynamic is clearly at work in *Wuthering Heights* when Heathcliff laments that he is failing in invention in his experiments on Isabella, as to what she 'could endure, and still creep shamefully cringing back'. Similarly, he experiments in child rearing, tracking how far noble material might be debased, and celebrating his success. He exults that Hareton, despite his enforced training in brutishness, 'is damnably fond of me! You'll own that I have outmatched Hindley there!' (p. 193)

The SPCA prize also led to the publication of two other works at the same period, W. H. Drummond, *The Rights of Animals, and Man's Obligation to Treat them with Humanity* (1838) which has a chapter on 'Love of Science Perverted – Vivisection' (which also drew on the

Elliotson)[16] and David Mushet's *The Wrongs of the Animal World* (1839). Mushet similarly invokes the atrocities of Magendie and his school, which he fears have taken secret root in England. His portrait is one of an uncontrollable psychological compulsion. There is, he remarks, 'no greater mark of vice than that of unsatiated appetite tantalized for ever with a prospect of satisfaction, but increasing like a devouring fire, can never cease to crave so long as any spot of soundness remains unpreyed upon'.[17] Vivisection is here clearly linked to depraved sexual appetite – a vice that can never be satisfied, but instead consumes the practitioner. Preying upon others, he is himself consumed: again there are parallels with Heathcliff who, despite his self-image of a cool, detached experimenter, is nonetheless entrapped by his own desires to extract revenge and to demonstrate his power of making others suffer.

Vivisection comes to the explicit attention of the national press in 1843 with a report in the *Times,* prominently titled, 'Vivisection', on the Earl of Carnarvon's address to the annual meeting of the RSPCA in which he returns to the topic of vivisection in the training of medical students and the infamous example of the iron bar through the dog's head. According to the Earl, this experiment, the *Times* reports, was 'to discover whether this long-tortured animal would retain any vestige of his old affection for the master – or should I rather say, the monster that had inflicted such sufferings upon him'.[18] Interestingly, the experiment thus singled out is not on mere reflex responses, but specifically on the brain, and the affections, and hence on the qualities which might demarcate human from animal life. It is impossible to know, given the paucity of sources on Emily Brontë's reading, where she might have encountered the term vivisection; she might have picked up a pamphlet from the SPCA for example, but we know that the family read the *Times,* and as this is one of the first usages in the press (outside medical journals), it is possible it was her source.[19] Certainly, the experimentation on the dog and its affections is suggestive of Heathcliff's own experimentation on Isabella and her dog (the two being interchangeable), while it also helps illuminate Heathcliff's expressed desire, with reference to the young Cathy and Linton to 'treat myself to a slow vivisection of those two, as an evening's amusement' (p. 238). As with the symbolic declaration of the hanging dog, Heathcliff self-consciously aligns himself with the figure of the cold-hearted scientist, taking his pleasure from animal experimentation conducted in the name of science. Heathcliff's object, in thus indulging in vivisection, is, like that of the Edinburgh scientists, to track the physiological sources of the affections which had sprung up between his two live subjects, thus reducing spirit and emotion to a material base of blood and nerve.

In his speech, the Earl of Carnarvon had worried about the effects of such 'contaminating scenes' on young medical students. What would you think, he asks, 'of the youth who could view such scenes of torture with eye and heart unmoved?'[20] Brontë tracks the cycles of oppression, as Heathcliff's own childhood subjection gives rise to a chain of violence. Thus, Isabella herself takes on Heathcliff's own mantle. Fleeing from Heathcliff, with marks of his abuse on her body, a bleeding cut under her ear, and her face 'scratched and bruised' (p. 151), she nonetheless rejoices that she has overturned his 'fiendish prudence' and awoken his 'murderous violence'. She has, in other words, disrupted his role of cool experimenter: 'Pulling out the nerves with red hot pincers requires more coolness than knocking on the head' (p. 153). The language once again evokes vivisection. The ancient mode of torture, using red hot pincers, is linked to the very modern practice of physiological experimentation on the nerves, as associated in particular with the 'foreign experimentalist' and vivisector, François Magendie. Isabella denies Heathcliff's human status to Nelly – 'He's not a human being' – and attributes her escape to the arousal of pleasure in her power to exasperate him which 'woke my instinct of self-preservation' (pp. 152–3). Far from pitying Heathcliff in his grief for the loss of Cathy, she 'taste[s] the delight of paying wrong for wrong', her Christian principles now wedded to a vindictive measure of justice, 'for every wrench of agony return a wrench, reduce him to my level' (p. 159). As with Heathcliff's debasement of Hareton, so that 'I've got him faster than his scoundrel of a father secured me, and lower' (p. 193), there is a preoccupation in the text with levels – both levels of suffering, and also positionality on a scale of being. The practice of vivisection brought into sharp focus the relationship between human and animal, raising questions about the nature of pleasure and pain, the ethics of animal experimentation for human gain, and more fundamentally, the viability of hierarchical structures of animal/human life.

Child Cruelty to Animals

Hierarchy was also implicit in the association between children and cruelty to animals: beings who were themselves low in the human scale were deemed to be particularly prone to exercising cruelty on those below. This preoccupation with child cruelty to animals was not simply a Victorian creation. For example, John Locke argued in *Some Thoughts*

Concerning Education (1693) that children should be brought up with 'an abhorrence of *killing* or tormenting any living creature' since 'they who delight in the suffering and destruction of inferior creatures, will not be apt to be very compassionate or benign to those of their own kind'.[21] Such sentiments clearly lie behind Hogarth's series of *Four Stages of Cruelty*, but it is at the end of the eighteenth century that these ideas take on powerful life in the huge outpouring of moral and educational texts for children which preached, with wearisome repetition, the necessity of kindness to animals. Such works were produced by writers from a diverse range of political and religious views, from Thomas Day's *Sandford and Merton* (1783–9), to Sarah Trimmer's *Fabulous Histories* (1786) and Mary Wollstonecraft's *Original Stories from Real Life* (1788) where, in interesting parallel to the scene in *Agnes Grey*, the governess kills a lark, which had been left in 'exquisite pain' after being shot by an 'idle boy'.[22]

One of the key texts of this genre was *Pity's Gift: A Collection of Interesting Tales to Excite the Compassion of Youth for the Animal Creation* by Samuel Pratt, an actor, writer, lapsed Anglican clergyman, and vigorous campaigner for animal rights. One of the most startling traits of Pratt's work, and indeed of many of these works, is the intensity of violence envisaged as a form of revenge on the perpetrators. Thus, in 'The Nightingale', the bird whose 'husband' and children have been shot, envisages the killer having his own children killed, and then will 'justice whisper – this is child for child'.[23] 'The Robin' offers an even more gruesome vision of retributive justice. The son of a shepherd, who had always counselled kindness to animals, finds a robin's nest and brings it home to his siblings, giving a bird to each. While the girls try to care for their birds, the boys inflict various forms of tortures, from pulling the bird around on a string and feeding it to a cat, to sticking pins in the eyes and taking 'a delight in seeing it bleed to death'. On discovering their crimes, the father in fury metes out appropriate punishment to each, scratching one with pins 'until his hands were all over blood' and setting his dog on the one who had given his bird to the cat.[24] The eldest son dies six months later, and animals and birds unite to attack his body as it is laid in the grave, thus establishing a union between the animal kingdom and patriarchal authority.

With the rise of the illustrated book for children in the early nineteenth century, there were innumerable tales on a similar theme. Thus in Mary Elliott's *The History of a Goldfinch: Intended to Excite in the Mind of Youth Humanity towards Brute Creation*, a boy who plucks the feathers off a goldfinch has his hair pulled out by his father in punishment: 'Go from my presence, you cruel, wicked boy, and never let me see your

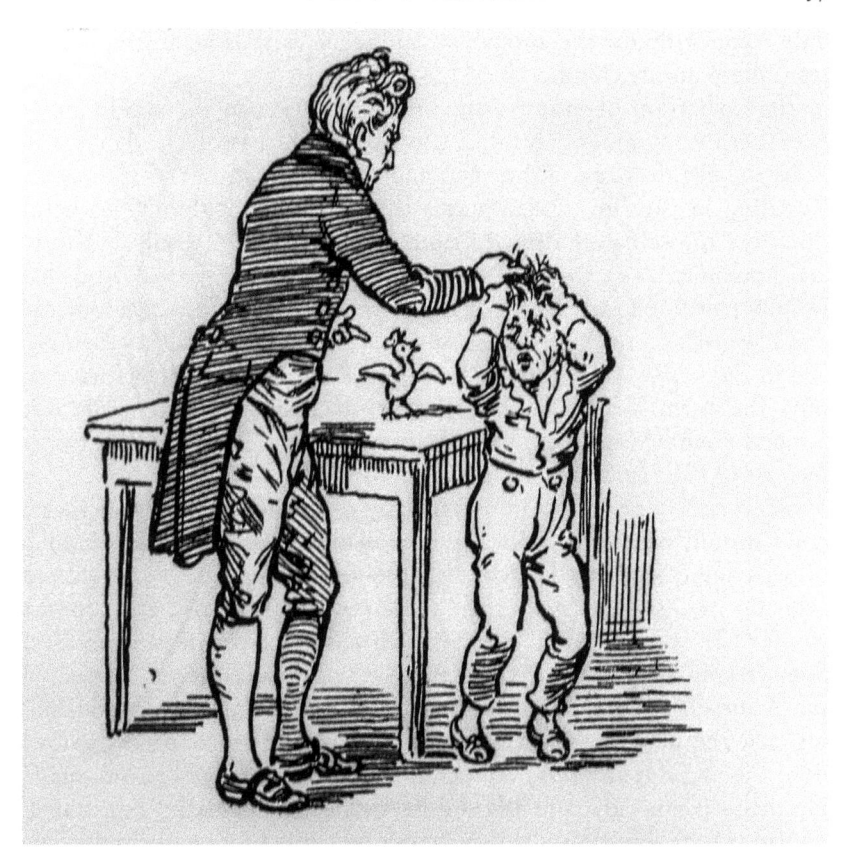

Figure 1.2 [Mary Elliott] 'The History of a Goldfinch', reproduced in Andrew W. Tuer, *Stories from Old-Fashioned Children's Books* (London: Leadenhall Press, 1899–1900), p. 211. Image courtesy of Oliver Christie.

face till you are sensible of your monstrous crime!'[25] In one reading, the tale instructs the child in his duties towards the lower creation, but on another, paternal violence (as depicted in the illustration to the 1823 version of the tale, Figure 1.2) is a reflection of that of the child: the tendencies that have to be controlled within the child are given license and social authority in the adult. Significantly, early American versions of the tale alter the subtitle to read: 'Addressed to those children who are dutiful to their parents, and humane to their fellow creatures'.[26] The subtext is made clear: concerns are focused primarily on disciplining children, and only secondarily on the animal, which becomes thus an instrument in controlling unruly childhood.

The invocation of monstrosity, in the depiction of the child's 'monstrous' crime, summons to mind another figure, that of Heathcliff who is, as Isabella declares, 'a monster, and not a human being' (p. 134).[27] Revelling in the sub-human status thus thrust upon him, Heathcliff distances himself very directly from the Victorian discourses of pity and humanity, as exemplified for him in the figure of Edgar: 'And that insipid, paltry creature attending her from *duty* and *humanity*! From *pity* and *charity*!' (p. 135). Inverting entirely the moral lessons as exemplified in *Pity's Gift*, and similar tales, he declares: 'I have no pity! I have no pity! The worms writhe, the more I yearn to crush out their entrails! It is a moral teething, and I grind with greater energy, in proportion to the increase of pain' (p. 134).

One of the first instructions in many of the moral tales was that a child should never tread on a worm. William Cowper, a poet much beloved and quoted by the Brontës, observed in 'The Task,' 'I would not enter on my list of friends ... the man who needlessly sets foot upon a worm'.[28] This section of the poem (which was much reprinted, with these lines serving as an epigraph for many RSPCA publications) speaks of the 'luxuriant growth' of cruelty if unchecked in a child, before concluding that the mercy of Heaven would be withheld from 'he that shows none – being ripe in years,/ And conscious of the outrage he commits'.[29] Heathcliff is not only conscious of the 'outrage' he commits, but dramatises in graphic form his alienation from prescribed forms of child development. His 'moral teething' increases rather than subdues his desire to inflict pain – both on others but also on himself, in this illuminating self-analysis of his masochistic and sadistic tendencies. He is both child and man, aware of the conventions of normative moral development, but trapped within an early 'teething' phase he is powerless to transcend.

It is a common move in criticism of the novel to associate Heathcliff with an amoral realm of nature, but this displaced form of romanticism, which merely preserves and inverts a binary opposition between the natural and the social, does not do justice to the complex engagement within the novel. *Wuthering Heights* is not a proto-Darwinian novel, celebrating a natural, asocial world of violence and passion; rather it tracks the intersection of the animal and the human, and the forms of violence that emerge if patterns of growth are perverted. It certainly does not defend violence to animals, but rather exposes the self-justifying language of a society which deploys the rhetoric of kindness to animals to control and coerce.

'The Training of Boys' and *Agnes Grey*

The theme of kindness to animals as a form of moral training was also taken up in the domestic conduct books of the early Victorian era. In a fascinating section on 'The Training of Boys' in *The Mothers of England, their Influence and Responsibility* (1843), the arch conservative Sarah Stickney Ellis argued fiercely for the doctrine of separate spheres and woman's duty to serve the 'nobler sex'. What emerges in the text, however, is rather a real dislike of boys and men, particularly in the sections on mistreatment of animals where she traces adult domestic cruelty towards the wife from the husband's treatment of animals when young. Thus, 'the strong taking advantage of the weak, and exulting in the suffering inflicted, and the mastery obtained, may begin with the little boy in the nursery, when he snatches his sister's kitten, and throws it into the nearest pond'.[30] Ellis portrays the mother as semi-powerless in controlling her boy's cruelty to animals. She cannot prevent it by direct authority, 'for that would make some boys prolong the amusement, for the purpose of showing their power', nor by entreaty, 'for that might possibly excite a laugh almost as exulting as that which is awakened by the sufferings of the tortured animal'. The answer, for this poor beset mother, is to start early so that the boy's innate 'love of power' and 'sense of mastery' are directed to protecting, rather than torturing, the weak – here defined as his sisters and animals.[31] For all its talk of male superiority, the male in Ellis's work is a frightening beast – able to exult in the torture of others, if not controlled from an early age by a wise female.

Ellis's scenario is re-enacted in Anne Brontë's *Agnes Grey*, which appears to offer a more straightforward response to contemporary ideas of animal/human relations and child upbringing than *Wuthering Heights*.[32] Agnes, in her position as governess, is in the hapless situation of Ellis's harassed mother; unable to rule by command or entreaty, for fear of exacerbating the cruelties of her young charge, master Tom Bloomfield. The chilling scene with the birds' nest has been prepared for by earlier events – when Agnes first meets Tom she is shown his garden which is covered in bird traps. Asked what he does with them he replies, 'Sometimes I give them to the cat, sometimes I cut them in pieces with my penknife; but the next, I mean to roast alive'. His recitation follows the structure of 'The Robin' and other moral tales, with an escalation of violence for each bird. Agnes takes over the script, informing him of his wickedness, and enjoining him to remember that 'birds can feel as well as you'. Tom's dismissive, 'Oh, that's nothing – I am not a bird' attracts the full invocation of deserved punishment in hell: 'you have heard where wicked people go to die; and if you don't leave off torturing innocent birds, remember, you will have to go when they there, and suffer just what you have made them suffer' (p. 20). Retributive justice will continue beyond the grave.

Brontë's depiction of Tom's response to the nest, however, is far removed from anything to be found in the didactic literature. Tom stands exultantly over the nest, 'with his legs wide apart, his hands thrust into his breeches-pockets, his body bent forward, and his face twisted into all manner of contortions in the ecstasy of his delight' (p. 42). The scene is inescapably sexualised: Tom, with his hands in his breeches-pockets, is at once child and man (in a reversal of Heathcliff's adult 'moral teething'), prefiguring the sexual tormentor he will undoubtedly become. Where Heathcliff was almost clinical in his vivisection, or self-torturing in his violence, Tom exhibits a more disturbing form of emotion: intense pleasure, reaching indeed to ecstasy. Tormented by Tom with her own powerlessness, Agnes reaches for a flat stone. She does not simply drop it, however:

> [H]aving once more vainly endeavoured to persuade the little tyrant to let the birds be carried back, I asked what he intended to do with them. With fiendish glee he commenced a list of torments, and while he was busied in the relation, I dropped the stone upon his intended victims, and crushed them flat beneath it. (p. 43)

To borrow from Michel Foucault, there is here an incitement to discourse, as Agnes virtually urges Tom to recite his proposed list of torments, suggesting a certain level of complicity in this sexualised display of violence. The scene shocks by its very starkness: there are no sentimental trappings about the sufferings of the poor birds, or any agonies of remorse on her part afterwards. Agnes, like Wollstonecraft's governess, depicts her act as one of humanity and duty; but as Heathcliff makes clear in *Wuthering Heights*, humanity and duty are concepts which can offer a convenient cloak of moral respectability for more murky motivations.

It is worth remembering that Agnes has already shown herself capable of violence in dealing with the children. When Mary Ann would not learn her lesson, 'I would shake her violently by the shoulders, or pull her long hair' (an act that recalls that of the father in the *Goldfinch*) (p. 29). Agnes's pupils are to her 'unimpressible, incomprehensible creatures' who need humanising, inhabiting the wrong side of the animal/human divide (p. 49). She despises them in much the same way that Heathcliff despises the Lintons. While there is clearly not a direct parallel between Heathcliff's hanging of Isabella's dog and Agnes's crushing of the birds, since her act can undoubtedly be seen as a humane one, there are nonetheless similarities in the ways in which each becomes involved in animal violence while seeking to blame the despised other party – Isabella or Tom. Agnes might crush the birds, but it is clear she would rather have crushed Tom with her stone; the birds become the symbolic replacement for the boy, who is also given agency for her violence. Inhabiting neither the community of family, nor of the servants' hall, Agnes, like Heathcliff, is a liminal figure. Although her violence and frustration only flash out at moments, they are located at the troubled boundary between animal and human, and also between child and adult.

Son and Dog in *The Professor*

The final scene I would like to consider is the shooting by William Crimsworth of his son's beloved dog at the end of *The Professor*, a novel which was written at the same time as *Wuthering Heights* and *Agnes Grey*, although published posthumously. The scene startles by its apparent gratuitousness. William and Frances are finally happily and comfortably settled; the novel could easily have ended with a quick depiction of wedded bliss and satisfied achievement. Instead, out of nowhere, we are introduced not only to the obligatory son, but also to the shooting of his dog. As in *Agnes Grey*, there is an ostensible humane reason for this violence – the animal has possibly been bitten by a rabid dog – but in narrative terms it functions symbolically to introduce the fraught relation between child and animal, and father and son. Victor, we learn, is an odd child; immersed in books, he rarely smiles, although he has 'a susceptibility to pleasurable sensations almost too keen', while his feeling for a mastiff-cub, given to him by his father's disruptive friend, Yorke Hunsden, 'strengthens almost to a passion' (p. 242). The mastiff, named Yorke in Hunsden's honour, is taken by him to a nearby town where it is bitten by a dog 'in a rabid state' (p. 243). William, on hearing the news, instantly takes out his gun and shoots the dog, without hesitation or consultation.

Rabies was, of course, a significant threat in the Victorian period, but as Ritvo, and more recently Neil Pemberton and Michael Worboys have shown, it played an even larger role in the cultural imaginary of the era.[33] As a disease, it broke down bodily boundaries between animal and human, introducing fluids and life forms from the dog into the human body. Conversely, the very phrase 'mad dog' suggested the conferment of human-like qualities onto the dog, if madness is understood in traditional terms as the loss of rationality, an attribute which is customarily only associated with human life. In *Wuthering Heights*, Heathcliff is figured as moving into a state of rabidity. Nelly describes his meeting with the dying Cathy in sensationalised terms: 'he gnashed at me, and foamed like a mad dog, and gathered her to him with greedy jealousy. I did not feel as if I were in the company of a creature of my own species' (p. 141). Passion, which transgresses social bounds, is figured as the crossing of a species barrier. Victor, with his 'glittering' dark eyes, unfortunate susceptibility to passion, and 'grinding of his teeth' (another form of 'moral teething') is a proto-Heathcliff (p. 245). The dog, which succumbs to disease when taken to the streets of the local town, is an ominous warning of what might lie ahead for his master.

Crimsworth, like Agnes, presents his action of shooting the dog as the only humane thing to do. His son, however, challenges this verdict: 'He might have been cured – you should have tried – you should have burnt the wound with hot iron, or covered it with caustic' (p. 243). These are, according to Elizabeth Gaskell, precisely the actions followed by Emily Brontë in her own life when bitten by a suspect dog, and also those attributed to Shirley Keeldar in Charlotte Brontë's later novel.[34] Victor is given a surprisingly knowledgeable and indeed persuasive reply which highlights the weakness of Crimsworth's defence and turns us rather to the symbolic function of the scene. It is placed in the context of discussions of Victor's upbringing which anticipate the debates about child-rearing in *The Tenant of Wildfell Hall*. Victor is the subject of a tripartite struggle: Hunsden fears that Frances's maternal caresses will 'make a milksop' of the boy; while Crimsworth, in turn, fears that Hunsden will encourage the 'ominous sparks' in his nature which must be crushed out of him:

> [T]he lad will some day get blows instead of blandishments – kicks instead of kisses – then for the fit of mute fury which will sicken his body and madden his soul – then for the ordeal of merited and salutary suffering – out of which he will come (I trust) a wiser and a better man. (p. 245)

Victor, the only well-loved son of a middle-class family, is predicted the life of a Heathcliff; beaten and kicked, and raging with an inner fury which, in sickening his body and maddening his soul, is portrayed as a form of psychological and spiritual rabies, destroying not only body and mind, but the ultimate mark of humanity, the soul. The terror of infection runs deep. Crimsworth resolves to send his son to Eton, knowing that he will find it hard to endure, in the hopes that it will 'ground him radically in the art of self-control' (p. 245). Crimsworth's platitudes are undermined by his own language. The prim assertion of the values of self-control is undercut by the weight of hesitancy in that bracketed phrase, 'I trust', and the text's evident desire that this 'mute fury' be given expression, unleashed upon the world, in Heathcliff-like scorn, rather than turned inward to sicken soul and body.

The three scenes of violence against animals I have examined take very different forms. The latter two are alike in claiming high-minded (if questionable) motivations, and are also the only scenes where the animals actually die. Heathcliff, significantly, is not directly responsible for animal murder – rather, his hanging of Isabella's dog signals his crossing over to inhabit, in self-disgust, the civilised world where people enjoy animal torture. Of the three texts, *Agnes Grey* is the most conventional. Agnes is the mouthpiece for many of the moral and religious pieties of the era with regard to animals, yet the violence of the crushing of the birds exposes the duality at the heart of this discourse: kindness to animals is enjoined as a means of crushing the feared 'animality' of the children themselves. The term animal in this context thus denotes both lower forms of creation which need to be protected in their helplessness, and also the attributes of children which need to be ruthlessly crushed. Agnes's act of 'humanity', in this case, is fuelled not by higher ethical motivations, but by anger, mounting to hatred, and fear. In *The Professor*, the dynamics of the scene are reversed, and its symbolic implications writ large. Charlotte Brontë makes clear that the shooting of the dog is a metonymic projection of the violence to be meted out to Victor as his 'electrical ardour' is curbed, constrained and excised. Passion in a child, like rabies, is an infection that must be exterminated. In this case, however, sympathy lies not with the adult, but with the child and his canine alter ego.

Wuthering Heights is undoubtedly the most complex of the three texts. Neither a celebration of nature's fierce laws, nor an endorsement of Victorian sentimental views of the animal realm, it destabilises divisions between animal and human, while tearing up the cultural scripts of child development. By hanging Isabella's dog, Heathcliff mockingly places himself within the Hogarthian model, while signalling his ironic distance from this narrative. Far from sinking to the gallows and dissection table, he operates explicitly within the confines of the law to enact his revenge and rise to the status of a landowning gentleman. All the animal epithets heaped upon him are not so much indications of his 'nature' as expressions of the bafflement and fear of those caught in his track. Unlike the other two texts, *Wuthering Heights* actively traces a child's path of development. Neglected and spurned in childhood, Heathcliff takes on in adulthood the refined cruelties of civilised society, as encapsulated in the pleasure he anticipates in conducting a 'vivisection' of Cathy and Linton. Heathcliff is probably the first 'vivisector' in literature and, thus, the head of a long line of fictional representations stressing the active cruelty of experimental physiology as enacted on the animal body. By employing this technical and little-used term, Emily Brontë indicates her keen engagement with the ethical and social debates in relation to science emerging at this time. This 'wolfish' and 'monstrous' figure is both heroic and despicable: in actively contemplating vivisection (albeit figuratively), Heathcliff seals his ascent into civilised society, and the degradation of his humanity.

In all three texts, contemplation of what it is to be human is enacted on, and through, the body of the animal. Each of the scenes marks a moment of disruption in the text: violence to animals functions as a form of interrogation of the hierarchies of power, whether child over animal, adult over child, or male over female. They also explore the modes of pleasure which come with mastery. For all three characters, their act is an assertion of power, which places the perpetrator (particularly Heathcliff and Agnes) in a position they affect to despise: Heathcliff enters into bourgeois life, Agnes domineers over her employer's child, and Crimsworth confirms his ascent to social respectability by sacrificing his son to the halls and playing fields of Eton. The reader, participating in these scenes and their unfolding, is complicit in their violence; the texts explore the forms of sickened admiration which Sarah Stickney Ellis expresses for those animal-torturing, masterful boys who will go on to rule the domestic fireside, the shires, and the British empire.

Notes

1 There have been various excellent studies addressing aspects of animal life in the Brontës' works, including Barbara Munson Goff, 'Between Natural Theology and Natural Selection: Breeding the Human Animal in *Wuthering Heights*, *Victorian Studies*, 27:4 (1984), 477–508; Lisa Surridge, 'Animals and Violence in *Wuthering Heights*', *Brontë Society Transactions*, 24:2 (1999), 161–73, and Ivan Kreilkamp, 'Petted Things: *Wuthering Heights* and the Animal', *Yale Journal of Criticism*, 18:1 (2005), 87–110.
2 See www.tate.org.uk/whats-on/tate-britain/exhibition/hogarth/hogarth-hogarths-modern-moral-series/hogarth-hogarths-4. Christine Alexander and Jane Sellars record in *The Art of the Brontës* (Cambridge: Cambridge University Press, 1995) that Branwell had made a copy of Hogarth's 'Idle Apprentices' in 1829, which suggests that at least some of Hogarth's prints were available in the Parsonage.
3 For an excellent discussion of the founding and development of the RSPCA, see Harriet Ritvo, *The Animal Estate: The English and Other*

Creatures in the Victorian Age (Cambridge Mass; Harvard University Press, 1987), chapter 3; James Turner, *Reckoning with the Beast: Animals, Pain, and Humanity in the Victorian Mind* (Baltimore: Johns Hopkins University Press, 1980). For further details on the impact of the Hogarth see Stephen F. Eisenman, *The Cry of Nature: Art and the Making of Animal Rights* (London: Reaktion Books, 2013) and Piers Beirne, *Hogarth's Art of Animal Cruelty: Satire, Suffering and Pictorial Propaganda* (Basingstoke: Palgrave Macmillan, 2015).

4 In *Bleak Houses: Marital Violence in Victorian Fiction* (Athens, OH: Ohio University Press, 2005) Lisa Surridge highlights the parallels between 'Beaten Animals and Beaten Wives', and the role of the law of couverture, pp. 86–95.

5 Vivisection is also discussed by Matthew Beaumont in 'Heathcliff's Great Hunger: The Cannibal Other in *Wuthering Heights*', *Journal of Victorian Culture*, 9:2 (2004), 137–63, where he links it, interestingly, to cannibalism (p. 154), and also Kreilkamp – although neither picks up on the innovative usage.

6 In her delirium Cathy speaks of the lapwing's nest, 'full of little skeletons' because Heathcliff had set a trap over it and the parents dared not feed their young (p. 108). There is, as many critics have noted, significant symbolism here with regard to Heathcliff's actions in the social and domestic sphere, as a 'cuckoo', despoiling nests. He is never shown, however, deliberately torturing animals. See Joseph Carroll, 'The Cuckoo's History: Human Nature in *Wuthering Heights*', *Philosophy and Literature*, 32:2 (2008), 241–57.

7 Goff, p. 498.

8 Electronic resources have made possible extensive searches of the available literature of the time, although it is important to note that the digitised material is still very partial, and particularly when it comes to more ephemeral materials such as pamphlets, or more local publications. Google Books shows virtually no citations of vivisection before the 1840s, and then very limited indeed, until the 1860s. The accuracy of its metadata, however, is very poor, with the wrong date (often by decades) attached to a significant level of material, thus some attributions for the 1820s, for example, are actually from the 1870s. Searches have also been conducted on the Brontës' favourite periodicals, *Blackwood's Edinburgh Magazine*, and *Chambers's Journal* as well as in the Proquest collection, 'British Periodicals', the Gale 'British Newspapers' and 'Nineteenth Century UK Periodicals' collections more generally, and the *Times Archive*.

9 Ritvo, p. 131.

10 Ibid., at pp. 157–8. Richard French, in his detailed study, *Antivivisection and Medical Science in Victorian Society* (Princeton: Princeton University Press, 1975) similarly argues that, 'During the first thirty years of the Society's operations the issue of animal experimentation slowly emerged from a position of concern but little urgency to a relatively significant priority', p. 27. The public lectures and demonstrations on living animals of

the French physiologist, François Magendie, in London in 1824 had given rise to a discussion in parliament, led by Martin, the sponsor of the 1822 bill against animal cruelty, between 24 February and 11 March 1825, but as French notes, he could find no further debate on vivisection until the major discussions of 1875 (French, p. 25).

11 French notes that despite the extension in 1835 of the law against cruelty to animals to encompass cats and dogs and other smaller animals, there were no records of prosecutions in this area until 1874 (pp. 28–9).

12 Revd. John Styles, *The Animal Creation: Its Claims on our Humanity, Stated and Enforced* (London: Thomas Ward and Co, 1839), p. 89.

13 Ibid., at p. 99.

14 In his extensive annotations of his copy of Thomas John Graham's *Modern Domestic Medicine* (London: Simpkin and Marshall et al., 1826) Patrick Brontë frequently compares the advice being offered to that in Elliotson, which is most probably John Elliotson, *Human Physiology* (London: Longman, Rees, Orme, Brown, Green and Longman, 1835).

15 Quoted in Styles, p. 98. Elliotson notes that this research was awarded the physiological prize from the French Institute in 1826. If Brontë had read this section of the Elliotson, she would also have found that the discussion was embedded in a larger discussion of how Brachet had maimed and then masturbated a Tom cat in order to understand how a paraplegic man had sired children. As Elliotson comments, 'This was what some bluff John Bulls would call French taste' (p. 449).

16 William Hamilton Drummond, *The Rights of Animals, and Man's Obligation to Treat them with Humanity* (London: John Mardon, 1838), p. 154.

17 David Mushet, *The Wrongs of the Animal World. To Which is Subjoined the Speech of Lord Erskine on the Same Subject* (London: Hatchard and Son, 1839), p. 211. Lord Erskine had proposed the first bill on Cruelty to Animals, in the House of Commons on 15 May 1809. The bill was defeated on that occasion.

18 'Vivisection', *Times*, August 19, 1843, p. 7.

19 The *Times Archive* records only six uses of the word 'vivisection' before 1850, and this was the first to offer a substantive discussion of the issue, and the only time the term was used as an item heading.

20 'Vivisection', p. 7.

21 John Locke, *Some Thoughts Concerning Education*, J. W. Yolton and J. S. Yolton, eds. (Oxford: Clarendon Press, 1989), section 116.

22 Mary Wollstonecraft, *Original Stories from Real Life: With Conversations Calculated to Regulate the Affections, and Form the Mind to Truth and Goodness* (1788; new edn., London: J. Johnson, 1796), p. 6. For a discussion of this literature see Tess Cosslett, *Talking Animals in British Children's Fiction* (Aldershot: Ashgate, 2006); Elizabeth A. Dolan, *Seeing Suffering in Women's Literature of the Romantic Period* (Aldershot: Ashgate, 2008); David Perkins, *Romanticism and Animal Rights* (Cambridge:

Cambridge University Press, 2003); Moira Ferguson, *Animal Advocacy and Englishwomen, 1780–1900: Patriots, Nation and Empire* (Ann Arbor: University of Michigan Press, 1998).
23 Samuel J. Pratt, *Pity's Gift: A Collection of Interesting Tales to Excite the Compassion of Youth for the Animal Creation.* Selections from Pratt by a Lady (London, 1798, printed for T. N. Longman and E. Newbery), p. 61.
24 Pratt, *Pity's Gift*, p. 116.
25 Mary Elliott, *The History of a Goldfinch* (London: W. and T. Darton, 1807). Extracted in Andrew W. Tuer, *Stories from Old-Fashioned Children's Books* (London: Leadenhall Press, 1899–1900), p. 211. The text was frequently reprinted during the century, both in England and the United States.
26 Mary Elliott, *The History of a Goldfinch: Addressed to those Children who are Dutiful to their Parents, and Humane to their Fellow Creatures* (Philadelphia: B. and T. Kite, 1807).
27 The accusation is repeated on p. 152.
28 William Cowper, 'The Task' Bk VI in *Life and Works of William Cowper*, Revd. T. S. Grimshawe, ed. (London: Saunders and Otley, 1836), Vol. 7, p. 118. This section, for example, was printed in the *Saturday Magazine* (31 August 1833) under the title 'Cruelty to Animals' (p. 78). The lines on the worm serve as an epigraph to 'Humanity to Animals Recommended' in *Society for Prevention of Cruelty to Animals. Short Stories.* (London, 1837).
29 Cowper, 'The Task', p. 118.
30 Sarah Stickney Ellis, *Mothers of England, their Influence and Responsibility* (London: Fisher, Son and Co., 1843), chapter 10, p. 304.
31 Ibid., at pp. 302–4.
32 I have touched on this point briefly in my introduction to the Oxford World's Classics edition of *Agnes Grey*, which also places the preoccupation with the relation of the child and animal in the context of medical psychiatry of the period.
33 Ritvo, *Animal Estate*, pp. 167–202; Neil Pemberton and Michael Worboys, *Rabies in Britain: Dogs, Disease and Culture, 1830–2000* (Basingstoke: Palgrave Macmillan, 2007). See pp. 56–60 for a short section on the Brontës.
34 See Elizabeth Gaskell, *The Life of Charlotte Brontë* (London: Dent, 1958), p. 184, and Juliet Barker, *The Brontës* (London: Weidenfeld and Nicolson, 1994), p. 198.

The research leading to these results has received funding from the European Research Council under the European Union's Seventh Framework Programme ERC Grant Agreement number 340121.

CHAPTER 2

Learning to Imagine
The Brontës and Nineteenth-Century Educational Ideals
Dinah Birch

Educational Change

Changes in the theory and practice of education throughout the first half of the nineteenth century were intimately associated with contested interpretations of human progress and with divided views on the role of the imagination in the development of the individual. The roots of these educational conflicts ran deep, for the motives for establishing new models for learning were never separate from its political, economic and religious arguments of the period; and they were central to the varieties of late Romanticism that formed its literary culture. The debates find complex and often divergent expressions in the writings of the Brontë siblings, emerging from the experiences of a family where education was both an inspiring ideal and the family business. For the Brontës, education was above all the means to the self-determination they craved, in both practical and creative terms. Congratulating W. S. Williams on his daughter's admission to the pioneering Queen's College in 1849, Charlotte Brontë is particularly forthright on the question. 'Come what may afterwards, an education secured is an advantage gained – a priceless advantage. Come what may it is a step towards independency'.[1] The family was not, however, of one mind in their understanding of what a condition of 'independency' might mean, and their disagreements reflected some of the sharpest disputes of their generation.

The pressure to manage ruptures in class and gender in an industrialising economy was often what stimulated the drive for educational reform. Schooling was widely seen to be key to moral and intellectual self-improvement, but it was also the bedrock of social cohesion and a necessary basis for the development of an effective workforce. These were among the fundamental tensions that shaped the educational arguments of the turbulent 1830s and 1840s, at a time when there was increasing anxiety about the potential association between ignorance and insurrection, particularly among the working poor. An educated population was widely recognised as a necessary foundation for stability and progress, while teaching the ignorant was also understood to be an ethical or spiritual duty. Here the imperatives of economic and cultural politics overlapped with those of religion, as they often did in the Brontë family. Dr James Kay (later Sir James Kay-Shuttleworth), who became one of the most influential educational reformers of the nineteenth century, was convinced that withholding a serious education from the poor would increase the risk of revolution, a fear that was beginning to disturb the governing classes. His widely read pamphlet on 'The Moral and Physical Condition of the Working Classes Employed in the Cotton Manufacture in Manchester' (1832), later cited in Friedrich Engels's *The Condition of the Working-Class in England* (1845), urged far-reaching change on the grounds of both principle and pragmatism:

> If a period ever existed when public peace was secured by refusing knowledge to the population, that epoch has lapsed. The policy of governments may have been little able to bear the scrutiny of the people. This may be the reason why the fountains of English literature have been sealed – and the works of our reformers, our patriots, and our confessors – the exhaustless sources of all that is pure and holy, and of good report[2] amongst us – *have not been made accessible and familiar to the poor*. Yet literature of this order is destined to determine the structure of our social constitution, and to become the mould of our national character; and they who would dam up the flood of truth from a lower ground, cannot prevent its silent transudation. A little knowledge is thus inevitable, and it is proverbially a dangerous thing.[3] Alarming disturbances of social order generally commence with *a people only partially instructed*. The preservation of *internal peace*, not less than the improvement of our national institutions, depends on the education of the working classes.[4]

Many agreed with Kay, seeing the spread of education as a matter of both moral duty and social expediency. But there was no consensus as to the means by which this acknowledged good was to be achieved. Opinions on how an effective education was to be delivered were sharply divided. Quarrels about the conflicting responsibilities of church (or chapel) and state in supporting education were increasingly bitter. Reformers promoted widely divergent ideas as to what should be seen as a good school, or a competent teacher, or a worthwhile course of study. Kay's primary appeal in warning of the dangers represented by an ignorant population is to the self-interest of his middle-class readers, but his Biblical and cultural references to the 'fountains of English literature' and the 'exhaustless sources of all that is pure and holy' imply that the kind of education he assumes in his readers – and has in mind for the working classes – is not just the acquisition of literacy. 'The poor man will not be made a much better member of society, by being only taught to read and write'.[5] Should schooling be designed to enable individuals to develop to their fullest human potential? And how far was it right that this potential for a measure of intellectual autonomy would be defined by social class? Should its aim be to fit young people to make the most efficient contribution to national prosperity, or perhaps simply to earn a living? Or was the real objective of education to impose conformity and inculcate discipline among growing children of all classes, ensuring that they would be content to accept the roles allocated to them within a settled social hierarchy? Could it, properly managed, deliver more than one of these objectives, or perhaps all of them? These questions were particularly pointed in the case of the Brontë family, where the responsibilities and privileges of gentility co-existed with the pressures of poverty and with the pressing need to use education as the means to secure an income, while extensive reading in Romantic poetry and fiction created a strong belief in the sustaining power and autonomy of the life of the imagination.

The urgency of these arguments was, in part, a consequence of the rapidity of industrial and urban development in early nineteenth-century Britain. This was a process that the Brontës experienced first-hand, as Patrick and Maria Brontë left the rural or provincial communities of their childhood and early youth to bring up their family in Haworth, a small industrial mill town in Yorkshire where smoking chimneys stood against the background of wild moors. The history of Haworth in the first half of the nineteenth century, as the town struggled to respond to the challenges of industrial growth, reflects the transformational forces that were sweeping through the country. From the perspective of the settled land-owning classes, traditional patterns of agricultural labour among the rural poor, punctuated by weekly attendance at an Anglican village church, had seemed to call for only the most rudimentary level of education. But the arrival of factories, mines and mills might call for different and more exacting levels of preparation for a competitive adult life for both workers and the expanding middle classes who owned and managed these new industrial ventures. Old social structures and the educational assumptions that went with them were fragmenting and the form that their replacements should take was far from clear.

Educating Women

For Charlotte, Emily and Anne Brontë, these issues were further complicated by their status as young ladies. In the early years of the nineteenth century, working-class girls were thought to need even less in the way of serious education than their brothers. Anything beyond a basic instruction in religious principles – and perhaps in the essentials of literacy – would be redundant. But the Brontë girls were the daughters of a clergyman, and it had long been accepted that something more was necessary for young women with claims to gentility. As both the Anglican church and non-conformist sects began to take their social responsibilities more seriously in the early decades of the nineteenth century, the education of the wives and daughters of churchmen assumed a special importance – not only because relative poverty often meant that (as was true for the Brontë sisters) they had to be especially careful to safeguard their status as ladies, but also because they would be expected to be in a position to teach the poor.[6] However, the question of what might be required for the education of women with claims to gentility was a matter of persistent controversy. In this respect, the immediate demands of a society undergoing rapid and fundamental change collide with the philosophical and political arguments that had first taken shape in earlier decades.

Here too, arguments would often turn on the competing claims of an instrumentalist view of education and a more aspirational model. Protests against the moral and intellectual triviality and practical futility of the education conventionally provided for middle-class women were longstanding. The first publication of Mary Wollstonecraft, the most eloquent and influential of an early generation of feminists, was a conduct book focusing on the issue of women's education. *Thoughts on the Education of Daughters: With Reflections on Female Conduct, in the More Important Duties of Life* (1787) argued for rationality and a firm morality as the needful basis for the education of girls: 'Indolence and a thoughtless disregard of everything except the present indulgence make many mothers, who may have momentary starts of tenderness, neglect their children. They follow a pleasing impulse and never reflect that reason should cultivate and govern those instincts which are implanted in us to render the path of duty pleasant'.[7] Wollstonecraft went on to identify women's education as a central issue in her *A Vindication of the Rights of Woman: with Strictures on Political and Moral Subjects* (1792):

> The education of women has, of late, been more attended to than formerly; yet they are still reckoned a frivolous sex, and ridiculed or pitied

by the writers who endeavor by satire or instruction to improve them. It is acknowledged that they spend many of the first years of their lives in acquiring a smattering of accomplishments: meanwhile strength of body and mind are sacrificed to libertine notions of beauty, to the desire of establishing themselves—the only way women can rise in the world—by marriage.[8]

Wollstonecraft's arguments for a 'rational' education as an indispensable precondition for the progress of her sex were essential to her feminism and they continued to be important to the cause of progressive reform. But they were always accompanied by an insistence that reason would support the fulfilment of duty in a woman's life. She was no advocate for self-cultivation for its own sake. The exercise of reason would make women more responsible and productive as full and active members of society; it would not promote what she disapprovingly terms 'doting self-love'.[9]

The issue was not, however, the preserve of radical thinkers. More conservative or religious figures were often equally dissatisfied with the superficiality of the education routinely provided for middle-class girls. Jane Austen was scathing about the pretensions of schools that preyed on ambitious families – establishments which, as described by Austen in 1815, 'professed, in long sentences of refined nonsense, to combine liberal acquirements with elegant morality and new systems – and where young ladies for enormous pay might be screwed out of health and into vanity'.[10] In her *Strictures on the Modern System of Female Education, with a View of the Principles and Conduct Prevalent Among Women of Ranks and Fortune* (1799), the evangelical reformer Hannah More took an explicitly religious view of the need for a more rigorous approach to the education of young ladies. Like Wollstonecraft, she deplored the contradictions in a system that actively encouraged thoughtless behaviour in women, and then condemned them for their light-mindedness. 'It is a singular injustice which is often exercised against women, first to give them a very defective Education, and then to expect from them the most undeviating purity of conduct.'[11] More wrote with Christian purpose. Her views on women's education balance the nurturing activities that she defined as essential to womanhood with a recognition – essential to her evangelical principles – of the need for disciplined self-culture. Intellectual independence would enable women to carry out duties more effectively. For a woman, 'the great uses of study are to enable her to regulate her own mind, and to be useful to others'.[12] Like many among the legions of would-be educational authorities that followed her, she emphasised that her aim was not simply to theorise about women, but to suggest an approach to their education that would 'qualify them for the practical purposes of life'.[13] One of these 'practical purposes' might be a capacity to teach. More's *Moral Sketches of Prevailing Opinions and Manners, Foreign and Domestic: With Reflections on Prayer* (1819) found its way into the library of the parsonage at Haworth, where it was heavily annotated by Patrick Brontë. More argued that women whose education had not been confined to 'useless accomplishments' could and should become educators: 'Ladies, whose own education not having been limited to the harp and the sketch-book, though not unskilled in either, are competent to teach others what themselves have been taught.'[14]

Women might be active as teachers either in the family setting or elsewhere, but it was widely assumed that the destiny of middle-class girls, or young ladies, was primarily domestic. Wollstonecraft's tart observation that their only means of rising in the world was through marriage remained largely true. Once married, their work would be in the home, where their responsibilities as the lofty guardians of morality would be combined with the mundane day-to-day duties of running a household. The work of a female teacher seemed a natural extension of this double identity, though in practice its uncertain social status gave rise to many of the dislocating tensions within competing models for the proper schooling of women.

The Brontës as Teachers and Pupils

The transformative power of education lay at the heart of the lives and ambitions of the Brontë family. Born in 1777, Patrick Brontë escaped a background of rural poverty in Ireland by working as a teacher and then gaining the degree from the university of Cambridge that enabled him to launch his career as an Anglican clergyman. His commitment to education had provided him with intellectual and spiritual authority and a gentleman's profession. It was an impressive achievement. His work as an educator also won him a wife. He met Maria Branwell – the Cornish woman he was to marry – in 1812 while acting as an examiner at the Wesleyan Academy, a school for the sons of Methodist ministers in Yorkshire. Maria was the niece of the headmaster. Much of Patrick's attraction for the warm-hearted Maria lay in his firm but affectionate authority as a teacher. Maria revered him as a man who would assume the role of a 'guide and instructor'[15] in her life. He was dependable, but his story of determined success in the face of adversity also gave him a heroic glow in her eyes. Not only was Patrick a teacher and a clergyman, he also became a published author with a significant body of poetry and fiction to his name. The marriage was happy, but after giving birth to six children in quick succession Maria died (probably of uterine cancer) in 1821. After the loss of her mother and later of her sisters Maria and Elizabeth, Charlotte Brontë became the oldest survivor in the close family circle of four surviving siblings. She assumed, in part, the role of the children's lost mother. Like Maria Branwell, Charlotte came to see Patrick as a hero, who had built a life for himself and his family through his mastery of learning. She was determined to follow suit. Branwell, Emily and Anne were more ambivalent in their responses to their father's daunting example. These family dynamics figure largely in the Brontës' differing interpretations of the relation between the instrumental utility of educational discipline and the intrinsic value of imaginative autonomy.

Charlotte's admiring devotion to her father meant that the concept of educational success that she internalised in childhood reflected the experiences of both her mother and her father. Like Maria Branwell, she was inclined to romanticise the relation between the powerful and benevolent male teacher and his grateful and adoring female pupil, an impulse which caused her much suffering when she fell in love with Constantin Heger, a married teacher she encountered in Brussels. Like Patrick, she identified the determined pursuit of success in education with personal fulfilment and professional opportunity. In Charlotte's case, however, this was a pattern that was shadowed with loss. Her experiences at the austere Clergy Daughters' School in Cowan Bridge in Lancashire, unforgettably recalled in her descriptions of Lowood School in *Jane Eyre*, left both physical and mental scars. Her two older sisters, Maria and Elizabeth, did not survive their time at the school, both succumbing to tuberculosis – the family plague that was also, in later years, to kill her two younger sisters, Emily and Anne. Charlotte's later years at Margaret Wooler's school at Roe Head, followed by work as a private governess and by her difficult but life-changing experiences as a pupil-teacher at the Pensionnat Heger in Brussels, provided the material of much of her mature fiction. Her developing use of this material reflects her ambivalent understanding of the educational institutions that had shaped so much of her early life.

Jane Eyre (1847) traces its heroine's steady progress from her passive and often unhappy years as a schoolgirl to her final identity as a woman for whom teaching had provided access to 'independency', security and finally romantic fulfilment. In Lowood School, Charlotte Brontë draws on her memories of both the Clergy Daughters' School and Roe Head. Her representation of Jane Eyre's later experiences as a governess draws on her own largely unhappy and frustrating experiences as a teacher at Roe Head and the emotional intensity of her life at the Pensionnat Heger. But these were not her only experiences of learning. Her earliest formation as a pupil, like that of most middle-class girls of her generation, had been in the context of the family. As a daughter of the household, she was taught alongside her siblings by her father and aunt. This was as important to her adult identity – and that of her sisters – as her years in the classroom. The home education of the Brontë children was a conventional one in many respects, formed on traditional definitions of gender. Patrick Brontë had accepted the commonly held view that the rigours of a classical education were suited to boys rather than girls. In Patrick's didactic novel *The Maid of Killarney*, published in 1818, the pompous Dr O'Leary observed that 'The education of a female ought most assuredly, to be competent, in order that she might enjoy herself, and be a fit companion for man. But, believe me, lovely, delicate and sprightly woman, is not formed by nature, to pore over the musty pages of Grecian and Roman literature, or to plod through the windings of Mathematical Problems, nor has Providence assigned for her sphere of action, either the cabinet or the field. Her forte is softness, tenderness and grace'.[16] Dr O'Leary was not Patrick Brontë, and Patrick's girls received some tuition in Latin and ancient history. But he did choose to focus the home education of his son Branwell primarily on the classics, while his daughters concentrated on reading and writing, scripture, sewing, geography, history, mathematics and a little French.

Though he felt that girls should not be educated in the same way as their brothers, Patrick's approach to his daughters' education allowed them liberal access to resources often denied to girls of their generation. He used standard textbooks like the Revd J. Goldsmith's lively *A Grammar of General Geography for the Use of Schools and Young Persons*,[17] Oliver Goldsmith's *History of England*, Charles Rollin's *Ancient History* and Richmal Mangnall's ubiquitous *Historical and Miscellaneous Questions*, but he also permitted all of his children to read Shakespeare and Byron in unabridged texts. As an accomplished poet with an enduring respect for the authority and vitality of poetry, he raised no objection to his daughters reading the poetry of Cowper, Campbell, Southey, Wordsworth and Scott. Periodicals – like the Tory journal *Blackwood's Edinburgh Magazine* (known as 'Maga') and *Fraser's Magazine* – gave the family access to provocative contemporary writing and acted as a source of information about the political controversies of the day. The cultural life of Haworth was active and the family frequently went to hear religious speakers, lecturers and musicians. The children's home education equipped them with much richer and more challenging intellectual experiences than they would have encountered in most girls' schools of the period. Later, Charlotte's friend Mary Taylor wrote to Elizabeth Gaskell about Charlotte's exceptionally well-stocked mind and her eager appetite for learning:

> She would confound us by knowing things that were out of our range altogether. She was acquainted with most of the short pieces of poetry that we had to learn by heart; would tell us the authors, the poems they were taken from, and sometimes repeat a page or two, and tell us the plot ... She picked up every scrap of information concerning painting, sculpture, poetry, music, &c., as if it were gold.[18]

Charlotte and Branwell, the two elder Brontë siblings, had a close and competitive relationship. They took the lead in the children's precociously active engagement with the issues of the day. It was Branwell who, at the age of eleven, initiated the miniature family journal, 'Branwell's Blackwood's Magazine', which first introduced the idea that writing for the public was an ambition that could become a reality. Later, Charlotte took over the production of the tiny journal, renaming it 'Blackwood's Young Men's Magazine' and providing it with a more imaginative range of subjects. The two younger girls, Emily and Anne, formed their own alliance within the intense family life of Haworth Parsonage. All four children participated in the construction of elaborate fictional worlds, where their reading in literature, history and politics was reflected in the narratives of Gondal and Angria. For the Brontë family, the boundary between the processes of education and creativity were always porous. They had absorbed the traditional values and practical skills communicated in the lessons provided by their father and aunt, but their early writing allowed them the freedom to challenge those values and create their own forms of narrative.

In this sense, education enabled the children to achieve their own imaginative autonomy. As the three Brontë girls grew, however, it became increasingly clear to their father that the nature of their education could not sensibly be dictated entirely by the needs of their personal development, or even their qualifications to succeed in a competitive marriage market. Like their brother Branwell, they would have to become financially independent. The family income, dependent on the salary that Patrick could earn as perpetual curate at Haworth, was scanty (£170 p.a.) and there could be no guarantee that the girls would acquire security through finding a prosperous husband. This was a widely shared concern among the rising middle classes in the early decades of the nineteenth century. Harriet Martineau's *Household Education* (1849) emphasised the need for young women to experience a wide-ranging and rigorous curriculum, pointing out that 'female redundancy'[19] was a growing threat to their future security:

> In former times, it was understood that every woman (except domestic servants) was maintained by her father, brother or husband; but it is not so now. The footing of women is changed, and it will change more ... What we have to think of is the necessity – in all justice, in all honour, in all prudence – that every girl's faculties should be made the most of, as carefully as boys'. While so many women are no longer sheltered, and protected, and supported, in safety from the world (as people used to say) every woman ought to be fitted to take care of herself.[20]

One of the reasons for Patrick's luckless decision to send his daughters to the Clergy Daughters' School was that the syllabus would equip the girls with the accomplishments expected of teachers, or governesses – the only profession that could reasonably be expected to provide them with a livelihood while maintaining their status as gentlewomen. The prospectus of the school made it clear that it was prepared to cater for this requirement, increasingly common among hard-pressed clerical families looking to equip their daughters to make their own way in the world: 'If a more liberal Education is required for any who may be sent to be Educated as Teachers and Governesses, an extra charge will probably be made.'[21] The school had a distinguished list of patrons – including Hannah More and William Wilberforce – and though the fees were exceptionally low at £14 per year, this was partly because they were subsidised by charitable donations. Patrick paid the extra required to educate the girls and prepare them to teach. Throughout their early lives, it was understood that Charlotte, Emily and Anne would be equipped to earn their living as teachers, either in schools or as private governesses working for families. Though Branwell was educated with different objectives in mind, he too followed his father in working as a tutor for a well-to-do family. For good or ill, it was understood that education, the family occupation, was the natural destiny and resource of the Brontë family.

These circumstances established significant tensions in the minds of the young Brontës. On the one hand, their faith in the values they had absorbed from their reading, the aspirational culture of Haworth and the example and teaching of their father was strong. These features were particularly marked in Charlotte's development as a young woman.

The disciplined acquisition of a carefully regulated and well-informed mind of the kind that a serious education could provide would be the key to their status in the world and their prospects for advancement. On the other hand, the material of their education, particularly when it took the form of independent reading and writing, argued for a more autonomous form of self-determination. The rebellious poetry of Byron and Scott, those two great favourites of the Brontë household, did not encourage the children to think of meek obedience and social conformity as the highest points of human achievement. A steady application to study was a potent model in their lives, but so too was Romantic self-determination.

These divided ambitions became particularly evident in the experiences of Charlotte and Emily as pupils in the Pensionnat Heger, a girls' school in Brussels. Charlotte was the driving force in the scheme to study overseas and her primary motivations were practical and professional. If she and her sisters were to establish a successful school for girls, as she hoped they might, they would need a competitive advantage; for, as Charlotte wrote to her aunt, 'schools in England are so very numerous, competition so great, that without some such step towards attaining superiority we shall probably have a very hard struggle, and may fail in the end'.[22] She and Emily would need the polish that fluent French and an understanding of Continental manners and culture could provide. She also saw the experiment as an emulation of her father's spirit and courage: 'When he left Ireland to go to Cambridge University, he was as ambitious as I am now.'[23]

Emily was a partner in the sisters' plan to found a school, but there is no evidence that she or Anne shared Charlotte's enthusiastically practical commitment to the project. Speculating on the family's future in her diary in 1841, Emily's imagined picture of the new school has the air of a rosy fantasy: 'we (i.e.) Charlotte, Anne and I – shall be all merrily seated in our own sitting-room in some pleasant and flourishing seminary, having just gathered in for the midsummer holydays.'[24] Neither Emily nor Anne seem to have been greatly disappointed when the proposed school failed to attract pupils (though Anne recalls the scheme in the successful family school for girls that she describes in *Agnes Grey*), and the plan was abandoned. Even Charlotte, who had initiated the scheme, was quickly resigned to its collapse. The school that might have provided the sisters with the independence they needed did not come into existence, but the legacy of the professional education that Charlotte and Emily acquired in Belgium remained a powerful force in their lives and writing.

The prospectus of the Athénée Royal, the fashionable boys' school in Brussels where Constantin Heger taught, described an establishment with values very different to those of the Clergy Daughters' School: 'The instructor must activate the thinking of his student, develop his intelligence, rectify his judgement, arm him with good principles, and set before those precious materials that provide him with the record of the human spirit: after that, it is up to the student to build the edifice.'[25] Heger's wife Zoë Parent-Heger was head teacher at the Pensionnat Heger, where middle-class girls were taught. Her school operated on more cautious principles, making the inculcation of good behaviour and religion the primary selling points for prospective parents, rather than the active self-determination advocated by the Athénée Royal. Constantin Heger combined his duties at the Athénée Royal with a post as professor of rhetoric and literature in his wife's school, where he helped to design the curriculum. He was able to import some of the innovative educational methods of the Athénée Royal into his practice at the generally more conservative Pensionnat Heger. He encouraged critical independence in his pupils, reading passages aloud from French texts to his pupils and then analysing 'the parts with them, pointing out in what such and such an author excelled, and where were the blemishes', following with an assignment in which they were to express 'their own thoughts in a somewhat similar manner'.[26] Both Emily and Charlotte benefited from this stimulating regime, but Charlotte found the disciplined self-examination that it encouraged, together with the expansion of her range of reading, particularly useful to the development of her writing. Heger took her work seriously, praising its strengths and berating her if an assignment was below par. For Charlotte, the fusion of personal interest and intellectual challenge represented in Heger's teaching was overwhelmingly attractive, and she became deeply attached to her teacher. Reflections of this unreciprocated passion emerge repeatedly in her fiction – in Frances Henri's devotion to William Crimsworth in *The Professor*, Jane Eyre's love for Edward Rochester in *Jane Eyre*, Shirley Keeldar's marriage to Louis Moore in *Shirley*, or Lucy Snowe's adulation of Paul Emmanuel in *Villette*. This recurrent motif amounts to more than the obsessive recollection of a frustrated love. In each case, personal growth within an educational context culminate in both emotional and social fulfilment. The converging elements of Charlotte Brontë's varied educational experiences, intended to make her a teacher, gave her the confidence and the skill to become a writer.

Emily, habitually self-contained and self-protective, was less driven by the urge to make a public mark in the world than Charlotte. The education that meant most to her lay in her private reading and in the close creative relationship that she developed with her younger sister Anne. It is hardly surprising that her own brief attempts to teach, in Miss Patchett's school at Halifax, were unsuccessful, for 'she could not easily associate with others'.[27] Her writing makes little reference to the experiences of formal schooling. Yet in *Wuthering Heights* Emily tacitly concedes that the social and imaginative identity of her characters is formed by their experiences as children. Their real education is of a kind that most often takes place outside the schoolroom. For her, as for Charlotte, social considerations sit alongside an imaginative imperative in her interpretation of educational priorities. Heathcliff is denied a gentleman's education by Hindley, and it is this deprivation that leads to Catherine Earnshaw's resistance to becoming his wife. 'It would degrade me to marry Heathcliff, now', she tells Nelly Dean, despite her conviction that 'he's more myself than I am'.[28] Later, after leaving the Heights and mysteriously acquiring something of a gentleman's polish, Heathcliff attempts to deprive Hindley's son Hareton of an education as he had once been deprived. When Nelly encounters Hareton as a small child, she is met with 'a string of curses'. Nelly offers the boy an orange:

> 'Who has taught you those fine words, my barn,' I inquired. 'The curate?'
> 'Damn the curate, and thee! Gie me that,' he replied.
> 'Tell us where you got your lessons, and you shall have it,' said I. 'Who's your master?'
> 'Devil daddy,' was his answer.
> 'And what do you learn from Daddy?' I continued.
> He jumped at the fruit; I raised it higher. 'What does he teach you?' I asked.
> 'Naught,' said he, 'but to keep out of his gait – Daddy cannot bide me, because I swear at him.'
> 'Ah! And the devil teaches you to swear at Daddy?' I observed.
> 'Aye – nay,' he drawled.
> 'Who then?'
> 'Heathcliff.' (p. 97)

Here, the association between Heathcliff's demonic rebellion and the rejection of the publicly sanctioned education identified with the social and spiritual values of the church could hardly be more apparent. Asked whether the curate teaches him to 'read and write', Hareton replies that 'the curate should have his – teeth dashed down his – throat, if he stepped over the threshold. Heathcliff had promised that!' (p. 98). Heathcliff is a grotesque reflection of everything that Patrick Brontë, the dutiful perpetual curate of Haworth, is not. Catherine's daughter, a less mutinous version of her mother, is later able to woo Hareton with a domestic education that, like Charlotte Brontë's numerous depictions of the relation between teacher and pupil, is heavily eroticised. This is a kind of private learning that does not take place in a classroom, but it is only through its benign agency that the ancient bitternesses can begin to resolve themselves. Again, Nelly is the observer: 'I perceived two such radiant faces bent over the page of the accepted book, that I did not doubt the treaty had been ratified on both sides, and the enemies were, thenceforth, sworn allies' (p. 280). Emily's model for education rests on an independent inwardness that defies the regulation of the publicly affirmed and often religious values that dominated its institutional delivery.

Anne Brontë, whose development as a writer was formed by complex interactions with her siblings, is consistently concerned with education as it is enacted both inside and outside the home. She was not involved in the bold experiment of acquiring a continental education in Brussels, and in some respects her perspectives on the processes of moral education that dominate both *Agnes Grey* and *The Tenant of Wildfell Hall* remain largely English in their emphasis. Like Emily, she is not interested in the schoolroom as a location for fiction. Her uncompromisingly didactic novels ('All true histories contain instruction'[29]) insist on the primacy of a home education that is built on values of honesty, sympathy and charity. Anne has no truck with the seductions of the ambiguously attractive and quasi-Byronic heroes who figure in the novels of her sisters. For her, the inner life of the imagination must be mediated through the committed practice of evangelical religion. Her protagonists, Agnes Grey and Helen Huntingdon, do not waver in the confident self-belief derived from faith, which carries them safely through a sea of oppression, injustice and misfortune.

In this sense, Anne's novels might seem to be less emotionally and formally complex than those of her sisters. What gives them their distinctive force is the extent to which Anne is prepared to attack the social and cultural context in which the domestic education she describes takes place. Anne's interpretation of the cynical values that high-minded young women encounter when they lose the protection of their childhood homes was influenced and informed by her own difficult experiences as a teacher and governess. Her fiction is scathing in its condemnation of the hypocrisy and cruelty that seem to her to dictate the behaviour of both the rising middle class and the established aristocracy – the groups endowed with the financial resources that enabled them to dominate the social hierarchy. This is particularly true of Anne's first novel, *Agnes Grey* (1847), where the inexperienced but resolute Agnes is exposed to a protracted demonstration of the coarse self-interest of those who suppose themselves to be her natural superiors. Agnes's situation as a poor clergyman's daughter working for her living as a governess, endowed with the status of gentility without the economic resources to support her position, is a familiar example of the ambivalent position of the governess, frequently explored in mid-Victorian fiction.[30] Agnes, however, is no meek victim of her circumstances. Despite the abuse she suffers at the hands of her coldly unsympathetic employers, her sturdy resilience is more than equal to the challenge of her position. She is finally rewarded with the hand of the worthy curate Edward Weston, who despite his modest circumstances is understood to be a far more desirable figure than the ineffectual curate who repeatedly fails to deliver an education to the Earnshaws and the Lintons in *Wuthering Heights*. Of all Patrick's four surviving children, it was Anne who had most actively assimilated the evangelical values of Haworth parsonage.

Agnes's triumphant vindication in *Agnes Grey* bears the marks of fictional wish-fulfilment, for Anne's own experiences of working as a governess had been unhappy and no consoling curate had appeared to offer an escape. Charlotte, who had also found the daily demands of a menial teaching post to be onerous, sardonically referred to her sister's work as 'the Land of Egypt and the House of Bondage'.[31] Anne's second novel, *The Tenant of Wildfell Hall*, a more ambitious and complex work than the relatively slight *Agnes Grey*, concedes the scale of the challenge that confronts principled young women as they attempt to find their way in an unprincipled world. Helen Huntingdon is, in her own account, culpably naïve and more than a little vain and priggish in her belief that the example of her own Christian life will be sufficient to turn her husband away from his dissolute habits. 'I shall consider my life well spent in saving from the consequences of his early errors, and striving to recall him to the path of virtue'.[32] After learning, painfully, that she can have no influence over her husband's debauched behaviour, Helen removes herself and her young son from the marital home. Refusing the supposed sexual duties of a wife, she affirms her primary identity as that of a mother, tasked with the moral responsibility of rescuing Arthur from the corrupting influence of his father: 'henceforth, we are husband and wife only in the name ... I am your child's mother, and *your* housekeeper – nothing more' (pp. 260–1).[33] Helen works to support herself through her talents as an artist after she has succeeded in escaping from her marital home. In this respect she becomes one of the very few women with claims to gentility in the fiction of the Brontë sisters who attempt to establish themselves in any profession other than teaching. Nevertheless, she considers her priority to be the education of her son, not making a name for herself as a painter. Having failed to teach her husband, she is determined to teach Arthur. Like Emily Brontë's young Hareton, Arthur seems likely to be degraded by the deliberately debasing instruction of his drunken and self-indulgent father, who taught him 'to tipple wine like papa, to swear like Mr. Hattersley, and to have his own way like a man, and sent mamma to the devil when she tried to prevent him' (p. 296). In *Wuthering Heights*, Hareton is rescued from his fallen condition, and educated, through his love for the younger Cathy. Little Arthur is also saved from the dismal fate of his father by a loving woman (both Hindley and Arthur Huntingdon are destroyed by alcohol), but in his case it is the patient tuition and example of his mother, and not that of a lover, that enables him to fulfil his human potential. Like Emily, Anne believed that the education that would finally matter most would take place in the context of a loving home and not in the institutional surroundings of the classroom.

Each of the three Brontë sisters was committed to the educational ideals that they had absorbed as children, but each arrived at a different interpretation of those ideals, reflecting their divergent aspirations and experiences as adults. The Brontës were not isolated from the political and cultural controversies that defined contemporary debates about the nature and purposes of education. Their family circumstances made them particularly sensitive to the tension between an externally focused instrumental view of education and the more introspective definitions of an education that would support the autonomy of the imagination in private, and sometimes secret, processes of creativity. These tensions were often painful, for each of the sisters knew from their own experience that the high principles of education were widely different from its uncomfortable practice for spirited young women compelled to earn a living in lowly positions as teachers. While teaching at Roe Head, Charlotte recorded her frustration in a fiery journal entry:

> I had been toiling for nearly an hour. I sat sinking from irritation and weariness into a kind of lethargy. The thought came over me: Am I to spend all the best part of my life in this wretched bondage, forcibly suppressing my rage at the idleness, the apathy and the hyperbolic and most asinine stupidity of these fat headed oafs and on compulsion assuming an air of kindness, patience and assiduity? ... Just then a dolt came up with a lesson. I thought I should have vomited.[34]

Charlotte, Emily and Anne had every reason to understand that the social status and financial security of young, middle-class women would often depend on their work as pupils and teachers and that this work could confirm and extend their influence in the world. But the deepest allegiance of the three sisters was given to the inward education of the self as it was nourished by solitary reading and reflection and expressed through their literary creativity. Though the sisters repeatedly wrote about differing patterns of education and their consequences, they finally found the fulfilment and success that meant most to them in their work as poets and novelists, and not as professional teachers. It was in their writing that they were able to develop the full measure of their human independence.

Notes

1 Letter to W. S. Williams, July 3, 1849; *The Letters of Charlotte Brontë with a selection of letters by her family and friends,* Margaret Smith, ed., 3 vols. (Oxford: Oxford University Press, 1995–2004), ii, p. 226; quoted in Drew Lamonica, *We Are Three Sisters: Self and Family in the Writing of the Brontës* (Columbia: University of Missouri Press, 2003), p. 35. Williams was literary adviser to Smith, Elder and Company. Queen's College was established in London in 1848, initially with the intention of providing a formal education and to award qualifications to governesses.
2 'Whatsoever things are true, whatsoever things are honest, whatsoever things are just, whatsoever things are pure, whatsoever things are lovely, whatsoever things are of good report; if there be any virtue, and if there be any praise, think on these things.' Philippians 4:8.
3 'A little learning is a dang'rous thing/Drink deep, or taste not the Pierian spring' Alexander Pope, *An Essay on Criticism* (1709), ll. 215–6.
4 James Phillips Kay, *The Moral and Physical Condition of the Working Classes Employed in the Cotton Manufacture in Manchester* (London: Ridgway, 2nd edn., 1832), p. 61.
5 Kay, *Condition of the Working Classes*, p. 63.
6 The point is developed in Midori Yamaguchi's thoughtful *Daughters of the Anglican Clergy: Religion, Gender and Identity in Victorian England* (Basingstoke: Palgrave Macmillan, 2014).
7 Mary Wollstonecraft, *Thoughts on the Education of Daughters: With Reflections on Female Conduct, in the More Important Duties of Life* (London: Joseph Johnson, 1787), p. 2.
8 Mary Wollstonecraft, *A Vindication of the Rights of Woman* and *A Vindication of the Rights of Men,* Janet Todd, ed. (Oxford: Oxford University Press, 1993), p. 74.
9 Wollstonecraft, *A Vindication*, p. 77.
10 Jane Austen, *Emma,* Richard Cronin and Dorothy McMillan, eds. (Cambridge: Cambridge University Press, 2005), p. 21.

11 Hannah More, *Strictures on the Modern System of Female Education, with a View of the Principles and Conduct Prevalent Among Women of Ranks and Fortune*, 2 vols. (London, 5th edn., 1799); ii, p. 27.
12 Ibid., at i, p. 2.
13 Ibid., at p. 1.
14 More, *Moral Sketches of Prevailing Opinions and Manners, Foreign and Domestic: With Reflections on Prayer* (London: Cadell & Davies, 1819), p. 22.
15 Maria Branwell, letter to Patrick Brontë, 18 September 1812, quoted in Juliet Barker, *The Brontës* (London: Weidenfeld and Nicolson, 1994), p. 53.
16 Patrick Brontë, *The Maid of Killarney*, in *Brontëana: The Rev. Patrick Brontë, A. B., His Collected Works and Life*, J. Horsfall Turner, ed. (Bingley: T. Harrison & Sons, 1898), p. 178; quoted in Barker, p. 117.
17 This attractive and widely read book was written by the radical politician Richard Phillips. The 'Revd. J. Goldsmith' was a pseudonym. For a detailed account of Charlotte's home schooling, see Sue Lonoff, 'The Education of Charlotte Brontë: A Pedagogical Case Study', *Pedagogy*, 1:3 (2001), 457–77.
18 Letter from Mary Taylor to Elizabeth Gaskell, 18 January 1856; in Jane Stevens, ed., *Mary Taylor: Friend of Charlotte Brontë: Letters from New Zealand & Elsewhere* (Oxford: Oxford University Press, 1972), pp. 158–9.
19 This was a phrase that became familiar through the conservative William Rathbone Greg's widely discussed article 'Why are Women Redundant?', *National Review*, 14 (April 1862), 434–60.
20 Harriet Martineau, *Household Education* (London: Edward Moxon, 1849), pp. 243–4. Charlotte Brontë later wrote to Harriet Martineau, recalling how she had 'read with astonishment those parts of 'Household Education' which relate my own experience'. Harriet Martineau, *Autobiography*, 2 vols. (London: Virago, 1983); ii, p. 324.
21 The advertisement appeared in the *Leeds Intelligencer*, 4 December 1823. Quoted in Barker, p. 118.
22 Letter from Charlotte Brontë to Elizabeth Branwell, 29 September 1841, *Letters,* i, p. 268.
23 Ibid., at p. 269.
24 July 30 1841; *Letters,* i, p. 262.
25 Quoted in Charlotte Brontë and Emily Brontë, *The Belgian Essays,* Sue Lonoff, ed. (New Haven, Conn.; Yale University Press, 1997), p. xx.
26 Elizabeth Gaskell, *The Life of Charlotte Brontë*, Angus Easson, ed. (Oxford: Oxford University Press, 1996), p. 178. Quoted in Lonoff, 'Education of Charlotte Brontë', 465.
27 Ellis H. Chadwick, *In the Footsteps of the Brontës* (London: Isaac Pitman, 1914), p. 124.
28 Emily Brontë, *Wuthering Heights,* Ian Jack, ed. (Oxford: Oxford University Press, 1976), p. 71.
29 Anne Brontë, *Agnes Grey*, Robert Inglesfield and Hilda Marsden, eds. (Oxford: Oxford University Press, 1988), p. 1.

30 Examples of treatments of the governess and her place in mid-Victorian fiction include Mary Martha Sherwood's *Caroline Mordaunt; or the Governess* (1835), Harriet Mordaunt's representation of Maria Young in *Deerbrook* (1839); Elizabeth Sewell's Emily Morton in *Amy Herbert* (1844); or Lady Blessington (Marguerite Gardiner) in *The Governess* (1839). Mary Poovey's seminal account of the uncertain position of a governess within a family, 'The anathematised race: the governess and *Jane Eyre*', in *Uneven Developments: The Ideological Work of Gender in Mid-Victorian England* (Chicago: Chicago University Press, 1988) remains influential; see also Cecilia Wadsö Lecaros, *The Governess Novel* (Lund: Lund University Press, 2001).

31 Letter from Charlotte Brontë to Ellen Nussey, 1 July 1841, *Letters*, i, p. 258.

32 Anne Brontë, *The Tenant of Wildfell Hall*, Herbert Rosengarten, ed. (Oxford: Oxford University Press, 1992), p.128.

33 The early twentieth-century feminist novelist May Sinclair famously noted the boldness of this moment: 'Thackeray, with the fear of Mrs Grundy before his eyes, would have shrunk from recording Mrs Huntingdon's ultimatum to her husband. The slamming of that bedroom door fairly resounds through the long emptiness of Anne's novel.' May Sinclair, *The Three Brontës* (London: Hutchinson, 1912), p. 54.

34 Roe Head Journal, entry dated 11 August 1836. See Christine Alexander, 'Charlotte Bronte at Roe Head', *Jane Eyre*, R. J. Dunn, ed., 3rd edn. (New York: W. W. Norton, 2001), p. 404.

CHAPTER 3

Charlotte Brontë and the Science of the Imagination
Janis McLarren Caldwell

To characterise an early nineteenth-century view of the human, one could do worse than to quote Byron's *Manfred*, with his complaint that:

> ...we,
> Half dust, half deity, alike unfit
> To sink or soar, with our mix'd essence make
> A conflict of its elements[1]

Manfred's view of the human is not only explicitly dualist – 'half dust, half deity' – it is also a protest against the awkwardness of the mixture, 'unfit' either to rise with pride or to fall to degradation. The young Charlotte Brontë, creating her Byronic hero Zamorna, doubtless encountered this Romantic conflict between body and soul. But Brontë's characters, also of Byronic 'mixed essence', do sink and soar – alternately. She plays out the 'conflict of ... elements' more fully than does Manfred, and, as I hope to show, she turns that perception of conflict into an engine for creative thought. Charlotte Brontë left a number of descriptions of her process of creative imagining – which are indebted to early psychology – and which, in turn, had an impact on later psychology, especially of the unconscious mind. As cognitive science today revisits the nature of 'the mind's eye', it is a good time to be reexamining how nineteenth-century authors conceived of the imagination. My hypothesis is that Charlotte Brontë was a good introspective observer herself, influencing – and being influenced by – scientific accounts of unconscious or semi-conscious states.

Writing about several Victorians, including Brontë, William Cohen has claimed that 'embodiment came to be the untranscendable horizon of the human'.[2] Rehearsing the evolution of psychology over the course of the century, he notes that as scientific naturalism took hold, the language of the 'soul' was eclipsed by the language of the 'mind' – and that mind and brain were increasingly identified with one another. Cohen quotes from Alexander Bain's *Mind and Body: The Theories of Their Relation* to demonstrate the growing materialism of the scientific community. Bain posits 'one substance, with two sets of properties, two sides, the physical and the mental – *a double-faced unity*'.[3] One might note that even Bain in his materialism thinks in terms of a dualism of properties. While I agree that Brontë was deeply interested in embodiment and in scientific accounts of the mind, her reading and writing is situated in an earlier phase of this transition. Brontë not only inherited a scriptural language of body and soul, she also specifically recorded experiences of disjunction between a soaring imagination and a sinking return to embodiment. For Brontë, the experience of imagining clashed with the experience of being embodied.

There has been some debate about the extent to which Brontë was influenced by materialist science. Noting Brontë's reading of *Blackwood's*, as well as the medical and scientific texts available to her through the Keighly Mechanics Institute, Sally Shuttleworth has demonstrated that Brontë uses terms from medical and scientific psychologies of her day, especially phrenology.[4] More recently, Louise Penner has argued that Shuttleworth gives a picture of Brontë unduly influenced by materialist psychology. Penner finds evidence in Brontë's letters that she found materialism abhorrent.[5] Anna Neill characterised Brontë's attitude as 'scientifically literate and phrenologically inclined', but notes that she was 'nonetheless enormously disturbed' by her friend Harriet Martineau's materialism.[6] This captures the undoubted influence of materialist thought on Brontë's work, but also her rejection of a thorough-going commitment to erasure of the soul. This is a position that Elaine Scarry, in another context, has called 'volitional materialism', when a Christian dualist upends the traditional hierarchy of soul over body, preferring to care especially for the body.[7]

Brontë's Roe Head writings give us an account of her early experiences of imagining, for which she reserves her most exalted language. But what did imagining mean for nineteenth-century writers and scientists? Helpfully, the primary Oxford English Dictionary (OED) definition of the imagination extends from the early modern period through the Victorian to the present: 'The power or capacity to form internal images or ideas of objects and situations not actually present to the senses.' But within that definition is an equivocation that reflects an ongoing debate about the imagination: does the imagination actually form mental pictures, or does it use more abstract linguistic concepts? Alan Richardson traces the history of pictorialism in his fascinating 'cognitive historicist' study *The Neural Sublime*,[8] an account which I will rehearse here because it is so relevant, not only to Richardson's Romantics, but also to the

Brontëan imagination. An appeal to common sense may insist that the Imagination 'images' – pictorialism is implicit in the very name. But centuries of thinkers have experimented with using their imaginations on demand – closing their eyes, say, and trying to call up a beloved face, only to find the mental image nothing but a grey, thin shadow that pales in comparison to the glorious reputation of the creative imagination. Mental imaging alone seems inadequate to the reputation of artistic imagination, so the argument goes – it must require higher thought, the complexities of language, working invisibly to produce the vividness of imagination. W. J. T. Mitchell characterises the canonical Romantic poets, Blake excepted, as leaning toward the 'anti-pictorialist' side of this debate. The Romantic tendency, according to Mitchell, was to disparage pictorialism as too closely associated with mere outward material visibility and to consider the deep truth as imageless, formed in the invisible, intangible powers of the mind.[9] Richardson argues convincingly, contra Mitchell's emphasis on Romantic transcendence, that even the canonical Romantic poets were deeply engaged with a more materialist science of the mind. The same could be said for Brontë, who mixed dualist language with scientific language. In comparison, modern cognitive science has, until recently, leaned toward anti-pictorialism, explaining consciousness largely on digital models that require no manipulation of spatial or analogue components. Of course, many literary theorists have enjoyed this exclusive emphasis on language at the expense of the image. But neuroscientist Stephen Kosslyn has been a strong advocate of mental imaging as a component of thought, collecting some impressive data with functional Magnetic Resonance Imaging (fMRI), demonstrating that many people use their visual cortex in ways consistent with the spatial manipulation of mental pictures to solve problems.[10] When imagining, then, most use the same neural pathways as those used in perception. Influenced by Kosslyn, Elaine Scarry has entered the pictorialist fray with her remarkable *Dreaming by the Book*, which investigates the verbal production of what she calls 'vivacity'. Vivacity, for Scarry, is the full, life-like imaginative vividness that readers report when immersed in successful literary art.

Where do we fit Charlotte Brontë into this history of the imagination, with the largely undervalued mental image? Charlotte Brontë's several accounts of imaginary travel to Angria while teaching at Roe Head are intensely sensate and primarily visual:

> Never shall I, Charlotte Brontë, forget what a voice of wild and wailing music now came thrillingly to my mind's, almost to my body's, ear; nor how distinctly I, sitting in the schoolroom at Roe Head, saw the Duke

of Zamorna leaning against the obelisk, with the mute marble Victory above him, the fern waving at his feet, his black horse turned loose grazing among the heather, the moonlight so mild and so exquisitely tranquil, sleeping upon that vast and vacant road, and the African sky quivering and shaking with stars expanded above all.[11]

Brontë claims 'distinct' vision, crisp in detail and vast in scope, including Zamorna, his horse, the landmarks, the foliage, the deserted road and the wide, starry, night sky. Her emphasis is on the sensate, the aural, and what 'I [Charlotte Brontë] saw'. In another entry at Roe Head, she describes her imaginary experience in primarily visual terms: 'verily … these eyes saw the defiled & violated Adrianopolis shedding its light on the river from lattices whence the invaders looked out & was not darkened'.[12] Here she insists on the evidence of what 'these eyes saw' and demonstrates again a vast, yet particular, visual experience – seeing all of Adrianopolis down to the detail of the light glimmering between the lattices. But her language also approaches the transcendent tone we typically associate with Romanticism: she is in an 'ecstasy' and her diction and rhythms are Biblical, 'verily' she writes, 'this foot trod'. She alternates between calling her experience a 'trance', which perhaps points towards a semi-conscious mental state, and an 'apparition', which implies a ghostly visitation.

Another Roe Head entry seems to draw even more from the language of the supernatural. Although it begins with a pictorial phrase in which she sees Angria 'almost in the vivid light of reality' she winds up to: 'Then came on me rushing impetuously, the mighty phantasm that this had conjured from nothing from nothing [sic] to a system strong as some religious creed'.[13] The 'phantasm' implies perhaps a visual illusion, or more probably a supernatural visitation, as it is 'conjured' and figured not only as religious belief, but indeed as divine creation *ex nihilo*. Thus, it is difficult to position Charlotte Brontë's Angrian imaginative experiences strictly alongside either transcendent or physiological theories of the creative imagination. She seems to bolster pictorialist accounts with her protestations that her imagining is visual, with the spatially holistic, instantly present and vividly detailed qualities of seeing, yet she is clearly indebted to a transcendental tradition that speaks of higher knowledge descending upon her from above, or at least from outside herself. In this way, she blurs the terms of Mitchell's debate.

As idealist as Brontë may sometimes sound when extolling the imagination, a good deal of her psychological thought was patently materialist. As an avid reader and imitator of *Blackwood's*, Brontë was exposed to the psychology of her day, as *Blackwood's* 'had a long tradition of publishing psychological studies and stories ... since the 1820s'.[14] Shuttleworth's work on Brontë and psychology has demonstrated how steeped Brontë was in the pseudo-science of phrenology. Brontë's novels are peppered with the language of phrenological 'organs' and 'propensities' signalling phrenology's radical program of an embodied self, capable of self-improvement and class mobility. The aspect of phrenology that I'd like to emphasise here is the sense that the brain is an imperfect whole, divided into discrete organs. Of a 'bumpy' consistency, the brain's organs were often figured as warring against one another. Franz Gall claimed that phrenology 'will explain the double man within you, and the reason why your propensities and your intellect, or your propensities and your reason are so often opposed to each other'.[15] When Brontë submitted her own head to one Dr Browne for phrenological evaluation, she was delighted by a reading that emphasised the tensions within her character. Importantly, she was said not only to be 'endowed ... with an exalted sense of the beautiful and ideal' expressed in an 'enthusiastic glow' but also to 'exhibit the presence of an intellect at once perspicacious and perspic[u]ous',[16] both shrewd and plain. My point is that, quite beyond the references to phrenology in her novels, Brontë considered her own psyche to be alternately excessive and restrained, like the duelling organs of Gall's phrenology. Perhaps this explains, to some extent, the sequence of passages of elevated imagination followed by those of ironic deflation so characteristic of Brontë's style.[17]

One of the most prominent features of Brontë's early accounts of her own creative process is the qualitative difference between spontaneous visual imaging and conscious sequential thought. In the Roe Head manuscripts, not voluminous in the first place, she describes several instances of trance-like visual imagining abruptly interrupted. I have read the most famous of these as suggesting an other-worldly state, but Brontë goes on to narrate an interruption:

> Then came on me rushing impetuously, all the mighty phantasm that this had conjured from nothing from nothing [sic] to a system strong as some religious creed. I felt as if I could have written gloriously – I longed to write ... But just then a dolt came up with a lesson. I thought I should have vomited.[18]

She soars here with mystical language, having 'conjured' a 'phantasm,' but the real world in the form of a doltish student brings her crashing down. Her glorious vision has been replaced by frustration so acute as to induce nausea, a revulsion felt deep within her body. This is the most dramatic of the Roe Head entries, but there are two other interruptions that are so similar in structure one could consider them as variations on a theme:

> I felt myself breathing quick and short as I beheld the Duke lifting up his sable crest, which undulated as the plume of a hearse waves to the wind … Miss Brontë what are you thinking about? said a voice that dissipated all the charm, and Miss Lister thrust her little, rough black head into my face! 'Sic transit' etc.[19]

> [L]ast night I did indeed … lean upon the thunder-wakening wings of such a stormy blast as I have seldom heard blow & it … whirled me away like heath in the wilderness for five seconds of ecstasy … while this apparition was before me the dining-room door opened and Miss W[ooler] came in with a plate of butter in her hand. 'A very stormy night my dear!' said she 'it is ma'am'. said I.[20]

The first displays impatience with Miss Lister's annoying 'little, rough black head', but also an attempt at superiority and mature resignation. Or perhaps she is attempting Byronic *ennui*, as in Latin she sighs, 'Sic transit' – loosely translated, 'Thus passes [the glory from the earth]'. It is ever so, she seems to say, 'etcetera', too bored to bother finishing the sentence.

The second quote repeats the pattern of vision, writing compulsion and interruption, but with a stylistic difference: '"A very stormy night my dear!" said she "it is ma'am". said I'. Here, Brontë's delicate, controlled little understatement confesses her stormy inner vision to the page but conceals it from the innocent Miss Wooler, with her pat of butter. It is a move familiar to the readers of *Jane Eyre*, this ladylike containment, this imaginative flight followed by an ironic correction of her own passions.

These passages form a distinct mini-genre of Charlotte Brontë's early writing, very like Samuel T. Coleridge's preface to 'Kubla Khan' in which the man from Porlock shatters his dream vision.[21] But where Coleridge produces a mythic fragment, Brontë narrates her way back to a quotidian reality. In each case, Brontë describes not only an imaginative transport, but also, crucially, the abrupt return to consciousness. And she finds in this disruption an opportunity for stylistic experimentation. In her famous 'Farewell to Angria', also written at Roe Head, the shift to realism is more pronounced, more intentional: 'The mind would cease from excitement and turn now to a cooler region, where the dawn breaks grey and sober and the coming day, for a time at least, is subdued in clouds'.[22] Brontë's brand of realism includes both imagining and a course correction, as if eschewing the loss-obsessed strand of romanticism.

Furthermore, in between imagining and sober realism, immediately upon leaving her visionary trance and coming back to ordinary consciousness,

Brontë embraces the possibility of translating her vision to words. She leaves a record that while at Roe Head her visual imagination was accompanied by a writing compulsion, a 'scribblemania' in which she writes furiously with eyes closed to better access her inner vision.[23] It seems as if this rapid writing were a catch-up game, with Brontë working hard to keep pace with the speed of unconscious visualisation. If mental imagining happens in a flash, then consecutive writing has to work hard to translate into words that visual whole, before one is interrupted, before it disappears forever.

Though she says farewell to Angria, Brontë's mature writing doesn't smooth over her characters' tendencies to both imaginative rapture and frustration. In *Jane Eyre*, Jane resorts several times to imaginative visions which she must then reassess. For instance, Jane is a visual artist and in explaining her art to Rochester, she appeals to her mind's eye. When Rochester assumes her pictures are copies, she replies that they came '[o]ut of my head'. She goes on to confess to her 'reader':

> The subjects had, indeed, risen vividly on my mind. As I saw them with the spiritual eye, before I attempted to embody them, they were striking; but my hand would not second my fancy, and in each case it had wrought out but a pale portrait of the thing I had conceived. (p. 125)

She uses the dualist language of body and spirit here, but as a metaphor, with the 'spiritual eye' standing in for the imagination and embodiment for the work of art. The force of the contrast falls on the difference between her 'vivid' imagination and her execution. Questioned about her state of mind when painting she replies, 'I was tormented by the contrast between my idea and my handiwork: in each case I had imagined something which I was quite powerless to realise' (p. 126). The gap between the 'imagined' and the 'realise[d]' is so great as to 'torment' her.

Presumably Jane is later more successful at writing her inner vision, given the autobiography which we read. She doesn't meditate on that process, but there is a point at which she reflects on the nature of her inward visions – and subsequent narration. At Thornfield, in her 'discontented' 'restlessness' pacing the leads, Jane 'allow[s] my mind's eye to dwell on whatever bright visions rose before it – and certainly they were many and glowing'. Significantly, these 'bright visions' become verbal. She longs to 'open my inward ear to a tale that was never ended – a tale my imagination created, and narrated continuously; quickened with all of incident, life, fire, feeling, that I desired and had not in my actual existence' (*JE*, p. 109). Jane Eyre's imagination, like Brontë's own, is both visual and verbal. She sees 'visions' with her 'mind's eye' and hears 'inward[ly]' a continuous mental narration.

This passage – on imagining – is immediately followed by the forthright objection to real-world limitations that sounds so fresh and passionate that it risks being read as Brontë's own manifesto rather than Jane Eyre's:

> It is vain to say human beings ought to be satisfied with tranquility: they must have action; and they will make it if they cannot find it … [W]omen feel just as men feel; they need exercise for their faculties and a field for their efforts as much as their brothers do; they suffer from too rigid a restraint, too absolute a stagnation, precisely as men would suffer; and it is narrow-minded in their more privileged fellow-creatures to say that they ought to confine themselves to making puddings and knitting stockings, to playing on the piano and embroidering bags. It is thoughtless to condemn them, or laugh at them, if they seek to do more or learn more than custom has pronounced necessary for their sex. (p. 109)

This paragraph seems so distinct from what precedes and follows it, that it invites quotation without reference to its context. A heightened state of consciousness is marked by rapid, present-tense narration and insistent parallel phrases, matching women's needs with men's. But we may forget that Jane Eyre is arguing for the right to imagine, to create an internal vision and narration 'quickened with all of incident, life, fire, feeling, that I desired and had not in my actual existence'.

Of course, Virginia Woolf famously denounces the 'jerk' after this outburst when the narration returns to Jane, alone on the leads.[24] 'What were they blaming Charlotte Brontë for, I wondered?' asks Woolf, attributing the discontinuity of the narrative to Brontë's intrusive anger. If we look at the celebration of the imagination that precedes the angry outburst and compare it to the Roe Head journals, we might attribute the 'jerk' in the narrative to a discontinuity between semi-conscious and conscious states. The manifesto becomes not just an angry tirade, but a wonderfully fluent articulation – more eloquent than in the juvenilia – of the frustration attendant upon reentry to the cool, sober world of the real.

In training her realist eye on her own imaginative experience, Brontë provides valuable introspective evidence of her mind's operation. In fact, Brontë's descriptions were instrumental in W. B. Carpenter's formulation of 'unconscious cerebration'. Coined in 1854, the term 'unconscious cerebration' indicated a kind of creative thought that takes place independent of one's volitional control. In his *Principles of Mental Physiology*, Carpenter quotes from Elizabeth Gaskell's *The Life of Charlotte Brontë*:

> She said that it was not every day that she could write. Sometimes weeks or even months elapsed before she felt that she had anything to add to

that portion of her story which was already written. Then, some morning she would waken [sic] up, and the progress of her tale lay clear and bright before her in distinct vision, its incidents and consequent thoughts being at such times more present to her mind than her actual life itself.[25]

Gaskell emphasises the distinct visual effect, presented whole to Brontë upon waking, more vivid than life itself. Carpenter cites this passage in evidence of the part that the unconscious mind plays in invention, whether 'Artistic, or Poetical, Scientific or Mechanical'.[26] His point is that, across many fields of endeavor, the interviewed artist or scientist will describe this phenomenon: when encountering a difficulty, '*the tangle will be more likely to unravel itself* (so to speak) *if the attention be completely withdrawn from it*, than by any amount of continued effort' (Carpenter's emphasis). This is the picture of the unconscious mind as a vital creative resource, operating outside of the reach of conscious attention.

Many other nineteenth-century theorists of the mind emphasised the productive work of the unconscious mind. Earlier in the century, Coleridge called such unconscious work 'latent'[27] thought and William Hamilton named it 'mental latency'.[28] E. S. Dallas, in a more literary vein, called it 'the hidden soul'[29] and Cobbe added her voice to Carpenter's term 'unconscious cerebration'.[30] Even more remarkable than this proliferation of names for the unconscious mind was the number of metaphors used: 'a palimpsest' (De Quincey),[31] a horse, as part of a horse and rider team (Carpenter and Cobbe),[32] a secretary or librarian (Cobbe),[33] a magpie or kleptomaniac (Dallas)[34] and a treasure-house (Winslow).[35] For instance, Cobbe writes, 'We have seen that we are not Centaurs, steed and rider in one, but horsemen, astride on roadsters which obey us when we guide them, and when we drop the reins, trot a little way of their own accord or canter off without our permission'.[36] As a Christian, Cobbe has a stake in asserting the separateness of conscious and unconscious minds and the control possible to the alert rider, but her illustration, like Carpenter's, figures the 'unconscious' as a horse, useful and tractable, harnessed by the rider's will. What is clear is that these first theories of the unconscious mind considered such latent thought a valuable asset – a sense that was later eclipsed by Freudian views of the unconscious, many of which have emphasised repression or pathology.

Though the unconscious was viewed as distinctly different from the conscious mind, many theorists were interested in the traffic between the two. Dallas writes,

> [W]e live in two concentric worlds of thought – an inner ring, of which we are conscious, and which may be described as illuminated; an outer one, of which we are unconscious, and which may be described as in the dark. Between the outer and the inner ring, between our unconscious and our conscious existence, there is a free and a constant but unobserved traffic for ever carried on.[37]

Dallas theorises not only 'traffic' between conscious and unconscious minds, but also 'free' and 'constant' communication, of which we are unaware. One example of this kind of communication is the epiphany, or so-called eureka moment. Cobbe theorised the eureka moment as a surfacing of the end-product of unconscious thought into the conscious mind. Before Freud, the unconscious was typically defined as performing involuntary functions so that it was possible to be fully aware of the results of unconscious processing. Cobbe describes the experience:

> For example; it is an every-day occurrence to most of us to forget a particular word, or a line of poetry, and to remember it some hours later, when we have ceased consciously to seek for it. We try, perhaps anxiously, at first to recover it, well aware that it lies somewhere hidden in our memory, but unable to seize it. As the saying is, we 'ransack our brains for it,' but, failing to find it, we at last turn our attention to other matters. By-and-by, when, so far as consciousness goes, our whole minds are absorbed in a different topic, we exclaim, 'Eureka! The word, or verse, is so-and-so'. So familiar is this phenomenon, that we are accustomed in similar straits to say, 'Never mind; I shall remember the missing word by-and-by, when I am not thinking of it'; and we deliberately turn away, not intending finally to abandon the pursuit, but precisely as if we were possessed of an obedient secretary or librarian, whom we could order to hunt up a missing document, or turn out a word in a dictionary, while we amused ourselves with something else.[38]

For Cobbe, unconscious cerebration is a hidden resource, working behind the scenes to retrieve memories and make connections – and it will do its job even more fully if we divert our attention from the task at hand.

Another feature of unconscious cerebration, similar to the eureka moment, is the holistic quality of imaginative productions. They appear as vivid, instantaneous and seamlessly whole, unlike the fragments that the conscious mind can generate in a similar amount of time. As Dallas writes, 'The most royal prerogative of imagination is its entireness, its love of wholes, its wonderful power of seeing the whole, or claiming the whole, of making whole, and – shall I add? – of swallowing whole.'[39] As a psychological literary critic, Dallas derived much of his psychology from Sir William Hamilton and his formulations demonstrate an integration of a literary tradition with the new science of the mind. In *The Gay Science*, Dallas gives the unconscious imagination a large place not only in artistic creativity, but in all of life. He writes,

> Thus nobody tells us what imagination really is, and how it happens that being, as some say, nothing at all, it plays an all-powerful part in human life. Driven to our own resources, we must see if we cannot give a clearer account of this wonder-working energy, and above all, cannot reconcile the philosophical analysis which reduces imagination to a shadow with the popular belief which gives it the empire of the mind.[40]

Dallas refers here to the pictorialism debate about whether the imagination thinks with ideas (or shadows) or whether it is vividly pictorial (ruling the 'empire of the mind'). But he settles it, I think, in a way that might have appealed to Charlotte Brontë – and to many of those Victorians interested in physical explanations of the mind – but equally invested in preserving the language of the soul. He writes that the 'imagination is but a name for the unknown, unconscious action of the mind – the whole mind or any of its faculties – for the *Hidden Soul* ... it is the entire mind in its secret working'.[41] This may seem nothing but an overly broad generalisation equating the imagination with the unconscious mind, but Dallas makes an interesting move by linking the imagination with a more primitive reflexive mind and then calling it by a name usually reserved for higher functions: the Soul.[42] In the earlier faculty psychology, the 'soul' included highest, most disembodied faculties, in contrast to the appetites, seated deep within bodily physiology.[43] But Dallas, by sinking the soul into the regions of the unconscious mind, by calling it 'hidden', has suggested a deep embodiment and automaticity of the soul. Thus, imagining becomes less a transcendental visitation or super-added artistic faculty, but instead an eruption from a core, primitive reflex. In this way, Dallas's 'hidden soul' serves as both the substratum of neural elaboration and the bedrock of the human psyche.

One of the reasons 'unconscious cerebration' is more than a Victorian curiosity is that cognitive scientists have picked up this Victorian thread in theorising the 'adaptive unconscious' in contrast to the Freudian unconscious. Psychologist Timothy Wilson makes this connection, calling William Hamilton, Thomas Laycock and William Carpenter 'prescient': 'Their description of nonconscious processes is remarkably similar to modern views; indeed quotations from some of their writings could easily be mistaken for entries in modern psychological journals'.[44] The adaptive unconscious enables swift judgements, and is responsible for what psychologist Gerd Gigerenzer calls 'fast and frugal thought'[45] – frugal because these fast judgements are made with little new information. Malcolm Gladwell has popularised these fast judgements as 'thin slicing' or 'thinking without thinking'.[46] In each case, fast thought is liable to stereotyping. But it is also highly educable, even to the point of providing expert judgement.

Introspective accounts of creativity, such as those provided by Charlotte Brontë, are once again being viewed as valuable. In *Rethinking Thought*, Laura Otis interviews creative people about their own thinking processes, wondering especially about verbal and visual cognition. Can minds be categorised as primarily visual or verbal? Otis writes,

> As I listened to innovative writers and scientists describe their minds, I realized that creativity doesn't come in visual and verbal flavors. It results more from collisions in which words and images come together and people skilled with one, the other, or both are forced to interconvert them.[47]

The translation back and forth between words and images is the important work, according to Otis. She nonetheless registers the difficulty: 'Trying to convey words through images and images through words can be exhausting, but there is no greater engine of creativity'.[48] That strenuous effort of crossing between words and images, or between conscious and unconscious cognition, can, despite conflict, produce a rich harvest of innovative thought.

And so, we return to the difference – for some the nauseating difference – between mental imaging and the conscious use of language itself. I began with the pictorialism debate as outlined by Mitchell and argued that Charlotte Brontë seems not to fall cleanly on either side. But the word/image dichotomy that Mitchell emphasises may not be so much a historical division as a physiological one, as some cognitive scientists suggest. Ellen Spolsky, in *Gaps of Nature: Literary Interpretation and the Modular Mind*, argues that cognitive scientists like Jerry Fodor, Gerard Edelman and Daniel Dennett have been elaborating a theory of consciousness that turns out to be quite consistent with post-structuralist critical theory. Spolsky writes, 'According to the modularity hypothesis proposed by Jerry Fodor, the mind/brain achieves its purposes, when it does, by a collection of relatively independent processing devices called modules that receive distinct kinds of information'.[49] The differences, sometimes the incommensurability between the information processed by these modules, results in 'gaps' in consciousness, which is more sparse than we usually imagine it to be. The mind achieves whatever cohesion it does through narrative, which is a sort of glue that fills in the gaps. One of Spolsky's primary examples of a mental 'gap' is that between visual and verbal modules. As Spolsky writes, Mitchell in his iconology 'comes close to the biology of the problem' when he asks why we have 'this compulsion to conceive of the relation between words and images in political terms, as a struggle for territory, a contest of rival ideologies'.[50] But Spolsky points to conflict on a cognitive level as well, rather than solely on an ideological level. And she claims that the collaboration and/ or incommensurability between such modules as the verbal and the visual may produce the kind of creativity and innovation that provoke change, rather than necessarily reinforcing received power relationships.

Some of the most dramatic examples of narratives that fill in the gaps come from neurology. Michael Gazzaniga, known primarily for his work on cerebral commissurotomy (or so-called 'split brain') patients, posits a 'left hemisphere interpreter' that, in the instance of right-brain strokes, manages to resolve conflicting pieces of information from the intact and damaged areas of the brain. The more conflict, the more elaborate the narrative, as in the following example. Gazzaniga writes:

> Patients with 'reduplicative paramnesia', because of damage to the brain, believe that there are copies of people or places. In short, they will remember another time and mix it with the present. As a result, they will create seemingly ridiculous, but masterful, stories to uphold what they know to be true due to the erroneous messages their damaged brain is sending their intact interpreter. One such patient believed the New York hospital where she was being treated was actually her home in Maine. When her doctor asked how this could be her home if there were elevators in the hallway, she said, 'Doctor, do you know how much it cost me to have those put in?' The [left hemisphere] interpreter will go to great lengths to make sure the inputs it receives are woven together to make sense – even when it must make great leaps to do so. Of course, these do not appear as 'great leaps' to the patient, but rather as clear evidence from the world around him or her.[51]

Neurologists call this fast, unconscious weaving of narrative 'confabulation'. This rapid 'filling in' is notoriously imperfect, but it is lively, with a seamless sense of real-world experience. My closing speculation is that the vivacity that Scarry describes is experienced by readers given a very incomplete set of authorial instructions. And that, if Charlotte Brontë can be read as a practiced confabulator, then she also knew the value to the reader of a deftly placed discontinuity which can, in turn, recruit the reader's 'filling in' functions. The famous 'jerk' after Jane Eyre's feminist manifesto may be attributable to anger, as per Virginia Woolf, or perhaps more accurately, to an imagination that plumbed some of the secrets of the unconscious psyche.

I have been arguing for some similarity and continuity between Charlotte Brontë's picture of the unconscious mind and a twenty-first-century neurologic description of the same. But how does Charlotte Brontë's idea of the 'human' compare to our own? Gazzaniga's patients demonstrate how wired the 'left brain interpreter' is for generating a coherent narrative from conflicting information and how unconscious that narrative construction turns out to be. The drive for the explanatory power of narrative seems to be one of the deeply embedded features of human brains. Humans, then, in a bare sense, are meaning-making animals (though perhaps not the only meaning-making animals).[52] What does this picture of a meaning-making, unconscious mind contribute to a twenty-first-century idea of the human? In making meaning about the self and the body, even materialists, it has been widely argued, tend in ordinary life to speak according to 'folk psychology,' which uses a different 'mind' language (in which one is an agent with intent) from 'body' language (in which one is a physical object). Western philosophy of mind still runs up against the mind/body problem, or put another way, the famous 'hard problem' of how brain becomes mind. Even monists, as are the majority of current philosophers and neuroscientists alike, use *two words*, with a hyphen or a slash – mind/body – to indicate a unitary entity. They run into the difficulty of accounting for the so-called 'qualia,' those beliefs, desires, intentions, and feelings – many of them unconscious – that require a vocabulary different from that of natural science. As twenty-first-century humans, if even the materialists among us indeed unconsciously resort to the language of 'natural dualism', we're not so different from Charlotte Brontë in encountering a schism between the vocabulary of experiencing an imagination and the vocabulary of being a physical object. Humans, then, are meaning-making animals that work with a disjunction between two radically different ways of considering the mind/body. Remarkable in Brontë's writing is the registered discontinuity between the ways in which we're at liberty to experience imaginary worlds and the ways in which we're distinctly limited by our bodies-as-objects. Bronte's skill in negotiating and narrating that untranslatable human difference goes some distance toward explaining the continued freshness and vivacity of her work.

Notes

1 George Gordon, Lord Byron, *Manfred* (Act 1, Scene 2) in *The Complete Poetical Works of Lord Byron* (Boston and New York: Houghton Mifflin, 1905), p. 482.
2 William A. Cohen, *Embodied: Victorian Literature and the Senses* (Minneapolis: University of Minnesota Press, 2008), p. xii.

3 Alexander Bain, *Mind and Body: The Theories of Their Relation* (New York: D. Appleton and Company, 1874) p. 196, quoted in Cohen, *Embodied*, p. 3.
4 Sally Shuttleworth, *Charlotte Brontë and Victorian Psychology* (Cambridge: Cambridge University Press, 1996).
5 Louise Penner, '"Not yet settled": Charlotte Brontë's Anti-Materialism', *Nineteenth-Century Gender Studies*, 4:1 (2008), 1–27.
6 Anna Neill, *Primitive Minds: Evolution and Spiritual Experience in the Victorian Novel* (Columbus: Ohio State University Press, 2013), p. 16
7 Elaine Scarry, 'Donne: "But yet the body is his booke"', in *Literature and the Body: Essays on Populations and Persons*, Elaine Scarry, ed. (Baltimore: Johns Hopkins University Press), pp. 70–105, p. 71.
8 Alan Richardson, *The Neural Sublime: Cognitive Theories and Romantic Texts* (Baltimore: Johns Hopkins University Press, 2010).
9 W. J. T. Mitchell, 'Visible Language: Blake's Wond'rous Art of Writing,' in *Romanticism and Contemporary Criticism*, Morris Eaves and Michael Fischer, eds. (Ithaca: Cornell University Press, 1986), pp. 46–95.
10 Stephen M. Kosslyn, *Image and Brain: The Resolution of the Imagery Debate* (Cambridge, Mass: MIT Press, 1994).
11 Charlotte Brontë, 'We wove a web in childhood', in *An Edition of the Early Writings of Charlotte Brontë*, Volume II: *The Rise of Angria*, 1833–1835, Part 2: 1834–1835, Christine Alexander, ed. (Oxford: Basil Blackwell, for the Shakespeare Head Press, 1991), p. 385. Juliet Barker notes that, though about Roe Head, this fragment may have been penned at home: 'Though the implication of both the poem and the prose continuation is that they were written actually at Roe Head, the manuscript is clearly signed and dated in Charlotte's longhand at the end "C. Bronte Decbr 19th Haworth 1835"' (Barker endnote 47, p. 884).
12 Charlotte Brontë, 'Well here I am at Roe-Head' [Roe Head Journal], 4 February 1836, quoted in Juliet Barker, *The Brontës* (New York: St. Martin's Press, 1994), p. 249.
13 Charlotte Brontë, 'All this day I have been in a dream', 11 August 1836, quoted in Barker, *The Brontës*, p. 255.
14 Jenny Bourne Taylor and Sally Shuttleworth, eds., *Embodied Selves: An Anthology of Psychological Texts, 1830–1890* (Oxford: Clarendon Press, 1998), p. 143.
15 Francois Joseph Gall, *On the Functions of the Brain*, trans. W. Lewis, 6 vols. (Boston: Marsh, Capen and Lyon, 1835), II, p. 43, quoted in Sally Shuttleworth, *Charlotte Brontë and Victorian Psychology* (Cambridge: Cambridge University Press, 1996), p. 62.
16 J. P. Browne, 'Phrenological Estimate of a Lady', 29 June 1851, quoted in Barker, *The Brontës*, p. 681.
17 For specific examples of this deflation of the supernatural, see Robert Heilman, 'Charlotte Brontë's New Gothic', in *Jane Austen to Joseph Conrad: Essays Collected in Memory of James T. Hillhouse*, Robert C. Rathburn and

Martin Steinmann Jr., eds. (Minneapolis: University of Minnesota Press, 1958) pp. 118–32, p. 120.
18 Brontë, 'All this day I have been in a dream', 11 August 1836, quoted in Barker, *The Brontës*, p. 255.
19 Brontë, 'We wove a web in childhood', in *An Edition of the Early Writings of Charlotte Brontë*, Volume II: *The Rise of Angria, 1833–1835*, Part 2: 1834–1835, Christine Alexander, ed., p. 385.
20 Brontë, 'Well here I am at Roe-Head' [Roe Head Journal], 4 February 1836, quoted in Barker, *The Brontës*, p. 249.
21 Samuel Taylor Coleridge, *The Poetical Works of Samuel Taylor Coleridge*, James Dykes Campbell, ed. (London: Macmillan, 1905), pp. 592–3.
22 Charlotte Brontë, 'Farewell to Angria', in *Jane Eyre: An Authoritative Text, Contexts, Criticism*, Richard J. Dunn, ed., 3rd edn. (New York: W.W. Norton & Co., 2001), p. 425.
23 Brontë, 'I'm just going to write because I cannot help it,' [ca. Oct. 1836], quoted in Barker, *The Brontës*, p. 255.
24 Virginia Woolf, *A Room of One's Own* (New York: Harcourt Brace Jovanovich, 1991), pp. 74–5.
25 Elizabeth Gaskell, *The Life of Charlotte Brontë*, 1857, quoted in William Benjamin Carpenter, *Principles of Mental Physiology: With Their Applications to the Training and Discipline of the Mind, and the Study of Its Morbid Conditions* (Cambridge University Press, 2011), p. 535.
26 Carpenter, *Principles*, p. 534.
27 Samuel Taylor Coleridge, *Biographia Literaria*, 2 vols. (New York: Kirk and Mercein, 1817), p. 74.
28 William Hamilton, *Lectures on Metaphysics and Logic*, Henry L. Mansel and John Veitch, eds., 2 vols. (Boston: Gould and Lincoln, 1871), vol. 1, p. 235.
29 E. S. Dallas, *The Gay Science*, 2 vols. (London: Chapman and Hall, 1866; New York: Johnson Reprint Corporation, 1969), p. 199.
30 Frances Power Cobbe, 'Unconscious Cerebration', in *Darwinism in Morals and Other Essays* (London: Williams and Norgate, 1872), pp. 305–34, p. 305.
31 Thomas De Quincey, 'Suspiria de Profundis: Being a Sequel to the Confessions of an English Opium-Eater', *Blackwood's Magazine*, 57 (June 1845), 739–51, p. 739.
32 Carpenter, *Principles*, p. 24, Cobbe, 'Unconscious Cerebration', p. 333.
33 Cobbe, 'Unconscious Cerebration', p. 308.
34 Dallas, *The Gay Science*, p. 216.
35 Forbes Winslow, *On Obscure Diseases of the Brain, and Disorders of the Mind* (Philadelphia: Blanchard & Lea, 1860), p. 292.
36 Cobbe, 'Unconscious Cerebration', p. 308.
37 Dallas, *The Gay Science*, p. 207.
38 Cobbe, 'Unconscious Cerebration', p. 308.
39 Dallas, *The Gay Science*, pp. 268–9.
40 Ibid., at p. 193.
41 Ibid., at p. 196.

42 A similar argument is made in Neill, *Primitive Minds*.
43 Rick Rylance, *Victorian Psychology and British Culture, 1850–1880* (Oxford: Oxford University Press, 2000), p. 27.
44 Timothy D. Wilson, *Strangers to Ourselves: Discovering the Adaptive Unconscious* (Cambridge, Mass.: Belknap Press of Harvard University Press, 2002), p. 10.
45 Gerd Gigerenzer and Daniel G. Goldstein, 'Reasoning the fast and frugal way: Models of bounded rationality', *Psychological Review*, 103 (1996), 650–69.
46 Malcolm Gladwell, *Blink: The Power of Thinking Without Thinking* (New York: Little, Brown and Company, 2005).
47 Laura Otis, *Rethinking Thought: Inside the Minds of Creative Scientists and Artists* (Oxford: Oxford University Press, 2015), p. 14.
48 Ibid.
49 Ellen Spolsky, *Gaps in Nature: Literary Interpretation and the Modular Mind* (Albany: State University of New York Press, 1993), p. 20.
50 Spolsky, *Gaps in Nature*, p. 29.
51 Michael S. Gazzaniga, *The Ethical Brain* (New York: Dana Press, 2005), pp. 149–50.
52 This definition is not necessarily exclusive, as other species could conceivably prove to have meaning-making systems of symbols. Nor is it comprehensive, excluding for instance some with serious brain damage, because of course even Gazzaniga's 'left brain interpreter' can be injured. In that case, being *homo sapiens* is enough to be considered human. But we still need a definition of the human that reaches beyond biology and seems to be valid across cultures.

CHAPTER 4

Being Human
De-Gendering Mental Anxiety; or Hysteria, Hypochondriasis, and Traumatic Memory in Charlotte Brontë's Villette

Alexandra Lewis

He who has escaped from the opening jaws of destruction, ought to pursue his onward course, without looking back upon the horrors from which he has been rescued. It not unfrequently happens, that a man, by too obstinately contemplating what he has left behind, remains for ever after fixed in the retrospective attitude. His foot is arrested in the path of life; his imagination feeds solely upon what has gone by … In the drama of the world his part is closed, although he may be still doomed to continue on the stage. It is in this way, that melancholy and madness are often generated by the permanent pangs of memory, by the continual presence of the past.

– John Reid, 'Essay V: Remorse', *Essays on Hypochondriasis, and Other Nervous Affections* (1821), 3rd edn, rev. (London: Longman, Hurst, Rees, Orme, and Brown, 1823), pp. 78–9

We are apt to be stunned by the blow of any great misfortune. Its immediate effect is a dimness of perception and a confusion of ideas. A calamity may be of such a magnitude, that, like any vast object of nature or of art, we must be removed to a certain distance from it before we can take into our view its entire dimensions. Time may obliterate fanciful and superficial sorrows, but it adds to those which are justified and nourished by reflection. In such a case, the current of melancholy thought deepens the channel in which it flows … We cannot be made invulnerably happy by any combination of events, however fortunate; but, on the other hand, a single event of a calamitous nature will often be found enough to darken the remainder of our days. The sunshine of life is continually liable to be interrupted by clouds, but there are clouds which no sunshine can disperse.

– John Reid, 'Essay XXVII: Real Evils, A Remedy for Those of the Imagination', *Essays on Hypochondriasis*, pp. 422–4

The art of memory is the art of forgetfulness.
— James Douglas, *On the Philosophy of the Mind*
(Edinburgh: Adam and Charles Black, 1839), p. 196

Charlotte Brontë's *Villette* is Lucy Snowe's autobiography, the self-conscious excursion into the workings of mind and memory of a fictional heroine plagued by traumatic losses and burdened with a self-confessed nervous disposition. Communicating states of mind both past (as the older Lucy recalls the thoughts and reactions of her younger self) and present (self-interpretation with the aid of retrospect), *Villette* explores the centrality of memory to ongoing experience. The form and content — and eloquent gaps and silences — of Lucy's life-writing unveil the possibilities inherent in remembrance, as well as delving imaginatively into that faculty's unreliability. *Villette* follows the mnemonic tradition of Emily Brontë's *Wuthering Heights*, which I have explored in terms of memory 'possessed' by disruptions and intrusions in the aftermath of terror and violent conflict.[1] Charlotte Brontë's final novel shares a concern with the scope for — and effect of — numerous pathologies of memory: both consciously willed and unconscious, uncontrollable. From the drama of total familial loss (reduced to symbolic stormy outlines), to the harrowing salty nightmare (said to repeat into old age yet infrequently mentioned), to the bizarre patterns of displacement in her recollections and recognition of an early beloved, Lucy Snowe not only suffers due to the inability to remember but also, as I have argued in detail elsewhere, attempts consciously 'to emulate the traumatic obliteration of memory as a kind of coping strategy'.[2] Charlotte Brontë's development of a particular discourse of traumatic memory is also, I have shown, bound up with her investigation of stagnation and miasma, leading to a new literary aesthetic of the diseased mental atmosphere.[3]

This chapter takes the complex life of the mind — and resulting mental and bodily markers of happiness and distress — as central to Charlotte Brontë's conception of what it means to be human. It demonstrates how, by negotiating between unconscious cerebration and ideologies of self-will in *Villette*, as well as by calling into play cultural conceptions of (male) hypochondriasis and (female) hysteria, Brontë devises a narrative of traumatic memory which not only considers the role of empathetic imagination in diagnosis and tests the possibilities of fictional autobiography (that creative form devoted to the exploration of the individual human subject) but also, vitally, complicates gendered notions of anxiety.

An analysis of the novels of the Brontës can teach us much about nineteenth-century understandings of mental health and psychic wounding, just as a fresh appraisal of nineteenth-century scientific and philosophical writings on the mind provides scope for newly detailed and nuanced ways of reading Charlotte Brontë's approach to the idea of what it means to be human. I contend that many responses to *Villette*, from the time of its publication to the present day, have overlooked crucial aspects of the text by accepting as the sole interpretive paradigm Brontë's use of the gendered language of nervous disease. In doing so they fall into reading the signs of human health and suffering through the limited understanding of the first-person fictional autobiographer and her doctor, rather than interrogating the language of the text as a whole. As this chapter shows, Brontë's text identifies and mediates between contradictions underpinning contemporary medical discourse on mental and mnemonic dysfunction, among them the distinct conceptions of the unsound and healthy mind in response to calamity, in order to reach a view of mental distress that is less gendered – and more broadly human – than the nineteenth-century interpretive categories of 'hysteria' and 'hypochondriasis' might seem to allow.

In a two-step fictional navigation of the prevailing cultural assumptions arising from a Bakhtinian 'collision and interaction' of scientific and medical languages (the heteroglossia which 'washes over' Brontë's literary language),[4] Brontë first subverts the gendered nature of (female) hysteria and (male) hypochondriasis through the deeply imaginative self-identifications of her narrator (with such sufferers as the King of Labassecour) and then troubles the application of hypochondriasis through her compelling focus on stunned memory. As such, building on Emily Brontë's delineation of uncontrollable, pathological remembrance in *Wuthering Heights*,[5] Charlotte Brontë facilitates a significant shift in the accentual register from internal weakness manifesting in bodily symptoms, to external shock, manifesting in traumatic interiority with dire effects for health both mental and physical.

Strong Will Maketh the (Hu)man: Gender, Memory, and Self-Control

Throughout the nineteenth century, across discourses ranging from the scientific to the theological, the supremacy of the individual will as both functional and moral force regulating behaviour was a social tenet reproduced if not with increasing certainty then, at least, with enduring centrality. In 1835, psychiatrist James Cowles Prichard's important work, *A Treatise on Insanity and Other Disorders Affecting the Mind*, set the tone for an insistence on personal responsibility in cases of mental disturbance, which would be upheld in both metaphysical and physiological analyses even as psychologists elucidated in greater detail the unconscious, latent, or reflex actions of the brain. While Prichard provides an early reference to severe mental shock occasioned by external events such as bereavement or 'some reverse of fortune',[6] he nonetheless aligns the causes of mental illness with pre-existing moral or mental deficiencies. John Conolly and John Gideon Millingen (physicians and asylum managers) each profess a slightly wider view, pointing always to the importance of self-control, yet describing with some trepidation the extent to which unchecked 'mental anxiety or disturbance' – '[l]ong before old age approaches', and separate from the onset of disease – could result in 'impairment of memory'.[7]

For Conolly, writing in 1830, the 'perfect man ... is one whose emotions, passions, and affections are [not only] unimpaired, but who governs them and directs them to good ends'.[8] However, of the 'Peculiarities of the Human Understanding, Which Do Not Amount to Insanity' – including those 'produced by various accidental causes' – the entrance of 'unbidden' thoughts, the experience of 'fear' strong enough to take over the mind, and the ability of 'sad recollection' to 'benumb'[9] mental activity all combine to challenge from within Conolly's text his desired certainty in the possibility for self-mastery by the sane individual. In Millingen's view, memory, 'the principal link in the intellectual chain', is similarly liable to disruption by 'fear'[10] or even by excess and obsessive focus in its own operation. A firm believer in the influence of remembered adversity on present thinking and development, Millingen cites with approval French physiologist Broussais's account of 'tyrannical memory': a state of mind in which uncontrollable mnemonic forces are capable of overcoming the will altogether, compelling the afflicted person to 'contemplate a crowd of images which he would most willingly dispel'.[11]

From around the time of *Villette*'s publication, writers on the mind sought to diminish the dangers of the facility for non-deliberate thought, which came increasingly to be appreciated as performing highly complex mental work, independent of conscious supervision. Physiological psychologists such as William B. Carpenter minimised the unconscious mind's potential for disruption, even as they emphasised its scope for action (making connections missed by the conscious, rational mind). In addition, they insinuated that the paradigmatic moral responsibility and conscious control so important earlier in the century could somehow be exerted over the uncontrollable domain through the habituated training and formation of reflexive action. This conceptual and rhetorical struggle continued to century's end. In 1876, George Eliot's *Daniel Deronda* would follow the orientation of Charlotte Brontë's *Villette* by exploring the limitations of the conscious will, challenging the rationalist, educable model of the unconscious propounded by Carpenter. Also published in 1876 was the fourth London edition of Carpenter's *Principles of Mental Physiology*, with an extended preface elaborating his mid-century position on strength in self-direction. Calling upon Thomas De Quincey's *Confessions* as an example of 'the *will*' too weak to carry positive '*desires*' for improvement 'into effect', Carpenter finds much both to question and to admire in the words of J. S. Mill, that 'what is really inspiriting and ennobling in the doctrine of Freewill, is the conviction that *we have real power over the formation of our own character*'.[12] Real power of will, certainly, but not entirely complete: the conviction wavers within Carpenter's own work with relation to memory, which 'constitutes the basis of our feeling of *personal identity*'.[13] Impressions are reproduced 'as Ideas when the appropriate *suggesting strings* are pulled', but Carpenter does not make a mere puppet of memory: character is often created and revealed as much through 'the accidental shining-in of a light upon some dark corner'[14] as it is through the controlled exercise of recollection. Consequently, the physical traces or grooves 'in the Cerebrum' of which memory for Carpenter is 'the psychological expression'[15] are as likely to be formed spontaneously as intentionally, and perhaps in undesirable patternings – as well as being open to the unexpected and, in Carpenter's terms, almost physiological penetration of an external shock.

Charlotte Brontë was writing on the cusp of major shifts, in which 'mid-century physiological psychologists were actively transforming earlier associationist and phrenological models of consciousness and self-control',[16] and her novels point up the problems (and contradictions) which beset both contemporaneous and later analyses such as those of Carpenter. Of course, as I will go on to explore in relation to hysteria and hypochondriasis, they also confront the way in which, for most Victorians, 'self-will' is gendered: men being viewed as innately more skilled in the science of self-discipline. The intellectual powers of women were routinely described as being both inferior to those of men and unable to withstand the greater flooding forces of female 'passion'.[17]

The Unsound and Healthy Mind in Calamity

As James Douglas remarked in *On the Philosophy of the Mind*, '[m]emory not only restores to us what we formerly knew, but is itself a great inlet of new thoughts', and, in linking various sensations and reflections through time, 'is even essential to perception itself',[18] a centrality Brontë's narrator comes on some level to appreciate as she seeks to make sense of her past, and future. Brontë's presentation of internal conflicts and nuances in Lucy's life-composition points to an awareness only beginning to be registered in the margins or between the lines of widely circulating psychological texts on the faculty of memory. Her writing takes up and puts into conversation the often separate lines of inquiry regarding, firstly, the centrality of memory in forming both present identity and future perceptions, and, secondly, the functioning of storage and recall. *Villette* remains an important nineteenth-century articulation of the ways memories could be lost, the ramifications of their disappearance, and the role of attention and emotion in establishing what kind of memories were retained, or would fade into seeming oblivion.

According to Douglas, memory 'frees us from the narrow boundary of present objects' by collecting 'scattered passages of former existence' into the landscape of awareness. However 'diminishing in dimness and distance',[19] Douglas's horizon of remembrance forms a history which is circular and enclosing rather than chronological, visible whichever direction the mind's eye may turn. Thomas Brown, in his earlier *Lectures on the Philosophy of the Human Mind*, imagined the present as 'a *bright point*' which at once keeps aglow and is enhanced by the '*twilight of the past*'.[20] Both perspectives establish the past as a point not only of departure but also of constant reference. In figuring remembrance as a light shining forth into unknown future territory, however, both imply that forgotten elements are of relative unimportance. For Douglas, the mind avoids 'inextricable chaos' by allowing 'gradual darkness [to] steal' over aspects of the past viewed neither with regularity nor 'abiding interest'.[21] Sharing John Reid's focus on remorse in the heightened, even unhealthy, recurrences of memory, Douglas acknowledges that '[t]he great remembrancer of the past is guilt', and that, separate from the conscious fastening or reversion of attention to past objects in normal mnemonic operations, the guilty memory appears, over time, in 'distinctness' from its surrounding details, growing 'not dimmer, but more vigorous, with the lapse of years'.[22] This may even occur as a 'sudden resuscitation ... of past existence' (a mnemonic capacity which Douglas interprets in moralistic terms, as proof of 'an immortality of retribution').[23] In general, though, for Douglas, as for many other early- to mid-nineteenth-century writers on memory, it is 'whatever we continually reiterate' which becomes 'indelibly engraved on the mind'.[24]

In *Villette*, forgetfulness is shown to be a far more complex phenomenon. The palpable presence of the past can be registered even, or especially, under cover of that unfathomable darkness of mind. In Brontë's final novel, mental chaos is shown to be produced not only by Douglas's overcrowding (nebulous facts and faces), but by under-population (blank spaces created by the inability to shine light upon passages of great importance). This results in overcrowding of another sort – jostling ghostly shapes of matters improperly or impartially remembered. For Brontë, it is precisely from these depths that 'sudden resuscitations of past existence' can spring, and spring unbidden. Lucy's habitual reflections on past occurrences are not always detailed or specific: she skirts the edges of remembrance. It is as if the narrator's access to important recollections has been blocked, and the memories themselves wrongly relegated to mental darkness, leaving dimly perceptible outlines which persist in troubling the mind while refusing to make themselves fully comprehensible. For Douglas, '[e]xperience is of the utmost value, but … only when the chaff is sifted out of it'.[25] Brontë is interested in a problem with experience which runs almost in the opposite direction. Too much of Lucy Snowe's personal history has been sifted out and must be regained in order successfully to link past with present and future and to enable that guiding light of assimilated memory to make clear the features of her social and mental landscape.

The two excerpts from Reid's *Essays on Hypochondriasis* which open this chapter provide a closer sense of the model of memory with which Charlotte Brontë is working. Examined together in order to highlight the internal contradictions operating in Reid's essays, the passages also catch up many of the ambiguities regarding the relationship between dysfunctions of memory and a spectrum of (often specifically female) nervous diseases which continued to enjoy critical currency at the time of *Villette*'s publication. Ambivalences and arguments surrounding the interaction between mind and body, volition and the unconscious, self-control and hysterical emotion, were in abundance throughout the nineteenth century, making Brontë's construction of a novel so deeply enmeshed in these divides a deliberate entrance into the wider debate. In the first extract, Reid impresses upon the convalescent reader that much of the past can and ought to be forgotten. He implicates a lack of self-control on the part of the nervous sufferer in any inability to rise above the 'permanent pangs of memory'. A retrospective attitude leading to 'melancholy and madness'[26] is, on this view, attributable not so much to the impact of the past but to the poor habit of obstinate contemplation.

In the second extract, which concludes the work, Reid has widened his view, positing instead the potentially inescapable effects of the past (volition notwithstanding), and even acknowledging that there are particular sorrows which are 'justified and nourished by reflection'[27] over time.

Reid positioned his work somewhere between medical tract and advice manual, sufficiently divested of 'technical phrases' as to appeal not only to the 'practitioner in physic' but also 'the community at large',[28] and the reason for this apparent disjunction between the two excerpts regarding the individual's ability and responsibility to transcend times of past trouble becomes clear on closer examination of the title of his final chapter: 'Real Evils, A Remedy for Those of the Imagination'. Reid is, in effect, describing the impact of remembered events on two classes of mind: one disordered, belonging to the self-indulgent hypochondriac, and that of a person in full mental health, who has suffered the misfortune of an external calamitous 'blow'.[29] It is intriguing that, while condemning melancholy in the one, he almost exonerates it in the other, accusing the hypochondriac of lacking a degree of agency which he hardly demands of – or perhaps takes for granted in – the healthier mind. In conceptualising an unexpected cloud which might be 'enough to darken the remainder of our days', too vast to be dispersed by sheer force of will, Reid's conclusion has much in common with modern understandings of trauma, and of the 'dimness of perception' accompanying a stunning 'blow'.[30] There is in Reid's analysis, however, the implication that, following this 'immediate effect', the 'entire dimensions'[31] of the sorrow can be taken into view and reflected upon in a conscious and deliberate manner. In *Villette*, Brontë questions the possibility ever of viewing the entire dimensions of the traumatic cloud. In addition, she mediates between and moves beyond Reid's distinct conceptions of the ill and the healthy mind in calamity.

At once partially acquiescing in and struggling to escape from the overarching authority of gendered nineteenth-century labels, Lucy figures herself as a hypochondriac with – and this is distinctly unusual for a Victorian hypochondriac – professed ailments which are primarily psychological. Brontë's use of fictional autobiography heightens the sense of urgency with which Lucy struggles to articulate the truth of her experience, in a medical language frustratingly ill-equipped to capture its reality. Through the obsessive repetitions (and self-generated points of intense focus) of Lucy's life-plot, as well as those aporias unable always to be seen by her heroine-narrator, Brontë develops the sense of a mind labouring under and stunted by the lurking presence of unassimilated memories.

Hysteria, Hypochondriasis, and the 'Extreme Boundary of Human Knowledge'

Brontë's *Villette* brings the language and ideas of contemporary physiological theory into a more potent and expansive metaphorical usage. In so doing, the novel challenges theories which reduce mind to body (claiming always a physical cause for psychic disturbance) and thus traverses what Henry Holland had referred to as 'the undiscovered and impassable space that lies beyond'[32] contemporary scientific understanding. Thomas Laycock observed that investigation of nervous phenomena, stretching 'into the dim regions of metaphysics' and physical sciences, was entangled 'with a hundred theological questions, themselves involved in doubt and obscurity':[33] for these, indeed, Lucy has no answer – 'no words' (p. 42). Holland's chapter 'On the Present State of Inquiry into the Nervous System' is primarily physiological in orientation, but he acknowledges that 'we are required to recognise the direct influence upon [organic functions] of all mental emotions, even of the simple act of attention of mind'.[34] For Holland, it is 'manifest' that:

> we are treading in this place on the *extreme boundary of human knowledge* ... We have powers before us for contemplation, which we cannot identify as the properties of any physical agents, nor interpret by any analogy beyond their own action.[35]

'Reasoning men in every age' have been 'perplexed' by the challenge adequately of comprehending, let alone communicating, the workings of the mind in either sickness or in health.[36] James Douglas criticised the work of earlier writers on memory (including Thomas Brown and Thomas Reid), complaining that 'the study of mind has retrograded', and 'teachers have become popular in proportion as they have substituted images for reflections'.[37] Douglas does, however, allow that perceived rhetorical imprecision serves as an indication that the scientific world was ill-equipped to attempt to describe a realm so firmly situated in the intricacies of imagery as well as in more precise, rational reflection: '[t]he different operations of our minds are so complicated, and so blended into each other, that it is difficult to use terms regarding them which are free from objections'.[38]

Through Lucy's autobiographical relation of her mental reality, as well as the implications of her silences, Brontë comes nearer than much of the scientific discourse to finding a language of mental machinery that makes way for an understanding of the effect of external blows or shocks upon the mind, and the enduring effect of stunned processes of memory upon present lived consciousness. Where Holland mobilises the discourse of self-control to establish the distinction between simple memory and willed recollection as 'one of the best marked lines of demarcation between the human intellect and that of other animals',[39] Brontë takes the problem of loss of control over deliberate recollection to penetrate to the heart of one of the most difficult – and least well understood – elements of human experience: the recurrence of traumatic images or flashbacks which are simultaneously unrecoverable (in any meaningful totality) and yet irrepressible. Lucy's struggles with what Holland described as mnemonic 'combinations' which 'come *unbidden and vaguely* into the mind'[40] paint as somewhat simplistic John Barlow's claims for each individual's absolute power 'to prevent or control' the most extreme forms of mental disturbance, whether arising from '[m]orbid affections of the nervous system and brain' or non-physiological '[m]orbid affections of the Intellectual force', including '*Inefficiency*' in control of 'the appetites or instinctive emotions'.[41] In Barlow's view, '[s]o nicely balanced indeed is the machine, that a grain can turn it to either side, but it is in the power of the will to cast that grain'[42] or, by implication, to re-cast or remove it. In Brontë's *Villette*, it is no straightforward matter even to locate the lodged grain of trauma, let alone perform its safe extraction.

As well as being linked to women's nervous system and reproductive organs, hysteria for the Victorians was clearly interwoven with the emotions and the passions. Thomas John Graham – author of *Modern Domestic Medicine* (a text well annotated by the Brontës' father Patrick) and namesake of the fictional doctor in *Villette*, Dr John Graham Bretton – aligned hysterical behaviour with both the thwarted expression of sexual desire and the under-use of women's reproductive organs (a meeting of the emotional and the physical). In terms which point up the fraught ambivalence of Lucy's 'happy' (p. 496) ending in *Villette*, with her ideal marriage left unconsummated, Graham's medical reference book directs that 'the surest remedy is a happy marriage'.[43]

Ilza Veith's *Hysteria* cites Robert Brudenell Carter as a perceptive writer on the subject (perhaps 'too embarrassingly perceptive for his Victorian compatriots'): as Carter elucidated, strong emotion will always 'manifest itself by the production of certain effects, either upon the intellect and will, or upon the physical organism'.[44] According to George Man Burrows, hysteria in females was often perceived to be the same as hypochondriasis in males, an intriguing consideration in light of Lucy's efforts to renounce hysterical symptoms in herself by observing them cogently in others, and instead identifying almost rapturously with the manifestations of hypochondriasis exhibited by that figurehead of male authority, the King of Labassecour. In Burrows's view, the 'respective phenomena' of hysteria and hypochondriasis 'prove that they are different diseases', though 'of the same family'.[45] Descriptions of hypochondriasis from a range of nineteenth-century writers show that it was often understood in markedly anti-hysteric terms: morbid over-thinking, with volition overtaken by fine-tuned egotistical concern, rather than the uncontained emotional excesses characterising popular conceptions of the phases of female hysteria.

Several recent commentators construct their analysis of *Villette* in terms of hysteria. While Athena Vrettos suggests that 'the hysterical first-person narrator' projects neurosis onto narrative form, she qualifies her use of 'hysteria' in an endnote, recognising that it was often used as a catch-all for female nervousness.[46] It is important to note, however, that not all analyses focus on, or even view Lucy as the sufferer of, mental illness in any form.[47] On my reading, the closest Lucy – and her narrative – comes to hysteria (aside from her agitated reactions to the mysterious nun) are those occasions when she reveals the alternately desirable and unbearable effects of the hysteria of others upon her own emotional world. Brontë's narrator unwittingly reveals that a 'hysterical cry' (p. 14) from Polly would ease her own indirectly disclosed anxiety, and conversely later admits that her severe 'impatience' with pupils who sob hysterically at M. Paul's departure masks the 'truth', that she 'could not bear' their 'gasping' emotion (p. 439). Crucially, by the time of Lucy's trance-like wanderings in the night fête scene, she is able to declare – with accuracy – that, '[t]empered by late incidents, my nerves disdained hysteria' (p. 470).

What then of Lucy's self-proclaimed hypochondriasis, and the interaction of this mode of understanding disordered mental experience with the impact of traumatic memory shown by Brontë to be at work throughout both Lucy's life and her later rendition of that life in (fictional) autobiographical form? Hypochondria first enters the text with Dr John as an all-encompassing diagnosis-by-default (as such, not unlike hysteria) for a range of physical and mental complaints with causes otherwise imperceptible or inexplicable. Brontë makes John's continued presence in *Villette* contingent upon the condition: he becomes the school's doctor only because 'the respectable Dr Pillule' has been 'summoned' by a 'rich old hypochondriac' (p. 96). Madame Beck's daughter, as John's patient, makes possible Lucy's explicit discussion of the 'simulation' of disease as a means of securing 'attentions and indulgences' (p. 97).

The obvious difference between Désirée's feigned low spirits and Lucy's 'dreadfully low-spirited' (p. 183) period of intense hopelessness – that one is false, the other experienced with real pains and sensations – points to an even more important distinction.

Hypochondria was and continues to be viewed as a 'morbid state of mind, characterized by general depression, melancholy, or low spirits, for which there is no real cause': this according to the *Oxford English Dictionary*, which, providing an extract from *Villette*, sagaciously chooses a passage describing the King's affliction (not the central character's) to demonstrate historical usage. In the nineteenth century, hypochondriasis – hypochondria in its pathological aspect, although the terms were often used interchangeably – was seen to be based in some genuine chronic complaint of the nerves (often accompanied by indigestion), but was similarly characterised by the sufferer's unfounded belief in, and anxiety concerning, more serious bodily disease: an unstable exchange between pure neurotic imagination and fretful exaggeration. In contrast, Brontë shows that Lucy's morbid mental suffering during the long vacation is firmly based in realities both physical and, significantly, emotional: the sudden disappearance of a social framework, which has provided her with a 'prop' (p. 156) in the avoidance of agonising remembrance. Dr John's diagnosis of 'Hypochondria' is followed by Lucy with '[a]cquiescence and a pause': his remarks that, in the absence of persistent physical illness, medicine cannot assist 'bore the safe sanction of custom, and the well-worn stamp of use' (p. 183). When next hypochondria is mentioned, it is Lucy's empathetic viewing of the King's face; Lucy later fosters a likeness with another 'imperial hypochondriac' (p. 274), Nebuchadnezzar, as she disclaims her ability to speak out, and crucially to be understood, regarding her experience of the effects of 'solitary confinement' (p. 273) upon the mind.

Although Lucy registers her frustration at society's unwillingness to understand 'privations' (p. 274) neither based nor evidenced in the corporeal, her adoption of the (vague yet fettering) diagnosis ultimately provides ironic justification for avoiding the kind of speech act – and self-knowledge – after which she appears to hanker. Envisaging herself as a hypochondriac allows Lucy to situate her understanding of pain in both the present and, to an extent, the realm of the imaginary, thus giving credence (and a name) to immediate suffering while obscuring the need for backward mental glances. Her re-definition by a condition – even a stigmatising condition predicated on unreal or inflated fears and involving deficiency of firmness – is shown to function, in Lucy's mind, as a shield which exonerates her from her often barely registered yet overriding fear of facing the past.

An 1850 review of the anonymous *Confessions of a Hypochondriac* offers a colourful description of the 'poor man's scourge and the rich man's plague': cowering 'beneath the cloak and cowl of self-loathing and disgust of life', the sneaky egotist Hypochondriasis is 'averse to everything except the bitter consciousness of its own all-absorbing sense of perpetual solitude and woe!'[48] As Holland observed in a new chapter 'On Hypochondriasis' in the third edition of *Medical Notes and Reflections*, the presence of this morbid state of mind is, like many others, a question of 'excess in degree, and in permanence, of those feelings which traverse at times the soundest and most equable minds'.[49] Delusive but not delirious, the hypochondriac will rationally describe the most illogical, 'unreasonable' feelings.[50] The feelings of which Brontë's narrator complains – feverish wretchedness, 'agonizing depression' (p. 159) – are certainly in excess, but they are hardly imaginary or unreasonable. Although her pain is often registered in violent physical terms, she does not invent bodily symptoms unrelated to her actual distress. Apart from her fever during the long vacation, the youthful Lucy is remarkably physically fit, perhaps sturdier than many nineteenth-century fictional heroines. She is certainly preoccupied by the manifold sensations of her unhappiness, yet far from eagerly 'looking in the glass' as does Burrows's hypochondriac,[51] Lucy studiously avoids it. Hers is no narcissistic fascination, and the legibility of misery on her visage causes only grim regret (p. 448, and see earlier p. 210).

In *Modern Domestic Medicine*, Graham advised that the physician should 'humour' the hypochondriac's 'foibles' and 'seem to fall in with his views'.[52] In what amounts to a Brontëan reversal of medical authority or at least an attempt to regain some of the patient's agency, it is rather Lucy in Brontë's text who, with control of the narrative viewpoint, humours Dr John's diagnoses – adopting them, but hesitantly, and expressing her own experiences of the labels he assigns her, often at variance with their official definitions.

Lucy declares herself too ignorant to give the reader her impressions of the concert held in the hall of the principal musical society, at which the royal family of Labassecour are present, but she speaks with great instinctive surety on the inner mental landscape of the King (despite her protestations regarding his unique physiognomical 'hieroglyphs' and the 'strain' on her 'powers of vision' given that this is her first encounter with royalty):

> Well do I recall that King – a man of fifty, a little bowed, a little gray: there was no face in all that assembly which resembled his. I had never

read, never been told anything of his nature or his habits; and at first the strong hieroglyphics graven as with iron stylet on his brow, round his eyes, beside his mouth, puzzled and baffled instinct. Ere long, however, if I did not *know*, at least I *felt*, the meaning of those characters written without hand. There sat a silent sufferer – a nervous, melancholy man. Those eyes had looked on the visits of a certain ghost – had long waited the comings and goings of that strangest spectre, Hypochondria. Perhaps he saw her now on that stage, over against him, amidst all that brilliant throng. Hypochondria has that wont, to rise in the midst of thousands – dark as Doom, pale as Malady, and well nigh strong as Death. Her comrade and victim thinks to be happy one moment – 'Not so,' says she; 'I come.' And she freezes the blood in his heart, and beclouds the light in his eye.

Some might say it was the foreign crown pressing the King's brows which bent them to that peculiar and painful fold; some might quote the effects of early bereavement. Something there might be of both of these; but these as embittered by that darkest foe of humanity – constitutional melancholy ... Full mournful and significant was that spectacle! Not the less so because, both for the aristocracy and the honest bourgeoisie of Labassecour, its peculiarity seemed to be wholly invisible: I could not discover that one soul present was either struck or touched. (pp. 213–4)

On observing the King, the admirable insight of Brontë's narrator (her recognition of his sadness) belies a simultaneous blindness (in her closed reading of its causes). Emphasising Lucy's marked eagerness to deny the impact of the past on the King's present condition – her strange certainty that while 'some might quote the effects' of 'early bereavement' (p. 213) as causal, they would be mistaken – Brontë troubles the reader's reception of Lucy's words, drawing attention to the unstated currency of the past in the central character's own despondency. Designing a scene of layered watching, Brontë ingeniously invites the reader to look beyond the face of Lucy's text, even as Lucy attempts to read the face of the King. The narrator's words reveal more than she knows. The startling impression of possibilities adamantly overlooked or closed off too abruptly within Lucy's analysis of another, raises again the 'spectre' (p. 213) of that which is missing from, or has been as determinedly consigned to the margins of, the narrator's own text of self.

Not only was hypochondriasis understood to present 'many varieties and shades, of which nostalgia perhaps is one'; '[v]iolent grief' was also acknowledged by Victorian commentators to produce 'the same train of feelings, and sometimes in an equal degree',[53] as did constitutional melancholy. Furthermore, hypochondriasis was seen in some instances specifically to arise in the aftermath of 'some powerful and afflicting moral event', whereby the hypochondriacal suffering may be aggravated precisely 'because memory perpetually recalls [the precursory event or cause], and opposes the effect of remedial aid'.[54] In any case, in the expectation of suffering to come, hypochondria creates its own traumatic history. The compounding memory of each visitation was frequently described as clouding hope for eventual cure or release. As Charles Lamb, cited by Millingen, put it: 'Black thoughts' are 'continually / Crowding my privacy', arriving 'unbidden, / like foes at a wedding' – intrusive memories which, materialising from a treacherous past, threaten to spoil the promise of future joy and attainment of any semblance of self-unity.[55]

Good Grief! Failures of Language and of Communication

Thomas Laycock voiced a cultural sensibility shared by many Victorian commentators when he remarked that '[g]rief is the most common and most injurious passion of humanity, with the exception of its sister emotion anxiety'.[56] Brontë's narrator is shown to suffer both in extreme degree, and her grief and anxiety are compulsive and compounding. By striving to subdue the memory of their source, Lucy entombs emotions and prevents their release – a mode of existence illustrated in her reaction to John Bretton's patchy correspondence. Having written poignantly of the pathetic 'blank' (p. 267) of an existence bereft of meaningful communication, Lucy inscribes also the attempted deletion of unsustainable yearning: by 'hiding my treasure' – thrusting his letters into an air-tight 'stoppered' bottle, then this into a tree-hollow which she covers with mortar – 'I meant also to bury a grief' (p. 296). Crucially, however, the grief itself is not deprived of life, merely concealed or 'interred' (p. 296).

Of the 'full mournful and significant ... spectacle' of the 'moody King' of Labassecour – a face overcast with the dread 'Doom' of Hypochondria; now distracted by the pleasant conversation of his smiling wife, now 'invariably relapsed' into gloom (p. 214) – Lucy feels herself to be a privileged observer. His sufferings, she insists, remain 'invisible' to others. This failure of reading (both her sense of the failed perception of others and, as I have suggested, the resonances concealed by Lucy's deeply empathetic though temporally and affectively superficial view) is repeated with variation in the wider failure of communication and interpretation of individual human trauma in *Villette* as a whole.

As with Lucy's confession to a Catholic priest, unheard by the reader and its bounds unknown, Brontë leaves open the question as to whether

Lucy fully embraces what her novelistic predecessor Jane Eyre referred to as the 'opportunity of relieving my grief by imparting it' (*JE*, p. 24). In *Villette*, just as Brontë's narrator feels profoundly the absence of language sufficient to explain the unquiet metal state that she awkwardly aligns with hypochondria (which at least moves it out of the limitingly gendered realm of 'hysteria'), she complains also of the lack of a language of trauma in which the rough expanses and 'perils' (p. 66) of witnessing and aftermath might be communicated to those unfamiliar with that 'region' (p. 181). When Mrs Bretton questions Lucy in markedly general terms about 'gone-by troubles' (p. 177), inability, embarrassment and stubborn refusal to answer coincide. In Lucy's image, the 'crew' of Mrs Bretton's 'stately ship ... could not conceive' of her dire alternate reality, 'so', as 'half-drowned life-boat man', she keeps her 'own counsel, and spins no yarns' (p. 181).

Discussing what he sees as Brontë's rejection of orality in favour of the 'material possibilities of print' and 'impersonal' intellectual labour for female authors, Ivan Kreilkamp insists that Brontë's final novel disallows her characters even the general possibility of 'finding satisfaction in free speaking'.[57] Contrary to this view, I suggest that Brontë does not simply hold 'that to give voice to personal emotions is to speak in a foreign and obsolete tongue'.[58] Rather, she shows her narrator's frustration when personal and cultural silences coalesce. It becomes apparent that Lucy's intense emotional experiences, arising from an unuttered traumatic past, are on one level unable to be put into words by the stunned, nervous individual; and, on another, interconnected level, have thus far historically been devoid of a language in which such experiences might be made comprehensible to others, across barriers of class, gender and comparative suffering. As Brontë's narrator explicitly acknowledges, '[s]ilence is of different kinds, and breathes different meanings' (p. 347). Her momentary powerlessness to proclaim that other emotional extreme, unequalled joy – speech 'dissolved or shivered in the effort' (p. 487) – is nowhere near as harmful as her ongoing inability to describe the sources of her agony.

The inadequate contemporary vocabulary available for attempted descriptions of the mental pain of trauma is one which Brontë, in *Villette*, works to develop and diversify. Of the numerous references to pain in the novel, the majority are of emotional and psychic origin, even if registered in bodily sensation: the 'pain or malady of sentiment' can go through one's 'whole system' (p. 221), winding itself 'wirily round [the] heart' (p. 231), and 'corroding' (p. 479) when 'long accumulating, long pent-up' (p. 162), causing Lucy to question of torturous despair 'what bodily illness was ever like this pain?' (p. 449).[59] Lucy's refusals to enunciate her distress to those populating her history are, after all, explained in highly evocative terms to the reader of her latterly composed narrative. In a sense, Lucy demonstrates the potential for crossing perceived limits of effective communication by registering, despite barriers of culture and translation, the baffling expressions of Nebuchadnezzar (p. 274) as antecedent to her own twin frustrations. While Lucy's 'descriptions of vocal repression are often so eloquent as to define a poetics of withheld speech', Brontë does not condone Lucy 'simply shutting up' as Ivan Kreilkamp suggests.[60] Each cry 'devoured', each 'start' (p. 465) forbidden, reveals rather the pain – and the seeping away of potential for self-understanding and for nurturing dialogue between individuals and cultures, doctors and patients – contained in the spaces where silence becomes sinister, a strangled surrender to trauma.

Lucy's silences are themselves points of 'crisis' (p. 444) which serve only to reinforce and perpetuate her ongoing traumatic reality. Towards the end of her record, she proves unable to make the slightest noise to alert M. Paul to her presence, even when driven by the strongest motivation to do so. Eclipsed from his view by Madame Beck, Lucy is 'struck' by 'the total default of self-assertion' (p. 444). Her inability to communicate present desire and past connection breeds grief and compounds inexpressibility, launching her into what 'seems, to my memory, an entire darkness' (p. 445). M. Paul's absence even when known to be alive and in the same city is 'unendurable' to Lucy, and she bears the devastations of memory 'alone' for those 'certain minutes' before the arrival of his secret missive: yet another cycle of traumatic repetition (and foreshadowing), set up here by Brontë to be carried through to the novel's close. In their heady exchange of love and assurance of intention, before M. Paul removes her to the Faubourg Clotilde, all is contained in brief words rooted in the present and future tense: 'My heart will break! ... Trust me!' (p. 481). When, finally, Lucy declares 'I want to tell you something ... I want to tell you all' (p. 490), it remains unclear just what that 'all', that something, might encompass. Released from imprisoning solitude and mental stagnation by the promise of a brighter future, she recalls that:

> I lacked not words now; fast I narrated; fluent I told my tale; it streamed on my tongue. ... All I had encountered I detailed, all I had recognized, heard, and seen ... the whole history, in brief ... rushed thither truthful, literal, ardent, bitter. (pp. 490–1)

No doubt it is truthful and bitter, but of which 'whole history' does she speak of speaking (briefly) here? Lucy's condensed report reveals that she relates the entire urgent history of Madame Beck's plot to keep them apart: a recent past that is readily available to her in the literal terms of conscious recall. M. Paul's intent, tender focus is a listening style which Lucy has never before experienced, and one she might hope will be replicated should the autobiography she inscribes reach her unidentified 'reader' (p. 371). Having found her ideal kinsman, partner, and audience in one; 'the next day – he sailed' (p. 492). Amidst all the fears ghosting this text, there stands the very real possibility that Lucy Snowe has never uttered, let alone captured or assimilated, the subjective truth of her traumatic past, either in writing or in speech. The final chapter may be entitled 'Finis', but Brontë's shaping of Lucy's autobiographical record refuses to assign closure either to romantic yearning or to the developing potential for narrative as healing in the prolonged life-aftermath of the traumatic experience of early familial loss.

The narrative's mnemonic sleights of hand are deeper-set than Lucy's conscious evasions and the resulting irony is imbued with the most severe pathos, because it is the cruel irony of a doubled and repeated experiential structure where the effect of trauma on memory is so shattering and total as to efface the narrator's awareness of the need for expression of that traumatic event at all. The remaining life of the fictional Lucy Snowe hangs like the dead weight of solitary imprisonment within an inescapable mental nightmare. The reader recoils from the 'perfect work' of destruction and from the narrator's fragmentation into 'a thousand weepers', straining for her beloved's voice amidst disturbingly dark sunlight and the mortal 'hush' (p. 495).

Exhorting the reader to 'scout the paradox' of initial happy 'torture' (p. 493) in absence, the narrator sets forth the possibility of a final suspended state, where her happiness might be located primarily in past memory, rather than in present joy. Foreseeing the way in which those powers of memory, so often avoided, might be harnessed to her advantage, Lucy had implored the 'White Angel' or sustaining light of a pristine moment in time with M. Paul to 'linger; leave its reflection on succeeding clouds; bequeath its cheer to that time which needs a ray in retrospect!' (p. 488). Even in her moment of supreme joy, she (not unreasonably) forecasts future storm – or perhaps it is the elderly Lucy, reaching that point in the narrative act, who imposes retrospectively this wish for sustaining memory, as well as the knowledge of impending darkness, on the lived minutiae. As was noted in *The Journal of Psychological Medicine and Mental Pathology* in 1850, joy 'is real and active, and centred in the present moment'; when joy springs from 'quiet and prospective' hope combined with 'bright memory' and 'present love', it is 'a state on earth little short of Elysium'.[61] If Lucy's remembered joy sustains her during the first three years, it is a flame, she admits, kept alive only by the 'full-handed, full-hearted' (p. 494) letters she receives from the absent traveller. We have already seen that, for Lucy, the revisiting of old letters is not enough to 'sustain' (p. 268) contentment. The possibility of love in the present (or future) fades as the flow of letters ceases. Lucy reverts to present tense narration – and memory is indeed 'bright' – positioning her expectant at the end of those three years; this may be blended with the muted hope in the vision of 'union' (p. 496) that their respective Protestant and Catholic heavens enable, but there can no longer be a 'real and active' joy, 'centred in the present moment'.

Might the narrator's silence here speak of the 'heroic composure' John Reid envisaged arising from '[t]ruly tragical misfortune', when the 'mind itself is enlarged by the magnitude of its misery'?[62] Perhaps. But Lucy Snowe is no mere hypochondriac replacing, at long last, imagined afflictions with 'real evils';[63] she is a survivor of trauma, plagued by sorrows which are at once 'justified and nourished by reflection', and yet set apart from full reflection and assimilation by the 'dimness of perception' and 'confusion of ideas' occasioned by what John Reid refers to as the overwhelming impact of their 'blow'.[64] Having taken most of her narrated youth to move towards mental recovery through the recovery of stunned memory (constructing spoken, though incomplete, testimonies of confession and release), Brontë's heroine endures, at novel's close, a second major wreck. M. Paul's death is steeped in the veiling metaphors of her first recurrent flashback of familial collapse – or, more accurately perhaps, in bringing to actuality the nightmare of stormy oceanic destruction, his loss informs the language which has been employed, in retrospect, to describe the earlier sorrow. These two intersecting traumas create a double aftermath, the effects of which, while reverberating backward through the life as remembered, pass unremarked by Lucy in a decades-long 'pause at once'. Unless Lucy's 'pause' contains the consolatory reflection of the bereaved, or a complete reversal of her relationship with mnemonic processes to enable a life 'nourished' (p. 494) by retrospective joy, it appears that in the blank space of her final twin silences Brontë's narrator figures herself as a traumatic sufferer to the very last, binding together willed repression and the unconscious blocking of memory to live out the remainder of her days under the shadow of that calamitous cloud which, in Reid's formulation, 'no sunshine can disperse'.[65]

As this chapter establishes, with reference to the work of John Barlow, Thomas Brown, George Man Burrows, William Carpenter, John Conolly, James Douglas, Thomas John Graham, Henry Holland, Thomas Laycock, J. G. Millingen, James Cowles Prichard, and John Reid, among others, Victorian hypochondriasis (or, for women, hysteria) and Victorian trauma have in common their involvement of both body and mind, the crucial distinction being that there is an internal source (whether corporeal or mental) for hypochondriasis, whereas a reaction to an overwhelming external event is central to trauma. For Charlotte Brontë, cultural understandings of the idea of the human are constrained by highly gendered medical categories which – through their delimited perspective – risk transforming sufferers into something less than human. For Brontë, it is not the case that to be human is to be traumatised – far from it. Nor is it the case that to be a woman is to be hysterical. Rather, Brontë's novel shows how the Victorian diagnostic categories 'hysteric' and 'hypochondriac' have been deeply gendered by cultural judgements – and her fiction provides a startlingly forceful original argument for understanding mental distress and mnemonic dysfunction in terms that take an appreciation of the 'human' beyond the limiting boundaries of gender stereotype. Brontë's *Villette* demonstrates – in a way that might be as instructive for twenty-first-century reading audiences as it was for those of the nineteenth century – that modes of insight which allow both for the centrality of memory and for communion between passion and reason, materialist views and intense emotion, body and mind, past and present, will be necessary to move toward a fuller understanding of human thought and, in particular, human response in the face of catastrophe.

Notes

1 Alexandra Lewis, 'Memory Possessed: Trauma and Pathologies of Remembrance in Emily Brontë's *Wuthering Heights*', in *Acts of Memory: The Victorians and Beyond*, Ryan Barnett and Serena Trowbridge, eds. (Newcastle upon Tyne: Cambridge Scholars Publishing, 2010), pp. 35–53.
2 Alexandra Lewis, 'Stagnation of Air and Mind: Picturing Trauma and Miasma in Charlotte Brontë's *Villette*', in *Picturing Women's Health*, Francesca Scott, Kate Scarth and Ji Won Chung, eds. (London: Pickering and Chatto, 2014), pp. 59–76, p. 61.
3 Lewis, 'Stagnation of Air and Mind', pp. 59–76.
4 M.M. Bakhtin, *The Dialogic Imagination: Four Essays*, Carl Emerson and Michael Holquist, trans. (Austin: University of Texas Press, 1981), p. 418.
5 Lewis, 'Memory Possessed', pp. 35–53.
6 James Cowles Prichard, *A Treatise on Insanity and Other Disorders Affecting the Mind* (London: Sherwood, Gilbert, and Piper, 1835), pp. 12–13.

7 John Conolly, *An Inquiry Concerning the Indications of Insanity* (London: John Taylor, 1830; repr. London: Dawsons, 1964), p. 143.
8 Ibid., at p. 60.
9 Ibid., at pp. 93, 174, 190, 226–7, 193.
10 J. G. Millingen, *The Passions; or, Mind and Matter. Illustrated by Considerations on Hereditary Insanity* (London: John and Daniel A. Darling, 1848), pp. 155, 138.
11 Millingen, p. 90. Quoting Broussais, *De l'Irritation et de la Folie* (1828).
12 William Carpenter, *Principles of Mental Physiology* (1874), 4th edn. (London: Henry S. King & Co., 1876; repr. 1877), pp. xliv–v. See J. S. Mill's *Autobiography* (1873; New York: Liberal Arts Press, 1957), p. 109. Carpenter's Prefaces to the first editions published in New York and London differ, with an additional passage of three to four pages appearing only in the first American edition yet entering in reworked form the fourth London edition of 1876.
13 Ibid., at p. 455.
14 Ibid., at pp. 434, 436 (citing the *Quarterly Review*, Oct. 1871, p. 318).
15 Ibid., at p. 436.
16 Jenny Bourne Taylor, 'Forms and Fallacies of Memory in 19th-Century Psychology: Henry Holland, William Carpenter and Frances Power Cobbe', *Endeavour*, 23:2 (1999), 60–4, p. 61.
17 Lewis, 'Stagnation of Air and Mind', p. 67, citing J. G. Millingen, *The Passions*, p. 160.
18 James Douglas, *On the Philosophy of the Mind* (Edinburgh: Adam and Charles Black; London: Longman, Orme, Brown, Green, & Longmans, 1839), p. 195.
19 Ibid.
20 Thomas Brown, *Lectures on the Philosophy of the Human Mind*, 4 vols. (Edinburgh: W. and C. Tait; London: Longman, Hurst, Rees, Orme, and Brown, 1820), 2: 366. Brown's popular work on mental philosophy (by 1851 it had reached its 19th edition) is one of a number of such texts that would have been available to the Brontës at the Keighley Mechanics' Institute Library: see Clifford Whone, 'Where the Brontës Borrowed Books: The Keighley Mechanics' Institute Library Catalogue of 1841', *Brontë Society Transactions*, 11 (1951), 344–58.
21 Douglas., at p. 196.
22 Ibid., at p. 197.
23 Ibid.
24 Ibid., at p. 202.
25 Ibid., at p. 196.
26 John Reid, *Essays on Hypochondriasis, and Other Nervous Affections* (1821), 3rd edn., rev. (London: Longman, Hurst, Rees, Orme, and Brown, 1823), p. 79.
27 Ibid., at p. 423.
28 Ibid., at pp. iii–iv.
29 Ibid., at p. 422.
30 Ibid., at pp. 422–23.

31 Ibid.
32 Henry Holland, *Medical Notes and Reflections* (1839), 2nd edn., rev. (London: Longman, Orme, Brown, Green, and Longmans, 1840), p. 600.
33 Thomas Laycock, *A Treatise on the Nervous Diseases of Women; Comprising an Inquiry into the Nature, Causes, and Treatment of Spinal and Hysterical Disorders* (London: Longman, Orme, Brown, Green, and Longmans, 1840), p. 86.
34 Holland, *Medical Notes*, pp. 599–638, p. 602.
35 Ibid., at pp. 602–3, emphasis added.
36 Ibid., at p. 603.
37 Douglas, p. 148.
38 Ibid., at p. 193.
39 Henry Holland, *Chapters on Mental Physiology*, 2nd edn., revised and enlarged (London: Longman, Brown, Green, Longmans, & Roberts, 1858), p. 154. Holland's thinking on lines of demarcation had developed since the first edition (1852), which grew from *Medical Notes and Reflections* (1839). The remark on human and animal intellect in the 1858 edition does not appear in that of 1852, although much of Holland's discussion of the distinction between simple and willed memory remains the same. It is worth remembering that Holland and Darwin were distant cousins and Darwin is known to have sought medical treatment from Holland in 1840, by which time Darwin's theory of evolution was forming (though his work towards definitive publication was still in progress).
40 Ibid., at p. 155, emphasis added.
41 John Barlow, *On Man's Power Over Himself to Prevent or Control Insanity* (1843), 2nd edn. (London: William Pickering, 1849), pp. 16–17. For Barlow (unlike Thomas Laycock and J. G. Millingen), and despite his title, women are equally able (and accountable) to exercise 'complete self-control in the midst of structural disease' (p. 36) and functional 'irrationality' (p. 41). Echoing the popular view that women are the 'high-minded purifier[s] of society' (p. 71), Barlow refreshingly attributes this not to refinement of emotion but superiority of thought and intellect: given her 'feeble muscular power...her strength must be that of knowledge' (p. 71). He attributes the over-representation of women in Asylums to unequal opportunities for mental fulfilment, entrenched in all forms of nineteenth-century culture (p. 69).
42 Ibid., at p. 65.
43 Thomas John Graham, *Modern Domestic Medicine; or, A Popular Treatise Illustrating the Character, Symptoms, Causes, Distinction, and Correct Treatment, of All Diseases Incident to the Human Frame; Embracing All the Modern Improvements in Medicine, with the Opinions and Practice of the Most Distinguished Physicians. The Whole Intended as a Medical Guide for the Use of Clergymen, Families, and Students in Medicine* (London: Simpkin and Marshall et al., 1826), pp. 350, 348.
44 Ilza Veith, *Hysteria: The History of a Disease* (Chicago: University of Chicago Press, 1965), pp. 200–1; quoting Carter, *On the Pathology and Treatment of Hysteria* (London: John Churchill, 1853).

45 George Man Burrows, *Commentaries on the Causes, Forms, Symptoms, and Treatment, Moral and Medical, of Insanity* (London: Thomas and George Underwood, 1828; repr. New York: Arno Press, 1976), p. 466.
46 Athena Vrettos, *Somatic Fictions: Imagining Illness in Victorian Culture* (Stanford: Stanford University Press, 1995), pp. 50, 197. See also Beth Torgerson, *Reading the Brontë Body: Disease, Desire and the Constraints of Culture* (New York: Palgrave Macmillan, 2005), pp. 59–88.
47 See Janice Carlisle, 'The Face in the Mirror: *Villette* and the Conventions of Autobiography', *ELH*, 46 (1979), 262–89, and Angela Hague, 'Charlotte Brontë and Intuitive Consciousness', *Texas Studies in Language and Literature*, 32:4 (1990), 584–601. Hague's interest is in the relationship between the conscious mind and 'the larger interconnected reality that lies beyond' – the way the 'nonrational dimension of experience becomes embedded in the human psyche' (p. 600). This provides a point of departure for my focus on the relationship between the conscious mind and the larger reality that lies *within*: not the supernatural but the unseen workings of memory, spurred by external traumatic events similarly 'embedded' in the mind.
48 Anon., 'On Hypochondriasis', *Journal of Psychological Medicine and Mental Pathology*, 3 (1850), 1–14, p. 1.
49 Henry Holland, *Medical Notes and Reflections*, 3rd edn. (London: Spottiswoode, 1855), p. 411.
50 Burrows, p. 471.
51 Ibid.
52 Graham, p. 346.
53 Burrows, pp. 466–7.
54 Ibid., at p. 482.
55 Millingen, p. 77, citing Lamb, 'Hypochondriacus', in *Miscellaneous Poems* (1841).
56 Laycock, p. 176.
57 Ivan Kreilkamp, *Voice and the Victorian Storyteller* (Cambridge: Cambridge University Press, 2005), pp. 125, 144.
58 Ibid., at p. 145.
59 See Sally Shuttleworth's discussion of the way Lucy's description of mental pain draws 'decisively on current theories of physiological psychology': *Charlotte Brontë and Victorian Psychology* (Cambridge: Cambridge University Press, 1996), p. 234.
60 Kreilkamp, pp. 152, 145.
61 Anon., 'The Passions', review of scientific works, *Journal of Psychological Medicine and Mental Pathology*, 3 (1850), 141–64, 159.
62 Reid, pp. 420–1.
63 Ibid., at p. 416.
64 Ibid., at pp. 422–3.
65 Ibid., at p. 424.

CHAPTER 5

Charlotte Brontë and the Listening Reader
Helen Groth

Charlotte Brontë's fictional worlds are filled with voices and sounds. Her narrators compel readers to listen in, attend and empathise, as many of her contemporary reviewers noted with some unease. Moving through a series of environments and scenarios where the limits of human responsibility for the other are continually tested, Brontë draws the reader into a dynamic process of physiological and psychological response. The reader is invited to hear and to feel alongside central characters whose minds are jarred by strange sounds, haunted by mysterious voices or estranged by the hubbub of alien, urban, natural or domestic environments. By attuning the ear of her readers to the subtle differences between sounds and voices, to how, for example, the sense of hearing is peaked when a character enters a new environment or encounters new people, Brontë creates a more contingent experientially determined understanding of how the mind of a young person is formed by circumstance. This acoustic stress correspondingly aligns with Victorian physiological models of how minds function and respond to environmental and social stimuli. Taking these parallels as a starting point, this chapter will initially trace the associations made between sound, silence and mind in the Victorian critical reception of Brontë's work and then progress on to a series of close analyses of scenes drawn from *Jane Eyre* and *Villette*, as well as selected verse, that exemplify Brontë's careful construction of a rich array of literary soundscapes that trained her readers to listen to otherwise silenced voices and unfamiliar sounds as prompts for thinking through the limits of human experience and cognition.

As Jonathan Sterne has observed, the early decades of the nineteenth century generated 'new ways and valorized new constructs of hearing and listening'.[1] Paralleling these shifts in understandings of auditory perception, the mass production of literature in the early decades of the nineteenth century is often narrated as a falling into silence of a once-vibrant, communal oral culture.[2] Countering this view, David Vincent has persuasively argued that the opposite was true. In contrast to the seminal plaints of Roland Barthes, Walter Benjamin and Walter J. Ong, to name just a few, Vincent insists that 'the sound of the human voice was magnified rather than quelled by the mass production and distribution of prose and verse'.[3] Vincent also challenges the simple opposition of 'the faceless publisher and soundless reader' at the centre of this version of the history of reading, arguing instead that this mythic silence was continuously 'disrupted by men and women reciting, singing, shouting, chanting, declaiming, and narrating'.[4] The history of the novel in particular, has proved the ideal medium for reflecting on how to render other worlds and multiple voices audible through a process of silent reading, as Ivan Kreilkamp, Garrett Stewart and others have argued.[5] Reading voices, as Stewart reminds us, always has an auditory dimension, evoked through a system of textual cues that generate voices in the reader's mind. In this context, it can be argued that Brontë sustained her exploration of literary vocality in the movement between verse and novelistic prose, by placing particular emphasis on the interplay between sound and silence in both forms. This acoustic stress required a particular attentiveness of the listening reader, who eavesdrops on conversations and encounters that take place in the resonant silence of literary form. Listening in this sense is a process that exposes the elusive, intangible and inexplicable fragility of human communication as it moves across and between media and minds.

Victorian Soundings

In contrast to Dickens and other contemporaries, Charlotte Brontë proved an elusive quarry for lionising fans, as Ann Thackeray Ritchie reports in her account of a party hosted in honour of the author of *Jane Eyre* by her father William Thackeray: 'It was a gloomy and silent evening. Every one waited for the brilliant conversation which never began at all. Miss Brontë retired to the sofa in the study and murmured a low word now and then to our kind governess'.[6] Thackeray's guests expected to hear and engage in impassioned and inspired conversation with their favourite heroine, instead they were rewarded with barely audible utterances that revealed nothing. Recounting Adelaide Proctor's version of the event on the following page, Ritchie continues:

> One day Mrs Proctor asked me if I knew what had happened once when my father invited a party to meet [the author of] *Jane Eyre* at his house. It was one of the dullest evenings she had ever spent in her life, she said. And then with a good deal of humour she described the situation – the

ladies who had all come expecting so much delightful conversation, and the gloom and the constraint, and how, finally, overwhelmed by the situation, my father had quietly left the room, left the house, and gone on to his club.[7]

Ivan Kreilkamp argues convincingly, in his study of voice and storytelling in Victorian fiction, that Brontë's silent withholding of the powerful speech of her novels from her disappointed readers in this scene is indicative of her interest in a different kind of writing that was not driven by the embodied charismatic speech of 'a personalized author figure'.[8] Instead Brontë's work opens up a very different kind of space, Kreilkamp argues, that generates a form of interiority belonging not to an individual but to 'a new kind of reader' who was not physically present and required 'a new narrative form suited to address it'.[9]

Brontë's reticence in the Thackeray scene undoubtedly typifies this new form of literary professionalism, which did not require the presence of the author-celebrity or the association of self-expressive speech with authorial achievement. But it also reflects the disciplined alignment between keeping silent and the hard work involved with earning the right to speak that characterises Charlotte Brontë's letters. As she writes to her editor W. S. Williams in early September 1848: 'unless I can have the courage to use the language of Truth in preference to the jargon of Conventionality, I ought to be silent'.[10] Conventional speech, formulaic expression or inauthentic jargon disrupts the mediation of truth – it is a form of noise that distracts rather than informs or enlightens. In the same letter, Brontë dismisses her own poetry in favour of her sister Emily's poems which 'stirred my heart like the sound of a trumpet when I read them alone and in secret' (2: 119). Contemptuous of the gendered dismissal of Emily's poetry as 'rhymes', Charlotte insists 'no woman that ever lived – ever wrote such poetry before – condensed energy, clearness, finish' (2: 119).

Brontë associates a similar clarity and truth with Thackeray in another letter to Williams a month later where she implicitly contrasts the optimism of Ralph Waldo Emerson's transcendental poetics to the scepticism of Thackeray: 'I feel forced to listen when a Thackeray speaks. I know the truth is delivering her oracles by his lips' (2: 129). Difficult as it may be to hear the truth, she urges Williams, it is preferable to the sophistry of the charmer 'charm he never so wisely' – a sentiment that alludes explicitly to Psalms 58.4: 'Even like the deaf adder that stoppeth her ears; / Which refuseth to hear the voice of the charmer: charm he never so wisely'. In this scenario, Brontë is the deaf adder who stops her ears from the naïve charms of Emerson and Williams' argument for religious freedom. While she agrees that 'thought and conscience are, or ought to be free', she insists that human nature cannot at this stage of its development do without 'laws and rules in social intercourse' (2: 129).

Brontë was also deaf to the charms of George Henry Lewes, particularly after his negative review of *Shirley* in the *Edinburgh Review* in January 1850. In contrast to Thackeray, to whom she felt compelled to listen, she associates Lewes with the transitory noise of celebrity in a letter, again to Williams, penned a month after the review was published:

> Lewes has many smart and some deserving points about him, but nothing truly great – and nothing truly great, I should think, will he ever produce. Yet he merits just such successes as the one you describe – triumphs public, brief and noisy – Notoriety suits Lewes – Fame – were it possible that he could achieve her – would be a thing uncongenial to him; he could not wait for the solemn blast of her trumpet – sounding long, and slowly waxing louder. (*Letters* 2:350)

True and enduring fame, such as her own, Brontë implies, requires subtle incremental volume increases, not short, sharp blasts of incendiary and derivative rhetoric. Lewes's fall from grace here is not without some caprice, however, given Brontë felt very differently in the wake of his positive review of *Jane Eyre* in *Fraser's Magazine* in December 1847 which elevated the novel to the level of sublime communication: 'it is soul speaking to soul; it is an utterance from the depths of a struggling, suffering, much-enduring spirit'.[11] Lewes stressed that 'a chord in the breast' of the 'most ignorant reader' 'vibrates sympathetically' in the presence of great literature.[12] If the novel is to rise from 'the poor level of street conjuring' to art, as *Jane Eyre* does, the author must discard 'the empty phantasmagoria of the library' for 'transcripts from the book of life'.[13] Integral to this transcription, he argues, is 'the psychological intuition of the artist', her power 'of connecting external appearances with internal effects – of representing the psychological interpretation of material phenomena'.[14]

This link between mind and sound can be traced throughout the Victorian reception of *Jane Eyre*. Eugene Forcade in the *Review des deux mondes* equates the novel's 'irresistible eloquence' with a modern, some claimed, revolutionary, understanding of mind and emotion.[15] An otherwise negative review in *The Spectator* also noted the intrusive dialogic narrative voice and 'minute anatomy of the mind' as symptoms of an unfortunate new trend in fiction, the favouring of character over plot.[16] Reviewers of *Shirley* continued this line of observation, but in the case of A. W. Fonblanque in the *Examiner* the over-reliance on an unrealistic intellectual dialogue and 'interviews detailed' conspired against psychological depth.[17] *Shirley* fared better at the hands of the reviewer at the

Weekly Chronicle who praised the novel, alongside *Jane Eyre*, as expressing 'the silent thoughts of the time'.[18] Both novels reveal, according to this reviewer, a 'new mind, unhackneyed and thoroughly original' which 'spoke its own thoughts, developed its own experience, delineated human character, with singular power and truth'.[19] Eugene Forcade was equally positive, writing that *Shirley* possessed 'scattered ... brilliant flashes of poetry and the utterances of a reflective mind'.[20] But the 'thundering' *Times*, as Bronte described it, was scathing about the infrequent nature of 'brilliancy' in *Shirley* and stale 'talk' barely mitigated by 'gems of rare thought'.[21] Reviews of *Villette* were equally mixed, but often returned to the theme of sound in their description of the text's reception. The reviewer in *The Examiner* complained that Lucy Snowe's 'needlessly tragical apostrophes' like 'some dirge to the burden of "I can't be happy"' that 'sounds from within' were excessive and should be 'expunged' from the next edition.[22] After his scathing assessment of *Shirley*, Lewes also placed particular emphasis on the acoustics of *Villette* in his review for *The Leader*. He praises Brontë's originality – 'Every page, every paragraph, is sharp with *individuality*. It is Currer Bell speaking too, not the Circulating Library reverberating echos [sic]'.[23] It is this originality that holds the reader 'spell-bound', transcending the formulaic commerce of popular fictions, he argues strenuously, and returns to this theme in a comparative review of Gaskell's *Ruth* and *Villette* for the *Westminster Review*: 'In this world, as Goethe tells us, "there are so few voices, and so many echoes" ... so few persons thinking and speaking for themselves, so many reverberating the vague noise of others.'[24] Noise, here, as with Brontë's criticism of Lewes for succumbing to the noisy attractions of transient celebrity and commerce, is contrasted with the elusive authentic sound of true literary inspiration that Brontë's work typifies.

William Caldwell Roscoe's summative review essay published almost a decade later in the *National Review* assesses all of the Brontës' work, as well as Gaskell's *The Life of Charlotte Brontë*.[25] Contrasting the rhetorical excesses of Gaskell's biography with Charlotte Brontë's subtle critical ear, Roscoe cites Brontë's considered assessment of *Wuthering Heights* at length:

> 'Having formed these beings,' says Charlotte, '[Emily] did not know what she had done. If the auditor of her work, when read in manuscript, shuddered under the grinding influence of natures so relentless and implacable, of spirits so lost and fallen; if it was complained that the mere hearing of certain vivid and fearful scenes banished sleep by night, and disturbed mental peace by day, Ellis Bell would wonder what was meant, and suspect the complainant of affectation'.[26]

Emily may have been unaware of the affective power of her narrative, but Charlotte was acutely conscious of how her sister's words sounded in readers' ears and is critical of affected auditors who might be distracted by the noise of Gothic affinities. While Roscoe quickly dismisses Brontë's defense of her sister as erring on the side of generosity, her implicit privileging of a more careful mode of literary audition unsettles his account of his own response to *Wuthering Heights* as a representative 'modern Englishman' who 'shudders as he reads'.[27]

The power of Charlotte Brontë's words is amplified by Roscoe's elevation of her so far above her siblings. He reproduces 'recorded conversations' from a friend as evidence of her struggle against the isolation of Haworth and emphasises her strong Irish accent as a symptom of her social dependence on her Irish-born father and estrangement from the local Yorkshire dialect; a deeply felt otherness that Roscoe argues explains her interest in pronunciation and the accurate transcription of her characters' voices. He also cites Gaskell's vivid portrait of Brontë in her biography: 'The usual expression was of quiet, listening intelligence; but now and then, on some just occasion for vivid interest, a light would shine out, as if some spiritual lamp had been kindled'.[28] He then parallels this citation with one of many acoustically charged moments in *Jane Eyre* as a palpable example of the way Brontë 'gives voice' to isolation. Longing to see beyond the sequestered domain of Thornfield to the 'busy world, towns, regions full of life I had heard of, but never seen', Jane climbs up to the attic and directly addresses the reader:

> Who blames me? Many, no doubt; and I shall be called discontented. I could not help it: the restlessness was in my nature; it agitated me to pain sometimes. Then my sole relief was to walk along the corridor of the third story, backwards and forwards, safe in the silence and solitude of the spot, and allow my mind's eye to dwell on whatever bright visions rose before it – and, certainly, they were many and glowing; to let my heart be heaved by the exultant movement, which, while it swelled it in trouble, expanded it with life; and best of all, to open my inward ear to a tale that has never ended – a tale my imagination created, and narrated continuously; quickened with all the incident, life, fire, feeling, that I desired, and had not in my actual existence.[29]

The inward ear in this passage, as with Charlotte's response to *Wuthering Heights*, evokes the continuous flow of an associative consciousness, a ceaseless involuntary movement 'quickened' by the external stimuli of events and sensation and the internal stimuli of emotion and desire. There is also an implicit parallel here between Jane's inward ear listening to a never-ending tale and readers hearing Jane's voice in their mind as they read, demanding complicity and attentive listening.

Listening/Sounding/Reading

Brontë's reliance on first-person narration in both *Jane Eyre* and *Villette* privileges the 'inward ear' and the alignment of reading with the involuntary flow of consciousness, a conception of mind that paralleled Lewes's own theory of the 'stream of consciousness' first formulated in *The Physiology of Common Life*.[30] Human consciousness was, for Lewes, a perpetually fluctuating process that he conceived of in spatial terms as a 'mass of *stationary waves* formed out of the individual waves of neural tremors'.[31] Both Jane Eyre and Lucy Snowe describe their interior lives in similarly resonant and spatial ways. Waves of sound and voices penetrate their thoughts as they encounter new people and environments, involuntarily triggering associations and unexpected actions. This interest in evoking the dynamics of an interior life aligns with the ethological theories of character that were being formulated by John Stuart Mill, Alexander Bain and, to some extent, Lewes, in the 1840s. Outlining his vision 'Of Ethology, or the Science of the Formation of Character' in the *A System of Logic*, Mill argued that moral character was forged out of a dynamic interaction between natural impulses and habit, a systemic approach to understanding the physiology of mind that both Lewes and later Bain pursued in *The Senses and the Intellect* and *The Emotions and the Will*.[32] While Bain resisted the automatism of Lewes, preferring to understand humans as agents operating within 'complex emotional and intellectual sequences', as Rick Rylance argues, his interest in hearing and aural sensation as integral to the dynamic relational processes of consciousness (i.e., the habitual mixing of feeling of thought, conscious and unconscious content) has clear affinities with both Mill and Lewes's materialist ethologies of character.[33] Furthermore, hearing is, according to Bain, integral to how we read and retain what we read: 'It is not necessary to read aloud in order to transfer the work from the eye to the voice, a mere whispered or muttered articulation, a mere *ideal* rehearsal, will take, and become coherent'.[34] The mind's retention of associative networks of resonant phrases is also integral to the reading experience for Bain. Striking words repeated may trigger multiple associations: 'I am uttering a connected series of words, and among these, one, two, or three, have by chance the echo of one of the falls of an old utterance; instantly I feel myself plunged in the entire current of the past', a past which may include a passage from John Milton, an extract from Alexander Pope or a 'piece from [Thomas] Campbell'.[35] Bain suggestively described the ear as a 'matrix for holding together our recollections of language' – when we listen we hear echoes of voices we have heard before, lines we have listened to, words we have read.[36]

The processes by which the ear records and recalls the voices of the past, drawing present reflections or reveries into 'the entire current of the past', is one of the recurring motifs in Brontë's narrative poems, published in 1846. Unlike her sisters' poems in the same volume, Charlotte's poems favour narrative and monologue forms. They reveal complex psychological mindscapes acutely attuned to environmental stimuli, to sounds and voices that surface in states of reverie. In 'The Teacher's Monologue', the speaker finds herself alone finally at the end of a long working day:

> THE room is quiet, thoughts alone
> People its mute tranquility;
> The yoke put on, the long task done, –
> I am, as it is bliss to be,
> Still and untroubled.[37]

Liberated from the noise, clatter and trouble of the classroom the speaker can finally hear her own thoughts in 'mute tranquility'. Memories of home flood back, as the poem develops into a lyrical meditation on the porous boundaries between dreaming, reverie and conscious nostalgia:

> Sometimes, I think 'tis but a dream
> I measure up so jealously,
> All the sweet thoughts I live on seem
> To vanish into vacancy:
> And then, this strange coarse world around
> Seems all that's palpable and true;
> And every sight, and every sound,
> Combines the spirit to subdue (p. 108)

Published only a year before *Jane Eyre*, the affinities between this speaker and Jane Eyre are striking. Both personae feel the determining oppression of everyday circumstances acutely, both are fully realised through the devices of interior monologue – apostrophe, first person narration, rhetorical questioning of an unnamed listening reader. To live only in 'this strange coarse world', to be limited by its petty relentless demands, is to live in a world of constant noise. Only in stolen moments of silence, alone in tranquil rooms, can both women imagine another way of being that connects them to other voices and possible versions of themselves.

Lewes, as I have noted, was quick to praise the psychological complexity of both Jane Eyre's and Lucy Snowe's internal monologues, and in terms that consistently aligned with contemporary discussions of the ethological and physiological aspects of character formation. Both novels exhibit a sustained concern with how the mind responds to voices and sounds, strange and familiar. In a climactic scene in *Villette*, Lucy Snowe wanders forth into the streets of Villette, after awakening from an extended period of solitary delirium spent in the 'ghastly' stillness of a 'long dormitory' tortured by images of white beds 'turning into spectres' (p. 160). Her senses peaked by the novelty of her surroundings, Lucy responds to the sounds of church bells and enters the hushed refuge of a Catholic Church: 'The bells of a church arrested me in passing; they seemed to call me in to the *salut*, and I went in'. Once inside she surrenders to the comforts of the habitual movements of the penitents waiting compliantly for their turn to take confession. 'Mechanically obedient,' she responds to the hushed voice of one of the penitents telling her to 'go now', a mechanical or involuntary action that, in turn, triggers an awakening from her reverie, as her mind speeds ahead to consider the ethical implications of what she is doing (p. 161). An English Protestant woman, in a Catholic European country, alienated and perplexed by her experiences, she nevertheless makes an ultimately conscious and pragmatic choice to survive and break her silence, as she informs her reader, as well as her confessor: 'I was perishing for a word of advice or an accent of comfort. I had been living for some weeks quite alone; I had been ill; I had a pressure of affliction on my mind of which it would hardly endure the weight' (p. 161). The following exchange moves beyond the formulaic constraints of ritualised confession, as the priest notes. Given Lucy is unfamiliar with the form, she finds much-needed solace in the 'the mere relief of communication in an ear which was human and sentient, yet consecrated' (p. 162). After reading 'the cast of physiognomy' of her confessor and determining he is French, not Belgian, she listens silently to his attempts to convert her before reverting to an interior monologue, to convince the reader that despite his 'sentimental French kindness' she was deaf to the appeal of his 'heretic narrative' (p. 163). So, while the eye in this scene remains a factor in assessing the moral character of her interlocutor, it is the ear that draws Lucy and the reader into the scene, attending to the ritualised sounds of the Catholic church and its potentially lulling effects on her Protestant ethos.

Brontë's stress on the ear in this scene and elsewhere, aligns with Jonathan Sterne's contention that sound slowly becomes a discrete object of inquiry in the nineteenth century, across a broad range of disciplines from aesthetics and philosophy, to physiology, physics and mechanics. Acoustic experimentation produced new techniques for analysing and reproducing the physiological process of auditory reception in the early decades of the nineteenth century, while philosophical inquiry attempted to harness what Sterne dubs 'techniques of listening' in the 'service of rationality'.[38] As this stress on reason might suggest, the importance of Sterne's work is that it shifts the focus from light and sight as the primary metaphors of Enlightenment conceptions of truth and understanding to the 'role of sound and hearing in modern life'.[39] What this shift reveals is an untold history of the ear that has been crowded out by more prevalent accounts of the domination of the visual in modern culture. It is notable that the term aural was coined in 1847, a significant year for Brontë, as meaning the physiological process of hearing. Previously the term auricular was associated with the ear and it various auditory processes. Auricular also had associations with an oral tradition of storytelling, whereas aural, to quote Sterne again, referred specifically to 'the middle ear, the inner ear, and the nerves that turn vibrations into what the brain perceives as sound'.[40] This context is particularly germane for the reviews cited earlier, most notably Lewes's but of others as well, which describe the reception of Brontë's work in terms of vibration. Although, it is also important to note in the context of the Victorian reviewers' rhetorical stress on sublime/subliminal communication between Brontë and her readers, that there is often an implied theological alignment of sight with the material world, the world of things, images and words and sound with spirit, transcendent voices and pure transparent mediation. Lewes's 1847 review of *Jane Eyre* in *Fraser's Magazine* is a striking instance of the multiple senses of perceptual vibration and communication in these reviews. On one level, communication happens in Brontë's extraordinary novel, to quote Lewes, at the level of 'soul speaking to soul'.[41] In Lewes's case, in line with the Victorian practice of using soul, mind and consciousness interchangeably, this allusion to sublime communication implies the subliminal process of the mind rather than an invocation of a theological sense of spirit or inspired speech. On a more unambiguously secular note in the same review, Lewes's description of the 'chord in the breast' of the 'most ignorant reader' that 'vibrates sympathetically' in the presence of great literature is explicitly psychological in tenor.[42] Inspired vibration is only possible, he argues, if it is sparked by an artist's superior 'psychological intuition' and her power 'of connecting external appearances with internal effects of representing the psychological interpretation of material phenomena'.[43]

Lewes, of course, is not alone in his praise of the communicative force of *Jane Eyre*. There is a long critical tradition that has analysed the formal and cultural implications of Jane as a demanding first-person narrator. As the reader quickly discovers, she speaks constantly and, as Virginia Woolf observed in 1916, 'never leaves us for a moment or allows us to forget her'.[44] Jane assiduously collects incorrect versions of her self and recounts them to the reader insisting all the while that we are listening to the true version. There are many striking scenes that I could focus upon. I will resist dwelling on the powerful dissonance between inner voices and the screams and shouting elicited by her vile cousins or the poignant terror Jane feels when she is condemned to the haunting silence of the red room and move on to her first impressions of Lowood when she arrives 'bewildered with the noise and motion of the coach' (p. 42). As Jane gradually becomes aware of her grim, new, environment she registers the transitions from the 'dreary silence' of the first compartments she encounters to the 'hum of many voices' as she emerges into a dimly lit, wide, long room full of girls of every age (p. 43). Attending closely to this hum of voices, she gradually realises that it was 'the combined result' of the girls' 'whispered repetitions' (p. 44). Against this backdrop Miss Miller's 'cries' and exclamatory commands jar abruptly, providing the first augury of the disciplinary excesses to come. Sensitised by exhaustion after this first encounter, Jane sleeps undisturbed by dreams waking only once 'to hear the wind rave in furious gusts' before 'a loud bell' sounded the beginning of the next day (p. 44).

Throughout the scenes that follow, Brontë uses Jane's heightened registration of sound to both signal her acute sense of her own difference from her environs and invite the reader to listen for further cues and signs of internal disquiet inspired by the unfamiliar hubbub of the school.

In response to yet another exclamatory command from Miss Miller to 'Form classes!' Jane observes, 'A great tumult succeeded for some minutes, during which Miss Miller repeatedly exclaimed, "Silence!" and "Order!"' (p. 45). Tumult seems excessive here, indicating the sensory overload Jane feels and her internal struggle to maintain some semblance of composure. At this stage she is still on the outside listening for cues that might help her make sense of this oddly ritualised behaviour: 'A pause of some seconds succeeded, filled up by the low, vague hum of numbers; Miss Miller walked from class to class, hushing this indefinite sound' (p. 45) until a mysterious 'distant bell tinkled' and, as Jane observes, 'Business now began' (p. 45). This is followed by yet another repetitive cycle of religious instruction before the 'indefatigable bell now sounded for the fourth time'. By this point Jane is weakened by hunger unsated by the miserable repast on offer and unsettled by the other girls' querulous 'whispered words' that elicit a swift rebuff: '"Silence!" ejaculated a voice' – that of a 'little dark personage' of morose aspect (p. 45). A 'dreary silence' results, broken by the shocked whisper of a sympathetic teacher that only Jane overhears, 'Abominable stuff! How shameful!' before she progresses towards the classroom. Once again Jane is confronted by another 'glorious tumult' of protests about breakfast, which Jane registers sympathetically – 'Poor things! It was the sole consolation they had' (p. 46) – before they are called to silence, a peremptory command which 'quelled the Babel clamour of tongues' (p. 46).

This sequence of sound-triggered observations distinguishes Jane as an acute observer of human behaviour, in the wake of the volcanic dissonance and unreliable testimony of the previous Gateshead scenes. Her careful listening to every fluctuating sound implicitly models a way of reading character in context that is, as the novel's Victorian reviewers observed, symptomatic of Brontë's commitment to two interrelated processes: the authentic capturing of her central character's complex interiority and a critical exploration of the relative value of speaking and listening as barometers of truth telling. Only by listening can the reader hear the subtle shift in register, the difference between voices, the varying rhythms and styles of speech, as well as between modes of hearing, for example, the unthinking habitual following of exclamatory commands versus the attentive listening for cues and signs of an unspoken or concealed truth in the above scene.

Sally Shuttleworth has argued that the self-regulation involved in modelling attentive listening to linguistic cues in Brontë's work is intimately tied to phrenological theories of mind that linked sanity with controlling bodily surges of energy. Shuttleworth contends that the measure of Jane's success as a narrator is that 'we as readers believe we are listening to the workings of "sane energy", rather than the ravings of delirium'; an impression amplified by Jane's syntax which often 'gives the impression of surges of energy which are yet restrained within legitimate social bounds'.[45] Few would contest the point that Brontë's representation of Jane's consciousness is shaped by the tension between reason and desire – one only needs to summon the familiar torrid and racially charged encounters with Bertha Mason in which self-interested passion overrides any ethical recognition of the other. But what I want to suggest in this final part of my analysis of the novel is that at climactic moments, such as Jane's early encounters with Thornfield Hall and it various inhabitants,

Brontë places particular stress on listening to non-verbal cues as a means of measuring and testing the limits of human perception and communication. In this context, careful listening requires unsparing acuity on the part of the reader and Jane herself, a cruel reality that gradually dawns as the true owner of the mysterious laughter is revealed.

Guided by the solicitous Mrs Fairfax on her first tour of Thornfield, Jane innocently revels in the gothic pleasures of the third story, wandering past dusty relics 'wrought by fingers that for two generations had been coffin-dust' (p. 105): 'I liked the hush, the gloom, the quaintness of these retreats in the day', she quietly observes (p. 106). Pacing softly along a passage that resembled 'a corridor in some Bluebeard's castle' Jane hears Bertha Mason's laugh for the first time: 'the last sound I expected to hear in so still a region, a laugh, struck my ears. It was a curious laugh – distinct, formal, mirthless' (p. 107). A few pages later, after the famous and much quoted passage in which Jane directly confronts the reader with the prospect of potential revolution of the 'millions [that] are in silent revolt against their lot' (p. 109) – and of women in particular – she returns to the sound of the laugh which she frequently heard when alone and listening with her 'inward ear to a tale that was never ended':

> When thus alone, I not unfrequently heard Grace Poole's laugh: the same peal, the same low, slow ha! ha! which, when first heard, had thrilled me: I heard, too, her eccentric murmurs; stranger than her laugh. There were days when she was quite silent; but there were others when I could not account for the sounds she made. (p. 110)

Tellingly, all of these sounds take place in the silence of Jane's retrospective, first-person narration; this, in turn, elevates her articulate introspection. The reader identifies throughout with Jane's 'inward ear' and listens along as she moves from the suppressed silence of a multitude of women, to re-playing the perplexing and various sounds of what she erroneously believed at the time to be 'Grace Poole's laugh'. Registering each nuance of the laugh, its alternating peal and 'low, slow ha!' as well as the 'eccentric murmurs' and equally perplexing silences, Brontë draws the reader into a complicit accumulation of acoustic evidence that belies the false information Jane has received at this stage in the narrative. At the same time, Jane's inner voice implicitly valorises silent writing and novel reading as the privileged form of communication, a medium that allows otherwise silenced voices to be heard and unspeakable truths to be articulated.

Villette reinforces this valorisation of resonant silence in Brontë's work. Like Jane Eyre, Lucy Snowe – as Kreilkamp has argued – revels in the power of 'withheld speech' and the value of silent reading, thinking and observation.[46] I would argue however, that the dynamics of listening is equally central to this novel. As in *Jane Eyre*, sound triggers intense involuntary responses – both cognitive and somatic – that are channelled into articulate streams of conscious thought, to invoke Lewes's terminology. Lucy's inward ear in varying states of dreaming, reverie and conscious reflection registers sensory or nervous stimuli as either resonant symbolic sound or unwelcome meaningless noise. Early in the novel, immured in the solitude of the invalided Miss Marchmont's bedchamber, Lucy struggles to 'stop my ears' against the 'subtle, searching cry' that she detects amid the violent chaos of the storm outside (p. 38). She describes herself as compelled to listen by something stronger than her will – 'compulsory observation had forced on me a theory as to what it boded' (p. 38). Involuntarily she remembers when she had heard this unearthly Banshee-like cry before and interprets it, from superstitious habit, as an augury of the invalid's immanent death. She listens and trembles at the thought of what this sound signifies: '"Our globe" I had said to myself, "seems at such periods torn and disordered; the feeble amongst us wither in her distempered breath, rushing hot from steaming volcanoes"' (pp. 38–9). Adrift in her own hallucinatory reverie, a state to which she returns at other climatic moments of introspective revelation in the novel, Lucy's language here echoes contemporary Victorian theories of unconscious thinking or daydreaming, which argued for the close proximity between the imperfect sleep where dreaming takes place and imperfect waking or reverie where the mind moves between conscious and unconscious thought.[47] Only sound remains as an associative trigger once we no longer, to quote Lewes, 'see objects, smell odours or taste flavours'.[48] Freed from the visual, the mind can wander across the globe at will, he continues, drawing its wild and various domains into the dreamer's unconscious orbit.

There are many scenes in *Villette* that use acoustic triggers to register Lucy's acuity as a listener and elicit a mirroring identification from the reader. Like Jane Eyre, Lucy Snowe often registers alien sounds and voices at the moment of first encounter with a new city or culture. Arriving in London for the first time she hears the 'strange speech of the cabmen and others waiting around' (p. 45). The sounds of street life where she alights from the coach, jar and alienate, but also peak her survival instincts. She must listen closely and make sense of speech that seemed 'to me odd as a foreign tongue' so she can find the inn where she is staying (p. 45). Only when morning comes and London's visual attractions are revealed, reaching a pinnacle when she scales the dome of St Paul's, that she truly absorbs the city's multi-sensorial appeal – 'its business, its rush, its roar, are such serious things, sights and sounds' (p. 49). But this relief proves transitory as Lucy sets sail for Boue-Marine soon after and again sound is used to evoke the existential upheaval of 'my homeless, anchorless, unsupported mind' (p. 51). The stewardess who 'talked all night' (p. 51) and the careless speech of Ginevra Fanshawe transcribed with caesura, exclamations and relentlessly repeated 'I's torments and exhausts her (pp. 54-6). Yet, repulsed as she is by Ginevra, Lucy listens and follows her instructions to find work with Madame Beck. Surrendering her will to an imagined Providence she moves in a virtually somnambulant state through the quiet streets of Villette until she stops with a start in front of the brass-plate bearing her future mistress's name. Throughout this scene unspoken thoughts 'volley' chaotically through Lucy's mind, loosening her perceptual hold on her surroundings. Mysteriously, she hears the voice of Providence who orders her to 'Stop here' and she does so obediently, surrendering to habit, as she will do later when the penitents instruct her to rise and take confession (p. 64). Unmoored and estranged, Lucy registers every unfamiliar sound and voice. Notably, and in conclusion, the first thing Lucy notes about her new mistress is – the 'unexpected' sound of her voice – an acoustic sign of the tortured miscommunications and misrecognitions to come (p. 65).

In 'Pilate's Wife's Dream', Charlotte Brontë crafts a monologue out of the silenced experience of Pilate's wife. From the first stanza the reader is drawn into a torrid first-person narration that begins with the crash of a bedside lamp struck by the anonymous speaker's arm, unconsciously flung out in the middle of a dream:

> I've quenched my lamp, I struck it in that start
> Which every limb convulsed, I heard it fall–
> The crash blent with my sleep, I saw depart
> Its lights, even as I woke, on yonder wall;[49]

Sound establishes the scene, the lamp is struck and crashes to the ground, the sleeper awakens, jarred into consciousness in a convulsive moment that sets the tone for the poem. The reader is drawn into the heightened emotion of Pilate's wife, who waits feverishly in the darkness for the death of Christ:

> Yet, Oh, for light! one ray would tranquilise
> My nerves, my pulses, more than effort can; (p. 2)

As dawn slowly comes, sounds 'invade my ears', the dull measured 'strokes of axe and hammer ring' through the streets, 'not loud' but

'Distinctly heard' (p. 2). Brontë uses sound effectively here to isolate the speaker. Pilate's wife's voice sounds out against these insistent background noises which also summon associations that Brontë assumes will resonate with her reader's repertoire of remembered biblical reading. Returning to Bain's observations about the ear as a 'matrix for holding together our recollections of language',[50] what is striking about this poem, as well as *Jane Eyre* and *Villette*, is how Brontë writes with this nexus of listening and reading seemingly in mind. 'Pilate's Wife's Dream' thus serves as a fitting point to end a discussion of the centrality of the ear to Charlotte's Brontë's imaginative exploration of the senses. So much has been said of the visual aspects of Brontë's work that her attention to sound and voice tends to be overlooked, and yet, when we return to familiar scenes as I have done in this chapter, such as Jane's first encounter at Lowood, it is impossible to ignore the sonic dimensions of her writing.

Notes

1 Jonathan Sterne, *The Audible Past: Cultural Origins of Sound Reproduction* (Durham: Duke University Press, 2003), p. 2.
2 Roland Barthes, 'The Death of the Author,' in *Image Music Text*, Stephen Heath, ed. (London: Fontana Press, 1977), pp. 142–8; Walter Benjamin, 'The Storyteller: Reflections on the Work of Nikolai Leskov', in *Illuminations*, Hannah Arendt, ed., Harry Zohn, trans. (New York: Schocken, 1968), pp. 83–109; Walter J. Ong, *Orality and Literacy* (London: Routledge, 1982).
3 David Vincent, *Literacy and Popular Culture: England 1750–1914* (Cambridge: Cambridge University Press, 1989), p. 201.
4 Ibid., at p. 3.
5 Ivan Kreilkamp, *Voice and the Victorian Storyteller* (Cambridge: Cambridge University Press, 2005); Garrett Stewart, *Reading Voices: Literature and the Phonotext* (Berkeley: University of California Press, 1990).
6 Anne Thackeray Ritchie cited in *The Shakespeare Head Brontë*, Thomas James Wise and John Alexander Symington, eds. (Oxford: Oxford University Press, 1990), vol. 3 (1849–1852), pp. 49–50.
7 Ritchie, cited in Wise and Symington, p. 50.
8 Kreilkamp, p. 123.
9 Ibid.
10 *The Letters of Charlotte Brontë, with a selection of letters by family and friends*, Margaret Smith, ed., 2 vols. (Oxford: Clarendon Press, 1995, 2000), 2: 118.
11 George Henry Lewes, 'Review of *Jane Eyre*', *Fraser's Magazine* (December 1847) in *The Brontës: The Critical Heritage*, Miriam Allott, ed. (London: Routledge, 1974), p. 156.
12 Ibid., at pp. 157–8.

13 Ibid., at p. 158.
14 Ibid., at p. 160.
15 Eugene Forcade, 'Review of *Jane Eyre*', *Review des deux mondes* (October 1848) in *The Brontës: The Critical Heritage*, p. 180.
16 Unsigned Review, *Spectator* (November 1847) in *The Brontës: The Critical Heritage*, p. 143.
17 A. W. Fonblanque, 'Review of *Shirley*', *Examiner* (November 1849) in *The Brontës: The Critical Heritage*, p. 219.
18 'Review of *Shirley*', *Weekly Chronicle* (November 1849) in *The Brontës: The Critical Heritage*, pp. 230–3, p. 232.
19 Ibid.
20 Eugene Forcade, 'Review of *Shirley*', *Revue des deux mondes* (November 1849) in *The Brontës: The Critical Heritage*, pp. 238–49, p. 238.
21 Bronte cited in *The Brontës: The Critical Heritage*, p. 230; 'Unsigned Review of *Shirley*' (7 December 1849) in *The Brontës: The Critical Heritage*, p. 254.
22 George Henry Lewes, 'Review of *Shirley*', *The Examiner* (February 1853) in *The Brontës: The Critical Heritage*, p. 292.
23 George Henry Lewes, 'Review of *Villette*', *The Leader* (February 1853) in *The Brontës: The Critical Heritage*, p. 305.
24 George Henry Lewes, 'Review of *Villette* and *Ruth*', *Westminster Review* (April 1853) in *The Brontës: The Critical Heritage*, p. 341.
25 William Caldwell Roscoe, 'Article VI – Miss Bronte', *The National Review* (July 1857), 127–64, p. 131.
26 Ibid., at p. 135.
27 Ibid.
28 Ibid., at p. 142.
29 Ibid., at p. 145.
30 George Henry Lewes, *The Physiology of Common Life*, 2 vols. (Edinburgh: Blackwood and Sons, 1859–1860), 2: 63.
31 George Henry Lewes, *Problems of Life and Mind*, 2 vols. (London: Trubner & Co., 1874–1879), 1: 150. Emphasis added.
32 John Stuart Mill, *A System of Logic: Ratiocinative and Inductive* (Cambridge: Cambridge University Press, 2011), 2: 511–30.
33 Rick Rylance, *Victorian Psychology and British Culture, 1850–1880* (Oxford: Oxford University Press, 2000), p. 193.
34 Alexander Bain, *The Senses and the Intellect* (London: John W. Parker and Son, 1855), p. 432.
35 Ibid.
36 Ibid., at p. 475.
37 Charlotte Brontë, 'The Teacher's Monologue' in *Poems by Currer, Ellis, and Acton Bell* (London: Aylott and Jones, 1846), pp. 107–10, p. 107.
38 Sterne, p. 2.
39 Ibid., at p. 3.
40 Ibid., at p. 10.
41 Lewes, 'Review of *Jane Eyre*', in Allott, ed., p.156.

42 Ibid., at pp. 157–8.
43 Ibid., at p. 160.
44 Virginia Woolf, '*Jane Eyre* and *Wuthering Heights*', *The Common Reader* (New York: Harcourt and Brace, 1925), p. 160.
45 Sally Shuttleworth, *Charlotte Bronte and Victorian Psychology* (Cambridge: Cambridge University Press, 1996), p. 152.
46 Kreilkamp, pp. 140–54.
47 Lewes, *Problems of Life and Mind*, vol. 2, p. 349.
48 Ibid., at p. 366.
49 Charlotte Brontë, 'Pilate's Wife's Dream', *Poems by Currer, Ellis, and Acton Bell*, p. 1.
50 Bain, p. 475.

CHAPTER 6

Burning Art and Political Resistance
Anne Brontë's Radical Imaginary of Wives, Enslaved People, and Animals in The Tenant of Wildfell Hall

Deborah Denenholz Morse

Early on in Anne Brontë's *The Tenant of Wildfell Hall* (1848), the rakish country gentleman Huntingdon removes the contents of the heroine Helen Graham's art portfolio in a symbolic disembowelment of her 'body' of art. Helen, in turn, burns the miniature she has drawn of her violent suitor rather than submit to its being forcibly taken by him after he enters unbidden into the library where she is painting. Huntingdon is, significantly, on his way to hunt in 'an expedition against the hapless partridges' (p. 134); he sets down his gun as he springs through the window. This scene represents male encroachment upon female artistic space – with its obvious aura of sexual violation – in connection with the profligate slaughter of birds by upper-class men. Helen is painting an allegorical picture of beautiful turtledoves, 'feathered lovers' (p. 135) whose intense absorption in each other contrasts sharply with the future bloody deaths of the partridges, just as her vision of true love is derided by Huntingdon as 'the wild extravagance of hope's imaginings' (p. 136). Upon his return from shooting, the interloper Huntingdon, 'all spattered and splashed as he was, and stained with the blood of his prey' (p. 137), again approaches Helen to ask about the burning of the miniature. His vanity does not allow him to comprehend that Helen will maintain authority over her art even if she has to destroy it.

Brontë's own art exposes the cultural and societal license that connects these acts of violence against women and animals. Her novel undermines masculine privilege, mastery and violence and reveals how dangerous women's art can be to the male-constructed social order. Moreover, there is another history that is imbricated with Brontë's critique but nevertheless is a significant context for *Tenant*: the history of chattel slavery and of abolitionist discourse, evoked both through Brontë's language and her plot. The novel's relation to women's rights has been much discussed;[1] in this chapter I am exploring much less familiar ground, for I will argue that Brontë consciously links legal, brutal control over animal (particularly canine) and enslaved human bodies to husbands' lawful control of their wives' bodies. As Helen says to her young friend Esther Hargrave, whose mother and brother attempt to force her into marrying for money and social position: 'You might as well sell yourself to slavery at once, as marry a man you dislike' (p. 317). Brontë connects violence toward women's bodies and the erased bodies of the enslaved that are the unrepresented, 'unnarrated hinterland'[2] of her novel, called forth by the shared, echoing language of feminist and abolitionist discourses. Animal bodies – in particular the bodies of dogs – are depicted both as animal bodies and as displaced women's (particularly wives') bodies. In *Tenant*, how a man treats his animals is indicative of how he will treat his wife.[3] All bodies – wife, dog and, by Helen's own metaphoric comparison, enslaved person – are under the country 'gentleman's' dominion and are of use to the master only insofar as they support him financially, sexually or emotionally.

Dogs in *The Tenant of Wildfell Hall*

Anne Brontë, like her sister Emily – and indeed, all the Brontës – was famously animal-loving. Anne represents male cruelty to animals as a form of oppression intimately connected to masculine viciousness to wives and enslaved people. The master's tyranny over his wife extends to his dog, which he kicks and abuses, and this treatment echoes abolitionist descriptions of the master's rule over an enslaved human's body. The oppression of wife, enslaved person and dog is imbricated – rights denied by English law, which recognises all of them as simply property.[4] In *The Tenant of Wildfell Hall*, Anne argues for the right of all God's creatures – including animals – to be liberated from abusive bondage. We might view Anne Brontë as in the spiritual company of the great anti-slavery activist and animal welfare advocate William Wilberforce, who a generation before the writing of *Tenant* acted from much the same religious ground when in 1824 he met in a London coffeehouse with the enlightened group that founded the Society for the Prevention of Cruelty to Animals (SPCA, after 1840 RSPCA). The Brontës wrote novels that depict many animals – in particular dogs – and they provided a home for several animals (including Emily's merlin Nero) in Haworth Parsonage as well. Emily's bull mastiff Keeper, the successor to the Irish terrier Grasper, is the most famous Brontëan dog, but Anne's spaniel Flossy, a gift from the Robinson girls, Anne's pupils at Thorp Green, was also a beloved family companion; Flossy was more readily associated with the domestic space than Emily's fierce Keeper. All three dogs were the subjects of drawings and paintings by the sisters; Emily drew a beautiful pencil drawing of Grasper[5] and painted a famous watercolour of Keeper that shows his tawny coat and massive neck, and Anne sketched and Emily painted delicate, quicksilver Flossy. Both Keeper and Flossy outlived Emily and Anne; Charlotte and Patrick resisted having blind old Keeper put to sleep and allowed him to die naturally. As Charlotte's most recent biographer Claire Harman states, 'neither had the heart to hasten by a single minute the departure of this last link to Emily'.[6] While I and other animal studies scholars have focused on dogs in Emily's *Wuthering Heights*,[7] and I have written on the Darwinian struggle of the human animal in *Jane Eyre*,[8] little work has been done on the representation of animals – in particular, dogs – in Anne's novels.[9] Moreover, although more notice has been given in recent years to what Beth Torgerson calls 'the Brontë body',[10] most of this attention has to do with representations of mental and physical illness in human bodies rather than with representations of animal bodies.

In both *Agnes Grey* and *The Tenant of Wildfell Hall*, Anne connects the confinement and abuse of animal bodies with human bodies. In *Agnes Grey*, boys and grown men abuse animals, most vividly represented by the cruel boy John Bloomfield's desire to torture fledgling birds, an impulse approved as manly by his elders (particularly his hunting Uncle Robson) but frustrated by his governess Agnes's merciful, quick killing of the birds with a large stone so that they will not suffer from her charge's cruelty.[11] *Agnes Grey* also includes a High Church, social-climbing clergyman, Hatfield, who kicks the cottager Nancy's cat, thus demonstrating his disdain of both lower-class Nancy and her beloved animal; Hatfield also canes the sweet wire-haired terrier Snap with 'a resounding thwack upon the animal's skull' which causes Snap to yelp piteously and run to Agnes, who 'stooped to caress the dog, with ostentatious pity to shew my disapproval of his [Hatfield's] severity' (p. 102). The treatment of animals – particularly Snap, rescued from a cruel rat-catcher by the minister Edward Weston, Agnes's future husband – is a mark of the true and godly man. Indeed, Agnes's joyful reunion with Snap immediately precedes her meeting once again with Weston – a coming together that results in their engagement. Thus, consideration toward animals as well as toward humans (particularly women) distinguishes the good man from the vicious one – and in Anne Brontë's work, the man who is good to animals is rewarded, for he will also be kind to his wife.

In contrast to Weston's rescue of Agnes's dog in *Agnes Grey*, in *The Tenant of Wildfell Hall*, Huntingdon abuses his dog and bloodies himself in the hunt while he violates his wife's trust after penetrating her body of art. In contrast, Helen's later suitor Gilbert Markham, her eventual second husband – introduced earlier in the novel than Huntingdon and, thus, a model against whom Huntingdon is measured – is represented with his dog at key moments in the text that lead to his connection with Helen. During her first marital quarrel with Huntingdon, Helen is 'determined to show him that my heart is not his slave' (p. 177) and she and her husband spent the night apart. The next day Huntingdon is in a foul mood; instead of striking his wife, he hits his dog:

> But his favourite cocker, Dash, that had been lying at my feet, took the liberty of jumping upon him and begging to lick his face. He struck it off with a smart blow; and the poor dog squeaked, and ran cowering back to me … he called it to him again; but Dash only looked sheepish and wagged the tip of his tail. He called again, more sharply, but Dash only clung the closer to me, and licked my hand as if imploring protection. Enraged at this, his master snatched up a heavy book and hurled it at his head. The poor dog set up a piteous outcry and ran to the door. I let him out, and then quietly took up the book. (p. 179)

Helen asks Huntingdon if the thrown book 'was intended for me' and he responds 'No, but I see you've got a taste of it' while looking at Helen's hand 'that had also been struck, and was rather severely grazed' (p. 180).

This passage clearly aligns the spaniel with Helen and makes Huntingdon's abuse of the dog a displacement for what he would like to do to his wife.[12] Both Dash's and Helen's bodies are injured in this disturbing scene. Dash is 'lying at [Helen's] feet' at the beginning of the episode and he retreats to her 'imploring protection.' Helen can only protect Dash by allowing him to escape from the room, as she will eventually flee Grassdale to be free of her tormentor. Both Dash, who 'took the liberty' and was punished, and Helen, who does not yet realise her marital bondage, are Huntingdon's property. Neither dog nor wife are free to do as they like; their bodies are owned by their master.

Gilbert Markham's relation to dogs suggests that he is of more promising material – although he must be educated through Helen's diary into Christian, feminist manliness.[13] He meets Helen after little Arthur gets caught in a tree while gazing at Markham's 'beautiful black and white setter', Sancho (a name that suggests Gilbert is aligned with Don Quixote) who becomes a kind of canine go-between for the couple, an excuse for Markham to keep visiting Helen, since 'between myself and my dog, her son derived a great deal of pleasure from the acquaintance' (p. 44). During their parting scene in the chapter 'Reconciliation', when Markham has read Helen's diary and knows she is still married, their passionate but restrained farewell occurs while Arthur plays outside with another of Markham's dogs, Rover. Markham does knock Helen's brother Lawrence (who he thinks is her lover) off his horse, severely injuring him in his jealous rage, but he is kind to horses themselves; there is a memorable early scene of him with little Arthur 'solemnly' riding a great draft horse, 'enthroned upon his monstrous steed' (p. 46) in his child's imagination. Animal bodies are safe around this Victorian man – and so is his beloved woman Helen's body, the body of his future wife. The language of slavery and abolition disappears from the novel after Helen's escape from Huntingdon and it does not reappear in relation to her second, much happier, marriage with Markham.

The Abolitionist Context of *The Tenant of Wildfell Hall*

Brontë scholars have already connected Emily's and Charlotte's novels with race discourses. During the past two decades, Maja-Lisa von Schneidern, Christopher Heywood, and most recently Beverly Taylor have analyzed the influence of the slave trade in Yorkshire upon the character of Heathcliff in *Wuthering Heights*.[14] The linking of Emily's novel with British Victorian slave history is also portrayed in Andrea Arnold's 2011 screen adaptation of *Wuthering Heights* and Adam Low's 2009 short film 'A Regular Black: The Hidden History of *Wuthering Heights*', which examines the connections between the Brontës and Yorkshire slave-owning families. At least one Haworth Parsonage-sponsored panel discussed the disturbing history represented in Low's film when famous Brontëan and Marxist literary theorist Terry Eagleton (*Myths of Power*) joined Caribbean-born playwright and novelist Caryl Phillips and journalist and Bronte Society President Bonnie Greer after a preview screening on June 9, 2012.[15] Through Jean Rhys's elegiac novel *Wide Sargasso Sea* (1966), which voices Bertha Mason/Antoinette Cosway's narrative, the imbrication of Charlotte's *Jane Eyre* with British West Indian slavery became more evident, and post-colonial scholars – most notably Gayatri Spivak, Susan Meyer, Carl Plasa, Sue Thomas and Carolyn Vallenga Berman[16] – have interpreted *Jane Eyre* in relation to that horrific history of enslavement, with disagreements as to the novel's complicity or resistance.[17] Professor David Richardson and Dr Nicholas Evans at The Wilberforce Institute for the Study of Slavery and Emancipation, Dr James R. T. E. Gregory and other scholars have done voluminous archival research into the connections between the slave trade, abolition, and the West Riding of Yorkshire.[18]

Like her more famous sisters, Anne Brontë would have been very familiar with abolitionist controversies. Their father, Patrick Brontë, was supported at St. John's College, Cambridge, by a sizarship paid for by sponsors, including the great abolitionist William Wilberforce who fought for twenty years to abolish the slave trade and then continued to work towards the abolition of slavery in the British West Indies.[19] As Beverly Taylor states:

> Even as they avoid speaking directly on the subject of chattel slavery, however, Charlotte and Emily (and less conspicuously, Anne) refer obliquely to race in ways that not only critique social oppression but also subversively refer to the exploitation of slave labor and other power imbalances that supported the colonial enterprise.[20]

The first clue to Anne Brontë's concern with the history of slavery, the slave trade and abolitionist thought is that she chooses to set her novel in the years 1821–8; as Olwyn Blouet reminds us in 'Slavery and Freedom', the 'second phase' of the political campaign against slavery 'began in earnest in 1823 with Thomas Fowell Buxton's founding of the Society for the Mitigation and Gradual Abolition of Slavery'.[21]

Brontë's critique of the British mastery over wives and slaves illustrates how the novel's heroine, Helen Huntingdon, has become the property of her husband under English law. Although this servitude has been related to eighteenth-century narratives of abduction and imprisonment such as Richardson's *Clarissa* or to the Regency masculine ideals Brontë so visibly exposes and critiques,[22] Helen's planned escape with her young son from her marital enslavement should also be contextualised by Brontë's choice to set her novel in a particularly charged historical moment of anti-slavery argument, the 1820s, when the abolition of slavery in all British colonies was fervently debated. This period also saw continued pervasive pro-slavery sentiment in Britain, as Paula E. Dumas states in her 2016 study, *Proslavery Britain: Fighting for Slavery in an Era of Abolition*:

> British abolitionists, it turns out, did not proceed unopposed, nor was abolition a universal goal among Britons ... Far from being passive, doomed onlookers watching from the sidelines on the road of abolition, politicians, writers, members of the West Indian interest, and their supporters actively fought to maintain colonial slavery and the prosperity of the colonies and Britain.[23]

Carol Senf writes that 'Brontë's novel could most accurately be described as the portrait of an age ... and the characters she paints represent almost every kind of individual who might have inhabited the English countryside during the third decade of the nineteenth century'.[24] Although Senf lists aristocrats, gentry, 'commercial newcomers' and servants, neither Senf nor other commentators on the novel mention that some Yorkshire rural landed gentry were supported by slave plantations in the British West Indies, plantations like the ones owned by Sir Thomas Bertram in Jane Austen's *Mansfield Park* (1815).[25]

Yorkshire and the West Riding had particular links to slavery and to abolitionism. As James Gregory writes, 'The Bradford region was connected with the transatlantic slave trade, though not in the obvious ways that a major slaving port like Bristol or Liverpool were, for like many other parts of Britain there were members of the landed classes resident in the area, who owned plantations where slaves labored'.[26] Gregory cites evidence that the Spencer Stanhope family of Horsforth near Leeds and the Skelton family of Little Horton owned enslaved persons. Despite the fact that Brontë does not specifically reference slavery in the colonies as the source of some country gentry's income, contemporary readers of the novel would have understood the 1820s context of intense abolitionist debate, and they would have known the revolutionary moment in which the novel was being published and the women's rights agitation in the decade previous to 1848.

Abolitionist discourse was fervent in the near neighbourhood of Haworth among both Anglican and dissenting clergy. Gregory quotes *The Leeds Mercury* of May 25, 1833 when he argues that 'Bradford abolitionists were conscious that they belonged to a "county, which, to its lasting honour, has done so much to effect the liberation of the slave"'.[27] In Bradford, as Gregory tells us, 1820s–30s abolitionism of the Baptist minister Samuel Godwin (1785–1871) was occurring in an environment conducive to change:

> Various events were signalling and contributing to a new phase in anti-slavery agitation. There was the establishment under Zachary Macaulay of The Antislavery Reporter (the organ of the British and Foreign Antislavery Society which had been established in 1823) from mid-1825, and the dissemination by the Antislavery Society of 'An Address to the Clergy of the Established Church and to Christian Ministers of All Denominations' the time was ripe for a revival of activity. Publications on the sugar trade, revelations of the public funds being used to support the plantation economy, news about repression of slaves in Jamaica and atrocities in Mauritius were brought before Parliament and public opinion. And it was the latter, public opinion, which the Antislavery Society appreciated as crucial, providing a spur to Parliamentary efforts.[28]

One man who is documented as a supporter of Godwin's abolitionist lectures in Bradford was the Revd. William Morgan, Patrick Brontë's close friend since their first curacies at All Saints' in Wellington, Shropshire – the man Patrick chose to baptise his daughters Charlotte, Emily and Anne as well as to officiate at the funerals of his wife, his daughters Maria and Elizabeth, and finally, his son Branwell, eighteen years after this 1830 Bradford event:[29] 'The lectures roused the town, and invigorated the anti-slavery committees of men and women. The Reverend George Stringer Bull, rector of Byerley, and soon to become famous as a factory reformer, gave his support, as did local worthies such as ... the Reverend W. Morgan'.[30]

Through her depiction of Helen's marital oppression and resistance within the pastoral-named Grassdale Manor, Anne Brontë recalls not only the despoiled Edens of the British West Indies in the consummate moment of 1820s abolitionist debate,[31] but also ongoing transatlantic abolitionism in the revolutionary time of the novel's writing. The year 1848 was the year of revolutions across Europe and a momentous year in the struggle for women's rights as well. In July 1848, the first convention on women's rights was held in Seneca, New York. The women's rights movement was imbricated with abolitionism at its outset; one of the convention's organisers, Elizabeth Cady Stanton, had met the Quaker women's rights orator Lucretia Mott at the first international anti-slavery convention conference, held in London in 1840. William Lloyd Garrison was among the abolitionists who encouraged women, including Stanton and Mott, to join the anti-slavery movement. Other scholars have also noted the revolutionary implications of the year Brontë published *Tenant*, but none have related that revolutionary consciousness to abolitionism.[32]

Brontë certainly does not equate the actual material culture surrounding an Englishwoman of the rural gentry with the wretched condition of an enslaved human being in the British West Indies, nor does Brontë compare the psychic distress of a well-fed and elegantly clothed gentlewoman to the despair of women held in bondage, forced into relentless and dangerous work and often sexually abused. Rather, Brontë critiques the patriarchal economic, social and cultural structures that ruled the English domestic space as well as British dominions abroad – colonies in which slavery was still sanctioned under British law in the 1820s. Carl Plasa, for instance, argues in a discussion of *Mansfield Park* and *Jane Eyre*: 'Yet in so far as both texts mobilize suspect analogies between colonial domination and forms of female oppression at home, they reveal themselves to share a textual politics whose assumptions solicit interrogation'.[33] Brontë's novel uses the language of master–slave relations as John Stuart Mill will use it two decades later to compare married women to slaves in *The Subjection of Women*. Language evoking slavery and abolitionist discourse is overt in *Tenant*, from the repeated use of words like 'master', 'slave', 'emancipation', and 'freedom', to representations of incarceration, abuse, rape, rebellion, burning, escape, concealment and disguise. It might be useful to recall how Plasa states that Charlotte's *Jane Eyre* recovers submerged colonial discourse and 'formulates its critique of gender and class relations by means to an habitual recourse to a metaphorical language of enslavement and mastery'.[34]

Female art and the female artist are a crucial element in this Brontëan imaginary. While the most-discussed scenes of incendiary female rebellion in the Brontë sisters' writings are in Charlotte's novels – the 'madwoman's' torching of Thornfield in *Jane Eyre* and the inflammatory performance of the actress Vashti in *Villette* – it is in *The Tenant of Wildfell Hall* that the rebellious female artist is most overtly a central figure, the heroine of the text.[35] Helen Graham is a painter whose art reflects her evolving consciousness from sentimental dilettante to authentic artist who must earn her own bread and that of her young son. Helen tells the brutal story of her marriage to the dissolute Arthur in a diary that she chooses to allow her future second husband, Gilbert Markham, to read, a document that educates Gilbert into a greater sympathy with women's sufferings and rights. This diary is a work of art, a fierce testament in which a woman's passionate experience of male cruelty and female anger and pain is recorded, a much more extended Brontëan 'testament' than Catherine Earnshaw's rebellious writing in the margins and between the lines of a male-authored Testament in Emily's *Wuthering Heights*. Helen's narrative and pictorial art depicting the truth of her oppression affronts her husband, the 'master' who burns her painting implements and riffles through her diary, an act that Stevie Davies calls 'a rape of Helen's spiritual world in the form of her diary-testament, as well as the temporary demolition of her means of securing liberty'.[36] This conflagration is – unbeknownst to Huntingdon – the smouldering fire that will lead to her escape from the enslavement of her marriage as Helen becomes more determined to flee her husband's tyranny, rising like a phoenix from the ashes to take flight from Grassdale. Helen's alliance with her old servant, Rachel, in order to effect her plans of escape is another element of resistance to upper-class male dominion as the two women of widely different social classes and ages successfully rebel against the master of the house in order to protect the child he is corrupting.

The wifely servitude that Helen enters into when she marries Huntingdon is soon marked by his abandonment of her when she is pregnant with little Arthur. Huntingdon insists upon penning Helen in while she is carrying his child; when she asks to go to London, he does not allow her to be 'contaminated' by the city. Meanwhile, he embarks upon a drunken, sexual binge with London prostitutes, exchanging the possession of one female body for another and, thereby, identifying 'owned' wife with 'rented' or 'bought' whore. Eventually, Huntingdon commits adultery with the married Annabella Wilmot, who becomes his mistress. When he tires of her, he takes another mistress whom he introduces into his household as a governess for little Arthur. Women are fungible commodities, easily exchanged or gotten rid of, like enslaved humans.

Long before this moment, Huntingdon has made clear that he considers his wife simply as his property and that he disdains his possession. In words that echo Othello's disclaimer of Desdemona – and therefore both introduce race consciousness and intimate that Helen is unchaste – Huntingdon has offered his wife to any man who desires her: '"My wife! What wife? I have no wife," replied Huntingdon … "or if I have, look you gentlemen, I value her so highly that any one among you, that can fancy her, may have her and welcome – you may, by Jove and [have] my blessing into the bargain!"' (p. 301).[37] Huntingdon's words precipitate the attempted rape of Helen by Huntingdon's crony Hargrave. The introduction of race through the allusion to *Othello* followed quickly by a near rape of female 'property' evokes chattel slavery. This event occurs in Huntingdon's library – unused by Helen's sottish husband, who does not read – a room that Helen has recreated as her art studio. In this space of female art created within the traditionally male space of the library, surrounded by male-authored texts, Helen is forced to defend herself with her palette knife against her would-be rapist. With an instrument of her art, she deters Hargrave's pseudo-chivalrous entreaties for Helen to 'let me protect you … You are my angel, my divinity' (pp. 302–3). She succeeds in preventing his sexual assault as well as in creating a lasting text in her diary that records the event.

Helen's diary painstakingly records her husband's degradation and her own sense of enslavement to his imperiously childish will. The abuse she sustains, the attempted rape she prevents and the escape in the night with her young child all seem to echo elements of slave narratives and prefigure scenes from *Uncle Tom's Cabin*, the mighty anti-slavery novel that would be published four years after Brontë's. Helen writes that she 'laboured hard all day' to hurry her escape, having decided that it is 'far better that he [little Arthur] should live in poverty and obscurity with a fugitive mother than in luxury and affluence with such a father' (p. 298). She goes on to say that 'I am looking forward to a speedy emancipation ... we shall leave him hours before the dawn, and it is not probable he will discover the loss of both, until the day is far advanced' (p. 307). Commentators on *The Tenant of Wildfell Hall* do not focus upon the previously cited *Othello* reference, the anti-slavery language of this passage ('escape', 'fugitive' 'emancipation') or Helen's initial plans for a more complete deliverance than her retreat to the sanctuary of her brother's property, Wildfell Hall. Helen dreams of an escape across the ocean and to the North that could be read within a context of both enslavement and the flight to freedom:

> Oh, I would take my precious charge at early dawn, take the coach to M—, flee to the port of —, cross the Atlantic, and seek a quiet, humble home in New England, where I would support myself and him by the labour of my hands. (p. 298)

Helen Graham Huntingdon's Art of Resistance

The 'labour of my hands' with which Helen plans to support little Arthur and herself is her painting: 'The palette and the easel, my darling playmates once, must be my sober toil-fellows now' (p. 298). Helen's paintings evolve from the decorative allegorical art and miniatures of her maidenhood through the loving portraits of her husband painted in her early married days, to the realistic landscapes she creates after her escape from Huntingdon and Grassdale, when she poses as a widow. Helen's diary and paintings – and even her acting the part of widow when she is still a wife – are representations of her increasing commitment to resisting the harsh realities of her patriarchal culture and to achieving freedom for both herself and her young son by means of her own creativity. The onerous and humiliating conditions of women's subjugation have been masked by romantic platitudes about love and marriage, and gender ideals that tell women that they are angels intended as the helpmeets and saviors of men, while men are assured that their 'manliness' consists of selfish brutishness. Anne Brontë engages these gender ideologies overtly in her novel;[38] in the early chapter 'A Controversy', Helen Graham, now living at Wildfell Hall after fleeing Grassdale, argues with the vicar's wife Mrs Millward about little Arthur's upbringing. While Helen has conditioned her son to dislike the taste of alcohol because of her husband's drunken habits, her older hostess Mrs Millward declares that 'you will treat him like a girl – you'll spoil his spirit, and make a mere Miss Nancy of him' (p. 29). To Gilbert's criticism that Arthur will be rendered too weak to be virtuous by his mother's close guidance, Helen insists: 'I will lead him by the hand, Mr. Markham, till he has strength to go alone' (p. 28).[39]

Helen's 'art of resistance' reveals the disturbing truth that married women were enslaved under the law, the property of their husbands, as were all children produced in the marriage.[40] Huntingdon increasingly will not allow his son – his property – to be a part of his mother's deeply Christian life. Instead, he tries to inculcate the lessons of mastery and 'manliness' into little Arthur while increasingly forcing his separation from his mother. In this control of his child, Huntingdon paradoxically enacts both the training of male progeny for the eventual role of master *and* the enslaved child's status as chattel, his history of arbitrary, enforced separation from the enslaved mother. Ultimately, by writing the narrative of marriage as a story of confinement, bondage and escape, Brontë might be seen as revising the romantic courtship plot as slave narrative – and ultimately allowing the enslaved mother and child their freedom and autonomy. Helen's insurrectionist art evolves from an examination of her soul that is not only Protestant, but explicitly Methodist (Brontë's Aunt Branwell's religion) as Lee A. Talley persuasively argues.[41] The document of her 'capture' and imprisonment in Grassdale becomes a fiery indictment of the injustices of her husband-owner, Huntingdon, whose sexual predation is signalled by his name. Helen at first ignores Huntingdon's dark side despite her Aunt Maxwell's warnings that he is 'a bit wildish' (p. 115). In another context I have argued that this term points to Huntingdon as a failed Romantic as well as to Anne's critique of Emily's aggrandised Heathcliff in *Wuthering Heights*.[42] Now I wonder whether Brontë was also criticising the portrayal of the enslaved human as primitive by depicting the master as the 'wild' man. This overturning of cultural/literary paradigms is joined by Brontë's refusal to let her dark-haired woman die, or even lose the man she loves (Gilbert Markham) to a lighter-haired rival (Eliza Millward), a pattern all the Brontës knew well from Sir Walter Scott's novels, particularly *Ivanhoe's* Rebecca/Rowena tension.

Helen's art begins as far from radical. Her diary is a painful history of her progress as a creative artist as well as the dissolution of her fantasised perfect marriage into a struggle between master and enslaved person. When she begins to write her personal narrative – or rather, the part of her diary that she gives to the man she now loves, Gilbert Markham – she has just returned to Staningley, her Aunt and Uncle Maxwell's home. A good deal has been written about this entry into Helen's private writing world through Gilbert Markham and about the structuring of *The Tenant of Wildfell Hall* as an embedded narrative; the woman's voice is at its centre, heard only as it is framed by Markham's letter to his brother-in-law Halford, a letter which draws upon Markham's own twenty-year-old journal.[43] I will not again enter upon this question except to state that the focus upon reading the novel is surely upon the writing voice of Helen's diary rather than upon the fact that Markham is reading her diary. There is no sense that he alters Helen's words or that he is an 'unreliable narrator'; he tells Halford that he knows he will not be 'satisfied with an abbreviation of the contents, and he shall have the whole, save, perhaps, a few passages here and there of merely temporal interest to the writer, or such that would serve to encumber the story rather than elucidate it' (p. 110). At most, Markham might possibly have slightly edited the diary, which might be interpreted as intrusive despite Helen's invitation for him to read it. That said, I can also sympathise with the critique that the diary is served up for male consumption as part of the homosocial bond.[44] My concern in this venue, however, is not with another go-round about the novel's structure, but with the art of Helen's diary itself.

That diary tells quite a story. Helen writes about herself initially not as she is in animated talk with her beloved Huntingdon, but as she is in conversation with her Aunt Maxwell, who warns Helen that her beauty may be a mixed blessing when it comes to choosing a husband. Helen, Aunt Maxwell warns, needs to 'first study; then approve; then love' (p. 112). From here Helen tells of her seeming rescue by the handsome, red-haired Huntingdon (later portrayed at times as a kind of diminished Milton's Satan in *Paradise Lost*) from the oppression of the terrible Mr. Boarham, who nearly lectures her to death. In the next diary entry, Helen writes of another release by the smile and touch of Huntingdon, this time from the clutches of the debauched old Mr. Wilmot: 'It was like turning from some purgatorial fiend to an angel of light' come to announce that 'the season of torment was past' (pp. 123-4). Perhaps not only the suggestion of Lucifer ('angel of light') and hence Milton's epic poem, but also the echo of *Hamlet* ('Hyperion to a Satyr') prepares us for tragedy – and might even suggest Helen's own misgivings – but at this point of her diary, Helen does not yet see that she is about to become enslaved. Indeed, she is influenced by the sentimental ideologies of femininity that urge her to 'save' erring men.

Helen persists in ignoring the warnings not only of her aunt's words, but also of Huntingdon's deeds, and she is married. Within eight weeks, Helen is writing with painful truthfulness:

> And do I regret the step I have taken? – No – though I must confess in my secret heart, that Arthur is not what I thought him at first, and if I

had known him in the beginning, as thoroughly as I do now, I probably never should have loved him, and if I had loved him first, and then made the discovery, I fear I should have thought it my duty not to have married him. (p. 171)

Within ten months, Helen is calling her diary a 'chronicle for the purpose of recording sayings and doings' (p. 196), a description suggesting Helen is recording history that on some level is not simply personal, but more broadly social. In this instance, Helen documents an evening in which her husband 'ardently' kisses the hand of Annabella Wilmot, after which Helen confronts him not only with his 'misdemeanour' but also with her disgust at his excessive drinking, which he uses as an excuse for his misbehaviour: '"You often take too much; and that is another practice I detest." He looked up astonished at my warmth' (p. 198). Within two months after this episode, Helen is a new mother, so we know that she is seven months' pregnant when her husband's evening dalliance with Annabella occurs.

It is the birth of little Arthur that brings forth the most passionate and moving sections of Helen's diary. She is deeply grateful for her son: 'and thank Heaven, I am a mother too. God has sent me a soul to educate for heaven, and given me a new and calmer bliss, and stronger hopes to comfort me' (p. 202). Understandably, Helen is fearful of Huntingdon's influence over the child: 'My little Arthur! There you lie in sweet, unconscious slumber, the tiny epitome of your father, but stainless yet as that pure snow, new-fallen from heaven – God shield thee from his errors! How will I watch and toil to guard thee from them?' (p. 203) Within five years, after experiencing her husband's flagrant drunkenness, his clandestine and then flaunted adultery with Annabella Wilmot, and his psychological cruelty to her, Helen writes of her small boy that she is:

> hoping nothing for his future life, and fervently wishing he had never been born – I feel the full extent of my calamity – and I feel it now. I know that day after day such feelings will return upon me. I am a slave, a prisoner – but that is nothing; if it were myself alone, I would not complain, but I am forbidden to rescue my son from ruin, and what was once my only consolation, is become the crowning source of my despair. (p. 312)

It is a lament that – although it is for a son's immortal soul rather than for his cruel life as an enslaved person – might echo that of an enslaved woman whose child is born into hopeless servitude. Indeed, the enslaved mother Cassy's misery over her enslaved children in Stowe's *Uncle Tom's Cabin* does sound a good deal like Helen's grief in Brontë's novel.[45] I hasten to state, again, that I am not equating the materially comfortable although spiritually bereft condition of the upper-class Englishwoman to the wretched condition of the enslaved mother, but rather suggesting that Brontë is critiquing the patriarchal economic and cultural structures that ruled both the English home and the British West Indies.

Helen's diary – but not the novel itself – ends with her tale of intrepid, secret escape from the confines of her husband's degraded ancestral home. As she is planning to flee, Helen thinks: 'the atmosphere of Grassdale seemed to stifle me, and I could only live by thinking of Wildfell Hall' (p. 325). Immediately after this, Helen acts upon the word of her trusted servant Rachel, who has evidence that the new 'governess' is actually Huntingdon's most recent mistress. The two women plan together, agreeing that, as Rachel declares, 'Only let us get shut of this wicked house, and we'll do right enough, you'll see' (p. 385). Once Helen has departed from Grassdale and has arrived at Wildfell Hall, she begins her diary entry: 'Thank Heaven, I am free and safe at last!' – language that voices the later words of American Civil War Negro spirituals most famously echoed in Martin Luther King's 1963 'I Have a Dream' speech.[46] After this beginning, she relates her actual flight in detail, which she effects not only in alliance with Rachel – 'my faithful friend beside me' (p. 390) – but also with the long-suffering old butler Benson; the debilitated Huntingdon has called the butler a 'brute' who could not possibly feel Huntingdon's exquisite sufferings because Benson is of the servant class. Huntingdon's view of Benson as not fully human is akin to the master's relegation of enslaved persons to animal or 'brute' status, a justification for their enslavement and vicious treatment. In response to pro-slavery apologias, late eighteenth-century Quaker abolitionists created the image of a kneeling African man asking, 'Am I not a man and a brother?' This image became the most famous and recognisable of the anti-slavery movement in the nineteenth century, with Josiah Wedgwood's cameo medallion as its most striking and influential early embodiment.[47]

After her daring escape, Helen returns to her mother's house and her own birthplace, Wildfell Hall, takes on her mother's name – 'Graham' – and puts on widow's weeds; in a sense, she is embodying her actual roles as daughter and mother and her imagined, quite possibly desired, role of widow rather than wife. In her disguise as a widow, Helen performs her own created role. In an anti-slavery context, Helen is 'passing' as something she is not: a woman free of her master. Although Helen, of course, is not an enslaved woman passing as white, she is a married woman, owned as property, who is disguising herself as a free woman – a working widow who receives pay for her labours. She denies her husband power over her and erases his existence and their history together. Helen is also claiming her maternal ancestry (her mother's name, her own birthplace, her mother's wedding ring) in resistance to patriarchy: her abusive husband, her absent, drunken father, Huntingdon's debauched cronies and the society that gives husbands legal claim to their wives' bodies.

Helen's art is ultimately more powerful than Huntingdon's oppression. At the outset of their courtship, Helen's drama with Huntingdon included pretending not to care about her favoured suitor when he has offended her – but as she then admitted, 'He was not, however, to be repelled by such acting as this' (p. 134). As much as he tries to quell the creative spirit of his wife thereafter, she ultimately controls his life, as his much later performance in affecting not to be able to sign over custody of his son illustrates. In this scene, Helen rewrites the script of domestic drama by demanding – and receiving – her husband's signature assigning legal custody of little Arthur to his wife. Helen's drama with Huntingdon ends as he dies in pain and misery, a degrading death that Helen records in detail in a letter to her brother Lawrence – a letter that Gilbert Markham, the man she now loves, reads. These two men, Helen's brother and future second husband – formerly attracted by shallow, flirtatious, pretty young women – are educated by Helen's words, as they have been by her presence. Lawrence, instead of wedding the social-climbing Jane Wilson, chooses as his wife the good, brilliant and lively Esther Hargrave, the young protégé of Helen. Gilbert Markham loses his attraction to the feline gossip and flirt Eliza Millward and comes to love and respect Helen instead. Both marriages promise to be partnerships of mutual love and respect rather than tyranny.

Helen's painting also evolves from imitative art to an authentic expression of her female experience. At first, she confesses in her diary that 'there is one face [Huntingdon's] I am always trying to paint or to sketch, and always without success, and that vexes me. As to the owner of that face, I cannot get him out of my mind – and indeed, I never try' (p. 110). Soon, Helen is drawing Huntingdon's face on the backs of her drawings even in public spaces, and when he discovers the recognizable traces of his countenance, Huntingdon cruelly uses the knowledge he gains of Helen's affection for him to make her jealous by flirting with Annabella Wilmot, with whom he later commits adultery.

In the later scene occurring during their courtship with which I open this chapter, Huntingdon jumps through an open window after hunting – visibly marked as a predator – and symbolically performs a kind of spiritual/artistic rape when he insists upon entering Helen's portfolio. Years later, when Helen is at Wildfell Hall, she has turned 'with its face to the wall' (p. 48) an adoring, erotically charged portrait of Huntingdon that she painted during the first year of their marriage. She comments tersely that 'the frame ... is handsome enough; it will serve for another painting' (p. 393). Helen's years of suffering have transformed her art, which now consists in realistic, detailed, representational landscapes – documents of truth as well as witnesses to God's creation. Even in her painting of landscapes, however, while she is in hiding from her husband, Helen must mask the place she is painting; she still fears his right to reclaim his lost spousal property.

In her writing and in her paintings, and finally, in her self-created role and performance as widowed mother, Helen claims the authority of the female artist to tell her truths from her own experience and to create her identity with her own imagination. Her story of the abuses engendered through male ownership of women's and animals' bodies is a powerful indictment not only of oppressive masculine prerogative, but of English law. Anne Brontë's imaginary in *The Tenant of Wildfell Hall* also evokes a related, displaced narrative: slavery and abolitionist discourse. Through echoing language of both women's rights and anti-slavery speeches and writings, Brontë recalls slavery as another dimension of the British Empire's oppression; although a dark upper-class English wife's travails are a far cry from an enslaved West Indian's trauma, the same male-authored-and-administered legal system views both figures as property rather than as persons. Anne Brontë writes in the belief that both English wives and enslaved West Indian people have the right to their freedom – and long for escape from their masters.

Notes

1 See especially Elizabeth Hollis Berry, *Anne Brontë's Radical Vision: Structures of Consciousness* (Victoria, BC: University of Victoria, English Literary Studies, 1994); Monika Hope Lee, '"A Mother Outlaw Vindicated": Social Critique in *The Tenant of Wildfell Hall*', *Nineteenth-Century Gender Studies*, 4:3 (2008); Rachel Carnell, 'Feminism and the Public Sphere in Anne Brontë's *The Tenant of Wildfell Hall*', *Nineteenth-Century Literature*, 53:1 (1998), 1–24; Elizabeth Langland, 'The Voicing of Feminine Desire in Anne Brontë's *The Tenant of Wildfell Hall*,' in *Gender and Discourse in Victorian Literature and Art*, Anthony H. Harrison and Beverly Taylor, eds. (DeKalb: Northern Illinois University Press, 1992), pp. 111–23; Stevie Davies, Introduction, Anne Brontë, *The Tenant of Wildfell Hall* (London: Penguin, 1996); and Andrea Westcott, '"A Matter of Strong Prejudice": Gilbert Markham's Self-Portrait', in *New Approaches to the Literary Art of Anne Brontë*, Julie Nash and Barbara A. Suess, eds. (Aldershot: Ashgate, 2001).

2 I use this term from John Sutherland, *Lives of the Novelists: A History of Fiction in 294 Lives* (New Haven: Yale University Press, 2012), p. 342, for all that is not narrated in a story but is implied as possible context, as other histories 'behind' and 'outside' the novel.

3 There is, of course, an expansive history of this in later Victorian novels, perhaps most famously in Leo Tolstoy's *Anna Karenina*, when Vronsky breaks the back of his thoroughbred mare Frou-Frou, who has to be shot, with the obvious foreshadowing of Anna's suicide as the end result of her adulterous liaison.

4 The first animal protection law passed in England was in 1822; Martin's Act was intended to protect cattle, horses and sheep and resulted in the prosecution of some lower-class abusers but none from the upper classes. The punishment was only a modest fine. The Act was sponsored by the Irish politician Richard Martin and supported by the abolitionists William Wilberforce and Thomas Fowell Buxton. The SPCA was formed in 1824 and became the RSPCA in 1840. See James Turner, *Reckoning with the Beast: Animals, Pain, and Humanity in the Victorian Mind* (Johns Hopkins University Press, 2000), chapter 3. See also Keridiana Chez, *Victorian Dogs, Victorian Men: Affect and Animals in Nineteenth-Century Literature and Culture* (Columbus: Ohio State University Press, 2017), for a fascinating analysis of dogs in relation to masculinity.

5 Christine Alexander and Jane Sellars call this drawing of Grasper 'a sensitive record' in Christine Alexander and Jane Sellars, *The Art of the Brontës* (Cambridge: Cambridge University Press, 1995).

6 Claire Harman, *Charlotte Brontë: A Fiery Heart* (New York: Alfred A. Knopf, 2016).

7 See especially Lisa Surridge, 'Animals and Violence in *Wuthering Heights*', *Brontë Studies*, 24:2 (1999), 161–73, published online 18 July 2013; Ivan Kreilkamp, 'Petted Things: *Wuthering Heights* and the Animal', *The Yale Journal of Criticism*, 18:1 (2005), 87–110; Deborah Denenholz Morse, 'The Mark of the Beast: Animals as Sites of Imperial Encounter' in *Victorian Animal Dreams*, Deborah Denenholz Morse and Martin Danahay, eds. (Aldershot, England: Ashgate Press 2007) and, more recently, Isabella Cooper, 'The Sinister Menagerie: Animality and Antipathy in *Wuthering Heights*', *Brontë Studies*, 40:1 (2015), 252–62, published online 15 August 2015 and Deborah Lutz, *The Brontë Cabinet: Three Lives in Nine Objects* (New York: Alfred A. Knopf, 2015) on Keeper as evoked through the brass dog collar at Haworth Parsonage Museum. Stories of Keeper go back in Brontë lore to Elizabeth Gaskell's *The Life of Charlotte Brontë* (1857) and the quite possibly apocryphal and certainly gothicised tale of Emily beating and then nursing Keeper because he got upon the clean counterpane on the bed.

8 See Deborah Denenholz Morse, 'The Forest Dell, the Attic, and the Moorland: Animal Places in *Jane Eyre*', in *Time, Space, and Place in Charlotte Brontë*, Diane Long Hoeveler and Deborah Denenholz Morse, eds. (New York: Routledge, 2016).

9 One notable exception is Maggie Berg, 'Hapless Dependents: Women and Animals in Anne Brontë's *Agnes Grey*', *Studies in the Novel*, 34:2 (2002), 177–97. Berg is interested in Anne's feminism as it equates society's treatment of animal and female human bodies. A second essay by Berg, '"Let me have its bowels then": Silence, Sacrificial Structure, and Anne Brontë's *The Tenant of Wildfell Hall*' (*LIT: Literature Interpretation Theory*, 21:1, 20–40, published online 1 March 2010), focuses upon male violence to animals and 'against certain human beings who are not regarded as fully human' (Abstract). Berg's Derridean analysis places Gilbert Markham within the "carno-phallogocentric" frame of violence; Janine Hornosty explicitly refutes Berg's views of Gilbert in 'Let's Not Have Its Bowels So Quickly Then: A Response to Maggie Berg', *Brontë Studies*, 39:2 (2014), 130–140, published online 9 April 2014.

10 See Beth Torgerson, *Reading the Brontë Body: Discourse, Desire, and the Constraints of Culture* (London: Palgrave Macmillan, 2005).

11 Discussed in this volume in relation to animals in Sally Shuttleworth, 'Hanging, Crushing, and Shooting'.

12 'Dash' is of course the name of Queen Victoria's beloved King Charles spaniel (1830–1840), which suggests a kind of trampling of all female authority and devotion. Victoria adored her spaniel, who was painted with her by Sir George Hayter in 1833 and with Victoria's other favorite pets by Edwin Henry Landseer in 1838. Dash was buried at Adelaide Cottage in Windsor Home Park, and a marble effigy with a glowing inscription was erected over his grave. For a capacious history of the alignment of lapdogs and ladies, see Laura Brown, *Homeless Dogs and Melancholy Apes: Humans and Other Animals in the Modern Literary Imagination* (Ithaca: Cornell University Press, 2010).

13 This issue has been much discussed. See Deborah Denenholz Morse, '"I Speak of Those I Do Know": Witnessing as Radical Gesture in Anne Brontë's *The Tenant of Wildfell Hall*', in *New Approaches to the Literary Art of Anne Brontë*, Julie Nash and Barbara A. Suess, eds. (Aldershot: Ashgate, 2001), pp. 103–26.

14 Maja-Lisa Von Schneidern, '*Wuthering Heights* and the Liverpool Slave Trade', *ELH*, 62:1 (1995), 171–96); Christopher Heywood, 'Introduction' in Emily Brontë, *Wuthering Heights* (Guelph, ON: Broadview Press, 2002); Beverly Taylor, 'Race, Slavery, and the Slave Trade', in *A Companion to the Brontës*, Diane Long Hoeveler and Deborah Denenholz Morse, eds. (London: Blackwell, 2016), pp. 339–53.

15 See The Brontë Society and Brontë Parsonage Museum Blog, Richard Wilcocks 'Why Does Heathcliff Have Only One Name?', June 13, 2013, http://bronteparsonage.blogspot.com/2012/06/why-does-heathcliff-have-only-one-name.html.

16 Gayatri Spivak, '*Jane Eyre* and Three Texts of Imperialism', *Critical Inquiry*, 12:1 (1985), 243–61; Susan L. Meyer, *Imperialism at Home: Race and Victorian Women's Fiction* (Ithaca, New York: Cornell University Press,

1996); Carl Plasa, *Charlotte Brontë* (London: Palgrave Macmillan, 2004); Sue Thomas, *Imperialism, Reform, and the Making of Englishness in Jane Eyre* (London: Palgrave Macmillan, 2008); Carolyn Vallenga Berman, *Creole Crossings: Domestic Fiction and the Reform of Colonial Slavery* (Ithaca: Cornell University Press, 2006). Other more recent work includes the suggestive essay by Sarah Fermi, 'A Question of Colour', *Brontë Studies*, 40:4 (2015), 334–342, published online 22 December 2015, that examines the possible connections between Cowan Bridge School and West Indian slave plantations.

17 See also my entries, 'Emily Brontë' and 'Charlotte Brontë' in *The Blackwell Encyclopedia of Victorian Literature*, 4 vols., Dino F. Felluga, Pamela K. Gilbert and Linda K. Hughes, eds. (Hoboken, NJ: Wiley-Blackwell, 2015) in which I discuss race and imperialism in relation to Anne's sisters' work.

18 See, for instance, Christine Verguson's 'Profiting from Slave Trade', available at www.bbc.co.uk/bradford/content/articles/2007/02/15/west_yorkshire_slave_trade_feature.shtml and "An 'avowedly repugnant' trade", available at www.bbc.co.uk/bradford/content/articles/2007/02/26/slavery_west_yorkshire_feature.shtml.

19 For a full account, see especially William Hague, *William Wilberforce: The Life of the Great Anti-Slave Trade Campaigner* (New York: Harcourt, 2008).

20 Taylor, 'Race, Slavery, and the Slave Trade', p. 340.

21 Olwyn Blouet, 'Slavery and Freedom in the British West Indies, 1823–1833: The Role of Education', *History of Education Quarterly*, 30:4 (1990), 625–43, p. 629.

22 See for instance Juliet McMaster, 'Imbecile laughter and desperate earnest in *The Tenant of Wildfell Hall*', *Modern Language Quarterly*, 43:4 (1982), 352–68 and Rachel Carnell, 'Feminism and the Public Sphere'.

23 Paula E. Dumas, *Proslavery Britain* (London: Palgrave Macmillan, 2016), pp. 1–2.

24 Carol Senf, '*The Tenant of Wildfell Hall*: Narrative Silences and Questions of Gender', *College English*, 52:4 (April 1990), 446–56, p. 450.

25 For the contemporary general public, this background surfaced once again with Patricia Rozema's explicit 1999 screen adaptation of *Mansfield Park*.

26 James Gregory, 'Historical Perspectives on the Transatlantic Slave Trade in Bradford, Yorkshire: Abolitionist Activity c. 1787–1865', available at www.academia.edu/3827577/Historical_perspectives_on_the_transatlantic_slave_trade_in_Bradford_Yorkshire_Abolitionist_activity_c.1787_-_1865.

27 Ibid., at p. 2.

28 Ibid., at p. 4.

29 See Juliet Barker, *The Brontës: Wild Genius on the Moors* (New York and London: Pegasus Books, 2013), pp. 31, 81–2, 89–90, 102, 119–20, 158–60, 672.

30 Gregory, 'Historical Perspectives', p. 5.

31 See especially Srividhya Swaminathan, *Debating the Slave Trade: Rhetoric of British National Identity, 1759–1815* (Aldershot: Ashgate, 2009) for the first phase of the abolitionist campaign preceding these debates.
32 See, for instance, Lee, 'A Mother Outlaw Vindicated', citing Berry, *Anne Brontë's Radical Vision*: 'While England saw no violent revolution, Brontë's novel, through its "radical vigour" and "searching reappraisal of orthodox" (Berry, p. 71), attempted a quiet sort of revolution by challenging the very foundations of English upper-class society through a scathing critique of laws and ideologies governing the family, marriage and mothering' (p. 1).
33 Carl Plasa, *Textual Politics from Slavery to Postcolonialism: Race and Identity* (London: Palgrave Macmillan, 2000), p. 6.
34 Plasa, *Charlotte Brontë*, pp. 81–2.
35 On Anne's and Helen's art, see especially Maggie Berg, '*The Tenant of Wildfell Hall*: Anne Brontë's *Jane Eyre*'; Julia Gergits, 'Women Artists at Home'; and Antonia Losano, 'Anne Brontë's Aesthetics: Painting in *The Tenant of Wildfell Hall*', all in *The Brontës in the World of the Arts*, Sandra Hagan and Juliette Wells, eds. (Aldershot: Ashgate 2008) and Jane Sellars, 'Art and the Artist as Heroine in the Novels of Charlotte, Emily, and Anne Brontë', *Brontë Studies*, 20:2 (1990), 57–76, published online 18 July 2013.
36 Stevie Davies, 'Notes to Anne Brontë', *The Tenant of Wildfell Hall* (London: Penguin, 1996), p. 526.
37 See *Othello* Act V, Scene 2, just after Othello has smothered Desdemona, as he hears Emilia at the door: 'My wife! my wife! what wife? I have no wife'. Right after Othello says these words, Desdemona revives for a moment and forgives him with her dying breath.
38 See for instance Davies, 'Introduction', *Tenant*, xiv–v: 'In a powerfully argued Miltonic debate about experience, choice, temptation, Helen contests the segregated education of male and female, with its over-protection of girls and over-exposure of boys'. See also Morse, 'Witnessing as Radical Gesture': 'Her speech ... witnesses to an ideal of ungendered nurturance' (113); and Lee, 'A Mother Outlaw Vindicated'.
39 One of the best discussions of childrearing and gender ideologies in *Tenant* is in Judith Pike, 'Breeching Boys: Milksops, Men's Clubs, and the Modelling of Masculinity in Anne Brontë's *Agnes Grey* and *The Tenant of Wildfell Hall*', *Brontë Studies*, 37:2 (2012), 112–24, published online 22 November 2013.
40 See Lee, 'A Mother Outlaw Vindicated' for the marriage and divorce law contexts to *Tenant*.
41 See Lee A. Talley, 'Anne Brontë's Method of Social Protest in *The Tenant of Wildfell Hall*', in *New Approaches to the Literary Art of Anne Brontë* (Aldershot: Ashgate, 2001), pp. 127–51.
42 See Morse, 'Witnessing as Radical Gesture', p. 109.
43 See especially Langland, 'The Voicing of Female Desire'.
44 See, in particular, Carnell, 'Feminism and the Public Sphere'.

45 Of course, the contemporary evocation of this story would be Toni Morrison's brilliant novel *Beloved*.
46 In Allison Argo's Emmy award-winning documentary 'The Urban Elephant' about the circus elephant Shirley, the African-American man who has been Shirley's loving zookeeper for 22 years tearfully says good-bye to Shirley at her new home, a Tennessee elephant sanctuary: 'I am going to miss her. But when I saw this place I told her that there will be no more chains. She is free now. I don't know who was the first to put a chain on her but I am glad to know I was the last to take it off. She is free at last.'
47 See the Smithsonian site for American History, available at http://american-history.si.edu/collections/search/object/nmah_596365

CHAPTER 7

Degraded Nature
Wuthering Heights *and the Last Poems of Emily Brontë*

Helen Small

The threat of degradation carries a potent charge in *Wuthering Heights*. It gears the narrative logic of the first generation's tragedy and the revenge drama played out by Heathcliff into the second generation. At the crux of that initial tragedy, a refusal to be degraded is the reason Catherine Earnshaw gives for not marrying Heathcliff, accepting instead Edgar Linton, 'handsome ... young ... cheerful ... rich' (a judgement on Heathcliff's condition that drives him back out into the world to make himself 'worthy of any one's regard' (p. 87)).[1] With her next breath Cathy repudiates the conventional social preference, but she will go on to gainsay herself again and choose Edgar Linton. The lines that relate to Heathcliff are among the most famous in nineteenth-century English literature, establishing the distinctively aggressive cast of Romantic love in this novel and defining the 'impersonal' egoism (in Leo Bersani's apt phrase) of its 'ontological drama'.[2] 'I dreamt once that I was [in heaven]', Cathy tells Nelly Dean,

> [But] heaven did not seem to be my home; and I broke my heart with weeping to come back to earth; and the angels were so angry that they flung me out into the middle of the heath on the top of Wuthering Heights; where I woke sobbing for joy ... I've no more business to marry Edgar Linton than I have to be in heaven; and if the wicked man in there had not brought Heathcliff so low, I shouldn't have thought of it. It would degrade me to marry Heathcliff, now; so he shall never know how I love him; and that, not because he's handsome, Nelly, but because he's more myself than I am. Whatever our souls are made of, his and mine are the same. (p. 71)

Expulsion from heaven, in the heretical eschatology of the dream, is no degradation, no 'fall' from grace. To be 'flung out', 'back to earth', is to come home. Compounding the spiritual with the natural and the material, her language endorses happiness as a physical condition – on earth, on the heath (by linguistic association, with Heathcliff) above Wuthering Heights, and in a body capable of weeping for joy at the familiar – only for her to choose, after all, security of social position at the price of alienation.

The rhetorical grandiosity of the dream conceals how much of the speech does not logically cohere, even if we grant Bersani's claim that there is an 'irrefutable logic' to elements of it.[3] Read as allegory, the dream is ill-fitted to its object: in what sense could life with Edgar Linton be a 'heaven', unless we are to hear the word as implicitly scornful, a conventional 'heaven' of social ease? (if this *is* satire, there is little to code us into it, beyond, perhaps, the dismissively pragmatic 'no business in heaven'). If the dream result of marriage with Edgar is a broken heart and desperate revulsion of the spirit, why is aversion to social debasement – the lesser threat to self in any normal scheme of things – decisive? Nelly may be out of sympathy, but she has cause to protest: 'I wish you would speak rationally'.

Identification with Heathcliff, we are to understand, is Cathy's true spiritual condition, but the religious lexicon – heaven, hell, angels, souls, wickedness – sounds at once potent and disconcertingly empty in light of the grounded naturalism of its application and the summary decision against the dream's conditions for happiness. The reference to 'Whatever our souls are made of' is a piece of metaphysical gesturalism: Catherine has no detectable interest in speculation of the theological kind: she co-opts the language of the Bible to her own ends and (Lockwood has already discovered) employs her copy of the Bible as a makeshift diary. But the allure of love as she first defines it is clear. Making each identical with the other, the claim 'I am Heathcliff' proposes Cathy and Heathcliff as the guarantors of one another's existence. Where one is strong, the other (by implication) is strong with and in them. The danger, equally, is that absolute identification makes each vulnerable to self-harm inflicted by the other. The elaboration that follows – 'He's *more myself* than I am' – is less straightforward. As Bersani reads it, it constructs a version of identity at once entranced and threatened by alterity: oriented towards an other whose existence supplements and completes one's own and potentially replaces it altogether. Narcissism and sado-masochism become two psychological manifestations of the one ontological outlook. Most puzzling of all, perhaps, is Cathy's exclusion of Heathcliff from knowledge of what she feels: 'He shall never know'. Why not? If they are one and the same, if he is more even than herself, surely he already knows?

If Cathy's choice against Edgar and for Heathcliff lacks cogency, it is less because the emotional imperatives of Romantic love are with

Heathcliff (Byron, Scott, Shelley reverberate through Lockwood's prose, but his possession of them alerts us to what is merely conventional, as well as positively familiar, about them by 1847); it is rather because talk of the soul has a privileged status which we are given no clear-enough warrant to overturn, even as the theological underpinnings are in doubt. We cannot dismiss the metaphysical stuff and simply focus on the narrative drama of degradation, rejection and revenge; but neither can we ignore the narrative direction of travel (how Cathy chooses to act) and accept without scepticism her claim to more-than-complete identity with Heathcliff (at least a suspicion that scepticism might be in order).

A common response from readers is to privilege the metaphysical claims and judge Cathy's actions harshly, rebuking her with her acknowledged inauthenticity. 'Having sold out her real self for the position of self-centred, spoiled wife of Edgar Linton', reads the *Encyclopedia of Feminist Literature's* entry on 'Catherine Earnshaw', more bluntly than most criticism but indicatively.[4] Isolating the transactional nature of her decision – acknowledgement of her love of Heathcliff traded for material self-interest – such readings discount as trivial her explanation that, if she had her way, her gain would also be Heathcliff's: 'If I marry Linton I can aid Heathcliff to rise, and place him out of my brother's power', she explains to an incredulous Nelly. Only in the light of this expectation that Edgar will allow her to serve another man's interests would degradation and alienation both be avoidable. Her position is to that extent logical (Bersani is right: marrying to further Heathcliff's interests follows from the claim of identity with him) but it goes without saying that her intentions are solipsistic (even as she gives solipsism an extended meaning) and socially scandalous.

What, and how much, should we take it Cathy means when she says that Heathcliff is degraded and capable of degrading her? The *Oxford English Dictionary* tells us that to degrade is: 'To reduce from a higher to a lower rank' (1); more broadly 'To lower in estimation; to bring into dishonour or contempt' (3); also, more intimately, 'To lower in character or quality; to debase.' (4a). Each of the definitions encompasses more and less objective ways of gauging the value of a person on the basis of external signs, with 1 tending more toward objectivity, 3 and 4a admitting more in the way of subjective social and ethical judgement. The plainest application of the term is in line with meaning 1, designating Heathcliff's visible drop in status from adoptive ward of her father to mere farm labourer under her brother's command (whether Hindley pays Heathcliff is not said). This loss of economic and social standing is accompanied by losses in access to cultural capital and the symbolic attributes of gentility that encroach upon definition, 3 and 4a – not that Heathcliff has ever cared much to cultivate the symbolic capital at his disposal.[5] We can probably assume (given her recently acquired love of fine dresses and fine living) that Cathy's revulsion against degradation takes account of such concerns. Conceivably, in admitting them she would be criticising herself for laying weight on outward appearances despite the priority Heathcliff has in her 'soul'. Such a reading could only be speculative. To move towards any more ethical reading of worth, whereby she would refuse association with a man she understands to be debased in his 'character or quality' seems much less compatible with her claim to *be* him, *more than* subsumed in him. Taking her at her word that she 'is' Heathcliff, his degradation is already hers and there should be nothing to be lost by association.

If searching after the extent of Cathy's meaning in this fashion seems an oddly logic-chopping approach to a work far more memorable for the rhetorical afflatus of its late Romanticism – and the vicious realism that subtends it – than for any investment in abstract reasons and reasoning, it is worth recalling that Emily Brontë had a head for logic. Her Brussels tutor, Constantin Héger, used just that phrase of her when he gave Elizabeth Gaskell his recollections of Emily and Charlotte: 'a head for logic, and a capability of argument, unusual in a man, and rare indeed in a woman'.[6] Others besides Nelly charge Cathy with incoherence. Heathcliff, her father and even Edgar all contest her 'senseless' passions (p. 105), though (with their diverse motives for opposing her) they all fail to bring her to sense. In practice, Cathy is exactingly logical when it serves her immediate purposes; quick to abandon logic when it does not.

Much the purer idealist than Cathy,[7] Heathcliff expresses scorn, as she does not, for the idea that Hindley or any other external party has the power to degrade him, or by extension, 'them': 'misery, and degradation, and death, and nothing that God or Satan could inflict would have parted us,' he rails against her as she lies dying: '*you*, of your own will, did it' (p. 142). Heathcliff does not deny the reality of injury and pain; he denies the ability of anyone but themselves to inflict it and, unlike her, he lays weight upon the real-world effects of her 'wilfulness': '*Why* did you despise me?' (p. 142), he asks – charging her with imposing a barrier between them that no one else had power to impose – and she does not answer. For him, her marriage to another man is decisive (it 'parts them'); for her, it is (or should have been) merely instrumental to her security of social standing and Heathcliff's vindication.

The revenge plot that Heathcliff will go on to pursue returns upon the idea that (his words to Cathy notwithstanding) degrading another human being is a powerful form of abuse: that, as he will later describe his own sufferings under Hindley, it constitutes a theft of one's property in oneself – 'I have made', he tells Nelly, 'wild endeavours to hold my right' (p. 288), where 'right' seems to comprehend everything he cares to possess: Cathy, property, Wuthering Heights and the respect of others if not their social regard. His sustained campaign against his oppressor and his oppressor's son is an amplification of a general pattern by which masters and servants, parents and children, siblings and neighbours test themselves against the existing hierarchies of power. The physical and moral humiliations exacted by the powerful on the relatively disempowered are pervasive in the novel, their violence the more brutal for being so casualised: Hindley, Cathy and Nelly pinching, punching and spitting at the newly arrived child Heathcliff; Hindley knocking the lad Heathcliff beneath the feet of the horse he has lent him under duress; Hindley thrusting the point of a knife (wet from herring cutting) between Nelly's teeth; Heathcliff dashing the drunken Hindley's head repeatedly against the flagstones (he has already slit the man's wrist open); Heathcliff forcing Joseph down on his knees in Hindley's blood; the young Linton grovelling at the young Catherine's feet, helpless with fear as his father discovers his failure to put on the required show of 'liveliness' (p. 237). Degradation is more often than not literalised by Brontë as physical abasement: it puts a person spatially 'below' their oppressor. If it does not actually cause physical harm, it threatens it and increases the subject's vulnerability to it.

The language of abuse is similarly a language of animalistic debasement: 'the slavering, shivering thing' (Heathcliff of Edgar, p. 102), 'puling chicken' (Heathcliff to Linton, p. 183); 'a centipede from the Indies' (Nelly channelling Heathcliff's view of Isabella, p. 93). By contrast with, say, D. H. Lawrence, who owed so much to reading her, Brontë has relatively little interest in abuse as a self-reflexive linguistic performance (a reinvigoration of the tongue) though she surely saw violence, verbal and dramatic, as a means of getting beneath the literary surfaces of politeness and sentiment and moralism to gauge orders of reality that escape conventional representation in the period. She treats verbal abuse as proximate to physical force, so that words, rather than substituting for violence, accompany it. Heathcliff pours scorn on Linton, then hauls the sick boy painfully to his feet; he warns Catherine to yield the door key or be 'knocked down' and duly drags her onto his knee and slaps her viciously around the head. Threats are rarely empty in *Wuthering Heights*.

And yet, for all this attention to degradation as an interpersonal drama with a social context, the novel consistently enjoins its reader to scorn the idea that violence, physical or verbal, does anything more or worse than expose the true nature of its object. So: Heathcliff stands up to his child persecutors; Joseph cussedly joins his hands and prays rather than clean up the blood; Nelly does not flinch against the knife. Hindley, on the other hand, is abject, having drunk himself 'to a point below irrationality' even before Heathcliff beats him unconscious; Linton is pathetic, grovelling – the more despicable because he tries to manipulate others by displaying his weakness ('the cockatrice' is Nelly's disgusted name for him when she discovers the extent of his desire for self-preservation [p. 243]).

Can an animal be degraded? Were one dealing with almost any other early nineteenth-century writer, the answer would make reference to the animal 'kingdom', where the 'higher' animals are deemed to be closer in capability to humans: more intelligent, more readily socialised and capable of accepting some conventional constraints on their natures. A dog or a horse or even a cat, for example, might be thought noble (or ignoble, as the case may be); a frog, probably, not. Ripe territory for speculation, but it may be doubted whether Brontë would have agreed with the notion of ranking species, nor with the (essentially moral) assumption that the degree of an animal's obedient socialisation will cause it to occupy a higher or lower place on a hierarchy. 'Degradation', for her, has to do with the damage done to the worth of a creature, judged not against any external taxonomy of worth but against the standard of what it should be in and of its own individual nature.

Brontë's much written-about devotion to animals was a distinctive feature of her personality, and the creatures she brought under her care were undoubtedly loved – cats, geese a merlin hawk found injured and nursed back to health, several dogs including her favourite, the 'tawny, strong-limbed' Keeper, whose massive brass collar is on display in the Haworth Parsonage museum ('so completely under her control, [that] she could quite easily make him spring and roar like a lion'[8]) – but she was not 'soft' about them. Charlotte's account of her sister curing Keeper of sleeping on the Parsonage beds describes a necessary assertion of authority over the dog, in an age when washing the sheets was a full day's hard labour. For some readers, Emily's actions verge on sadism.[9] She doesn't (as dog-owners have traditionally done) beat Keeper with a stick (having 'none to hand') or throw water on him or force his head to the ground. She squares up to him so that they face off in the manner of two unequally matched fist-fighters:

> [H]er bare clenched fist struck against his red fierce eyes, before he had time to make his spring, and, in the language of the turf, she 'punished him' till his eyes were swelled up, and the half-blind, stupefied beast was led to his accustomed lair, to have his swollen head fomented and cared for by the very Emily herself. The generous dog owed her no grudge.[10]

The anthropomorphism is Gaskell's and probably also Charlotte's. Again, one wonders how far Emily would have endorsed it. The 'bare fist' suggests a working-class fist-fight rather than a gentlemanly encounter in the ring – more aggressive than sporting; only once her authority has been established do more gentrified codes come into the description (she is caringly reparative; the dog is 'generous' in defeat). But there is, in reality, only one fighter here. The scenario depends upon her having confidence that the dog will not bite her; that he recognises his 'master' and will hesitate before springing.

There is no authoritatively normative standard for 'human' behaviour in *Wuthering Heights*, just as there is no standard for animal behaviour not grounded in apprehension of what is due to a particular nature. Heathcliff treats his son as a servilely malleable creature, his 'bad nature' so plain for all to see that Catherine can name it to his father without fear of antagonising him, though she is wrong about the degree to which Linton harbours the potential to be something better – 'I'm glad I've a better [nature], to forgive [his bad one] … I know he loves me and for that reason I love him' (p. 254). Her efforts to redeem her young husband are self-deluding (when secure of his own safety he is a sadist) but Heathcliff's efforts to degrade Hindley's son are, by the same token, quite futile: the boy has an innate worth that external 'degradation' cannot touch and that would-be oppressors necessarily discover. 'One [son] is gold put to the use of paving-stones,' Heathcliff will conclude, 'and the other is tin polished to ape a service of silver. *Mine* has nothing valuable about it' (p. 193). The boy Hareton is rude, rough and at times violent, but Catherine has only to look at the grown man when he comes in from washing in the yard, cheeks glowing and hair wet, to adjust her earlier scorn for him. It is almost the only point at which Brontë reads like Lawrence. Worth shows itself in physical health, native intelligence, the ability to take blows without being crushed and a generous heart. That the revenge motive can be so casually dispensed with at last ('where is the use?' asks Heathcliff, looking at Hareton) only confirms what has become clear by then: that the strong know their strength and will fight to preserve their priority; the weak live in fear, knowing themselves to be at risk.

Laying out the philosophical assumptions behind the complex ways in which we speak of 'nature', Kate Soper helps to define this ubiquitous, troubling aspect of Brontë's naturalism. '[T]he notion of our having a "nature"', Soper observes, 'carries with it something of that same necessity we attribute to animal and inorganic modes of being.' Human beings typically set themselves apart from other animals by understanding themselves to be 'capable of defying their "nature" in ways denied to other animals', but to speak of 'human nature' is to refute that exceptionalism:

> [It is] to imply that we are possessed of preordained features, and subject to their order of needs in the way that other creatures also are. These features may be supposed to be very different from those of animals, but in describing them as 'natural' to us we are imputing a similar determination and necessity to them.[11]

The psychological attraction of thinking about *individual* human beings as having 'good' or 'bad', 'strong' or 'weak' natures (we may speculate) is that, unlike talk of 'character', which tends to blur those features that are 'defined and delineated' circumstantially with those that are dispositionally 'set',[12] talk of 'nature' privileges the 'set': it posits an essential quality or value at the level of the individual over which the cultivation of character is a veneer. It, thus, provides for something in us that precedes our social and cultural experiences and offers a foothold against the threats to self-identity that come with constructivist versions of character, however weak or strong. Needless to say, it is an exceptionally restrictive view of self-identity.

The principal ethical challenge posed by natural determinism of the extreme sort expressed by *Wuthering Heights* – and a reason the novel is so resistant to conventional moral readings – is that it is not obvious what we owe to each other, and indeed to ourselves, in the way of nature. Do we owe anything beyond the correct recognition of our own and each other's intrinsic worth and a willingness to accept what we find? Do we owe even that much if 'worth will out' regardless? How does one recognise natural worth? How far does it go beyond physical and emotional strength, in Brontë's thinking, to encompass more Nietzschean qualities? How far does it anticipate the Nietzschean contempt for the values system of Christian humanism – compassion, equality of moral worth, the assumption of moral sympathy? More simply: what check is there on the power of the strong? What role, if any, may there be for education and culture? Emily Brontë, we might remember, was a reluctant teacher, said to have told her students at Roe

Head School that she preferred the family dog to any of them;[13] not a few readers have thought the scenes of Hareton's belated education by Catherine sentimental and dispensable.[14]

One of the difficulties in the way of posing ethical questions of Brontë's writing is that they are apt to sound, like Edgar and Isabella Linton, weakly uncomprehending of the forces they are contending with. One can, as Isobel Armstrong does in this volume, sensibly read *Jane Eyre* in company with Giorgio Agamben as a confrontation with the ethical problem: How much can be taken away from the human being before they cease to be a human being?[15] That same question has very little purchase on Emily Brontë's text, even as it constructs a plot almost entirely out of the concept of aggressive degradation. If we are what we are, by nature, degradation and elevation are 'merely' superficial, social alterations in value and the person who rates them more highly fails in the basic, quasi-instinctual business of knowing how we stand by our very natures in relation to others. When Cathy tells Isabella that she is a fool to romanticise Heathcliff – he is a 'fierce, pitiless, wolfish man' (p. 90) – the 'wolfishness' (which no doubt increases rather decreases Isabella's infatuation) is less the point than the necessity of his ferocity: he will do what it is in his nature to do, just as Cathy herself will pursue her desires without pity for her husband or sister-in-law. When we see Nelly trying to teach Catherine, with equal lack of success, not to romanticise Linton, we should conclude not that these headstrong, young women are painfully misguided, but that it is in their nature to fall prey to the wolf, having flattered themselves that they are exceptions to a rule that carries no favourable exceptions.

There is only one (important and negative) exception to natural necessity as it shapes the narrative action of *Wuthering Heights* and it marks the point at which a limited role for the ethical comes into operation. If altering a person's social standing does not affect their true value, cases of self-devaluation require one to say something more. The doctrine of 'human nature', as Brontë espouses it, holds that a thing is what it is *unless* physically or emotionally denatured – by the poison of drink, for example (Hindley), or sickness (Hindley's wife, Frances), or (the most difficult, but also the purest, case) wilful 'falseness' to itself (Cathy).

The devoir piece that Brontë wrote on 'Le Palais de la Mort' for Héger at the Pensionnat in Brussels is derived from a fable of Florian, so cannot be taken as indicative of independent thinking on this subject, but the way in which she interprets the victory of 'Intemperance' in the competition to be the first minister of Death suggests that she warmed early on to the theme of 'denaturing'. She had, by then, some experience of her brother Branwell's alcohol and drug abuse (and by the time she began writing *Wuthering Heights* much more):

> If once I make the acquaintance with the father [boasts Intemperance], my influence will extend to the son, and before men unite to banish me from their society, I will have changed their entire nature and made the whole species an easier prey for your Majesty.[16]

Intemperance does damage on a terrible scale because it is heritable, altering the quality not just of the individual but of generations thereafter. The ambitions attributed to it are imperial in this moral theatre, though Brontë's arch dramatization of Intemperance (a pompous braggart, like a drunk) brings the devoir piece precariously close to moral comedy. Her only recorded response to Branwell's dissolute unraveling indicates pragmatism and not moralism when confronted with the real thing: 'He is a hopeless being', she told Charlotte bluntly.[17] It is a statement of fact. She was contesting the embattled efforts of her father and sisters to reform him, but she wasn't washing her hands of him. She is said to have attempted on several occasions to protect him (though the evidence is suspect);[18] she was possibly the one who hauled him drunk up the stairs to bed on too many nights – the only one strong enough, perhaps.[19] These are the acts of someone who recognises Branwell for what he is: the damage has been done. He no longer has, if he ever had, the inner qualities that would allow for restoration. It was not, incidentally, Branwell's sense of himself. One of his more startling assertions in a characteristically self-glamorising letter to his friend Francis Grundy is that his rapid slide down the slippery slopes of alcoholism, drug-addiction and alleged adultery with the wife of his employer – or more scandalously 'amorous behaviour' towards her son[20] – was all in the manner of a deliberate experiment upon himself: '[I had a] determination to find how far mind could carry body without both being chucked into hell'.[21]

A view of 'human nature' at once so attentive to the individual, so conscious of external threats to flourishing, yet so regardless of the capacity for individuals to be influenced for the better, may seem at once admirable in the specificity of its regard for the other, and terrible in the rapidity, intuitiveness and closedness of its judgments. It is a philosophy that makes almost no reference to social agreements, or, as Hannah Arendt put it, the worldly 'in-between' of people[22] – only to their animal interaction and the strictly limited social-being that pertains between animals. It places itself in no recognisable public sphere beyond the immediate unit of the family, the primal field of conflict. It is (for good reasons) an ethical outlook liable, even now, to give offence, but it has this much, at least, due to it: it is coherent in its own terms.

The really intractable problems presented by *Wuthering Heights* arise from the co-existence of this natural necessitarianism focused on the individual with those other aspects of the novel, from which this chapter began, that ask us to imagine human identity as radically 'impersonal' and un-self-contained. The naturalism gears almost all the narrative action of the novel; the 'ontological drama' belongs to its rhetoric – to the lyric flights of Catherine's speech and the metaphysical possibilities that flow from it (though, as they shape plot, they are subject always to a controlling scepticism in the narrative point of view). *Wuthering Heights* presents, in its plot, a series of narrative agents who 'are what they are' – and who, in so far as they are strong, will brook no degrading action by others; at the level of its emotional pronouncements, it is a powerful fantasy of love as the ultimate form of self-escape – the self given over to another who is (at a level more metaphysical than psychological) greater than oneself.

John Bayley's brief, but thought-provoking, view of Emily Brontë is worth recalling here. As he saw it, *Wuthering Heights* is 'the most accurate and yet the most misleading children's book ever written'; it deals in 'basic matters' of the kind that preoccupy five- or six-year-olds – 'violence, food, pets, an obsession with being or merging with someone else, kitchen love and hate, a wish to disappear' – but misleads the reader into responding to them as if they were adult passions and obsessions. Like a child, it hides 'what it is up to' and that instinct for concealment makes it fundamentally unlike the 'whole-hearted' fiction of Brontë's siblings.[23] If it isn't conceptually coherent, it may be, he encourages one to suspect, because Emily had less of herself invested in this, her first (and only) novel than Charlotte, Anne or Branwell in theirs, and was nonchalant of, if not more actively opposed to, their confessional sincerities.

To read *Wuthering Heights* after this fashion is to circumvent (perhaps too completely) problems of interpretation that do indeed resist 'adult' resolution. It is not incompatible with Bayley's reading to suggest an alternative way through the difficulty. If *Wuthering Heights* exhibits many of the qualities of the child's unsupervised play, it also has the qualities of that other more disciplined childhood occupation, the 'devoir' piece or formal exercise. On this view, the novel in effect sets itself two tasks (independent of the case made by Edward Chitham for its originating as a single volume novel, expanded when circumstances required another volume to be filled[24]). Each task is coherent on its own terms, but they do not 'add up' to an integrated whole because the rules of engagement are incommensurate. *Wuthering Heights* is a love story about two powerful natures that react, differently, against the threat of degradation and so destroy their own happiness; it is also a lyric expression of love that has no narrative articulation, but is rather an intensely egoistic imagination of an escape from ego (degradation cannot touch it). Set together, these are irreconcilable and unreconciled. On that reading, there would be no 'dialectical relation' between Romantic individualism and social realism, no resolvable antagonism between worldliness and unworldliness, or between psychology and morality; simply a setting of two modes side-by-side, and the author (as Gustave Flaubert and James Joyce would have it) invisible and indifferent.

Wuthering Heights has no successor fictions that might serve to confirm or challenge an account derived from it of Brontë's distinctive voice, or style, or preoccupations. For that reason, it tends to carry a weight in excess of what it perhaps should bear as a statement of her philosophy – and I have been suggesting that she may have had no clearly articulated philosophy, for all her talent at logic. It is oddly rare for critics to look for comparative critical purposes to her poetry, some of which (on the evidence available) was written in close conjunction with the novel;[25] those who have done so have tended to concentrate on the great late lyric 'No Coward Soul', which just precedes *Wuthering Heights* in composition. The last two poems entered in Emily Brontë's 'Gondal Notebook' (Manuscript B of the poems) appear in the collected *Poems of Emily Brontë* (Derek Roper and Edward Chitham, eds.) as Poem 126 and Poem 127.[26] They are dated 14 September 1846 and 13 May 1848, which places their completion (or at least their transcription) respectively at some two months after the first version of *Wuthering Heights* went to Henry Colburn publishers; and around five months after the novel's revision and publication – four months before it became evident that Brontë was suffering from terminal consumption. So far as we know, these were the last two poems she deemed, in some sense, finished. The long gap between them indicates just how little she viewed in that light after Charlotte extracted her reluctant agreement to the publication of the *Poems* of Currer, Ellis and Acton Bell (1846). ('No Coward Soul' is the only other poem clearly thought of as complete after that point.[27])

'Why ask to know the date – the clime?' (Poem 126) and 'Why ask to know what date what clime' (Poem 127) are alternative takes on a single theme: the morally degrading effects of civil war. They are closely related to the dramatic and emotional terrain of *Wuthering Heights*, but (like the novel) they put normative ethics under considerable strain. Unlike the novel, they are oriented (Poem 126 especially) towards the 'timelessness' of romance rather than towards present-day realism. Poem 126 is a long narrative poem (264 lines of iambic, tetrameter couplets), reminiscent of Byron, Walter Scott and the Border ballads. Its Gondal origins are visible, though quite when the civil war occurs in the Gondal sagas is a matter for speculation (W. D. Paden's 1958 reconstruction suggests a republican uprising against the nobility after the apparent extinction of the royal line[28] – which makes some sense of a very incomplete picture). Poem 127 is a heavily stripped-down version of the same: twenty-five lines, without the political drama and without the appearance of a wild woman as a quasi-Coleridgean figure for poetic lyricism.

The speaker of Poem 126 is an unnamed combatant in a foreign civil war, drawn by his dislike of inaction and a theoretical enthusiasm for the republican cause (the cause of 'liberty') to participate in another country's internal conflicts. He is not quite a mercenary, but in the heat of battle he becomes rapidly brutalised. The republican unit to which he is attached captures and imprisons a young nobleman, who is then mocked, abused and perhaps tortured – only spared death so that he can be ransomed in due course. So 'gentle' is the nobleman that the degradation meted out to him by the speaker and his company symbolically degrades those who inflict it – but without empowering their victim. Unable to resist, the nobleman becomes symbolically infantilised: physically vulnerable and without apparent resources once the code of honour is broken.

> The gentle blood could ill sustain
> Degrading taunts, unhonoured pain.
> Bold had he shown himself to lead;
> Eager to smite and proud to bleed—
> A man, amid the battle's storm;
> An infant in the after calm. (p. 187: 106-11)

When the prisoner's daughter ('a wretched child ... / A shape of fear and misery' [p. 190: 202–4]) begs to see her father, the speaker rebuffs her contemptuously – only to find that the tide of battle turns. His own young son is taken captive, the nobleman is released and in command again, and now the speaker throws himself on the mercy of the man he had degraded. The dying prisoner again shows his gentility: his last act is to order that the monarchist troops spare his torturer's son. But when the speaker attempts restitution by caring for the dead man's daughter, she proves beyond rescue, an implacable figure, 'savage' beyond words.

Implicitly, she is, in the final lines of the poem, an avatar for the betrayal of lyric poetry itself, more Coleridge than Byron:

> she was full of anguish wild
> And hated us like blackest hell
> And weary with her savage woe
> One moonless night I let her go. (p. 192: 261–5)

That poetry might be subject to degradation from its high original status in classical lyric once it enters the marketplace of modern, increasingly democratic literature was a not uncommon claim in the Romantic period – the subject of one of the great theoretical disputes fought out in the literary press, between Thomas Love Peacock, Percy Shelley and Byron; and within philosophy between Rousseau and Hegel. If Brontë had any familiarity with those debates, they are most likely to have reached her via Byron (so often said to be responsible for a degraded life, a degraded poetry, a degraded audience) and/or through the writings of Peacock, whose account of modern poetry as a degraded 'age of brass'[29] was a provocation to his friend Shelley to compose a 'Defence of Poetry' fit for modern liberal political purpose.

Poem 126 is, in its own terms and in that wider cultural context, a very conservative-sounding production, tying the destruction of the lyric tongue to a debased libertarianism. The privacy of the production seems only to reinforce that conservatism, since the very conditions of writing look so inimical to the idea of publication, and indeed of audience. To a reader familiar with *Wuthering Heights* it is oddly as though Edgar had been made the hero of the drama, Heathcliff deprived of motive and Cathy sent 'savage' into the wings.

The second and final poem (Poem 127) is a much tighter redaction and revision of the same theme – and it is the version I am more interested in. It abandons the social hierarchical dramatization of degradation entirely and replaces it with something that looks much more like a dark meditation on the self-degrading consequences of all humanity in the pursuit of power:

> Why ask to know what date what clime
> There [i.e., in that place of oppression] dwelt our own humanity
> power-worshippers from earliest time
> Feet-kissers of triumphant crime
> crushers of helpless misery
> Crushing down Justice honouring Wrong
> If That be feeble this be strong
> Shedders of blood shedders of tears

Self-cursers avid of distress
yet mocking heaven with senseless prayers
For mercy on the merciless

It was the autumn of the year
Where grain grows yellow in the ear
Day after day from noon to noon
That August, sun blazed bright as June
But we with unregarding eyes
Saw panting earth and glowing skies
No hand the reaper's sickle held
Nor bound the ripe sheaves in the field
Our corn was garnered months before
Threshed out and kneaded up with gore
Ground when the ears were milky sweet
With furious toil of hoofs and feet
I doubly cursed on foriegn sod
Fought neither for my home nor God. (pp. 192–3)

There is little of the hectic Romantic action of 1846, here (although there is some). The poem retains the compounded tenor of fierce remorse, but turns what was the speaker's extended rebuke to himself (relinquished lyricism the price of his crimes against a nobler nature) into what sounds in its opening lines much more like an indictment of humankind – or, at least, of a generation. The opposition between gentle noble and brutalising republican that structured the earlier poem has disappeared and in its place is a near-allegorical confrontation with the nature of power. 'There dwelt our own humanity / power-worshippers from earliest time': it is not entirely clear whether the power-worshippers are one branch of humanity with whom we are asked to recognise reluctant kinship, or whether these lines describe our common condition in all times and all places (so that to distinguish one date and clime would be irrelevant). The sheer ferocity of denunciation in the first stanza may suggest a more fundamental cynicism: all humanity degraded by the love of power.

Much of the ferocity comes with the sudden intensification of diction in line four. 'Feet-kisser' is apparently a neologism: *OED* credits Robert Browning with its first use in 1869; but *OED* may require revision here. Such aggravated causticity in the diction is not, in any case, a sustained feature of the poem, which increasingly achieves its effects rather through abstraction and intensifying reduction of the range of terms: 'crushers of helpless misery / Crushing down Justice'; 'Shedders of blood shedders of tears'. The first is a relatively simple repetition, approaching what Joseph Slaughter describes as 'the apparent impasse of tautology': tautology as the rhetorical form given to an 'obviousness' that nonetheless merits expression.[30] Those who crush helpless misery crush Justice. The weeding out of the emotive 'helpless' brings the reiteration down to a basic acknowledgement of an injury that should require no special pleading. 'Shedders of blood shedders of tears' offers a more complex compaction by means of ambiguation: those who shed the blood of others also cause tears to be shed; or, those who shed blood may become, given the quick reversals of war or the coming of remorse, themselves shedders of tears.

By the final three lines of the stanza, there is ample room for doubt about where the agency of distress lies:

> Self-cursers avid of distress
> yet mocking heaven with senseless prayers
> For mercy on the merciless

The 'self-cursers' are presumably not avid of their own distress – rather they inflict distress on others – but they may themselves meet with distress in the course of a conflict that recognises no external standards of justice or mercy. This is, it seems, the incoherent condition of all humanity, turning against its own kind in what, from a longer (theological) perspective is a form of constant civil war on earth: the human race self-cursing in its merciless eagerness for power but still 'mock[ing] heaven' with prayers for mercy.

The second stanza shifts from moral castigation to an allegory of Nature that suggests the mellowness of John Keats's 'Ode to Autumn', rewritten in the harsher light of Milton's hell and Byron's 'Darkness'. The proper process of maturation and harvest has been perverted by war (implicitly, by man's corruption, but corruption is no longer the focus): too-young corn ('milky sweet') was pounded to destruction by war, mixed with gore upon the fields, where it should have been 'threashed' and 'ground' with sickle and mill in due time. The distinctively Brontëan aspect of the description is that it draws eye and ear in admiration to the brutal destruction it supposedly condemns, but on closer inspection doesn't condemn much, if at all. Like a John Martin painting, Poem 127 seems more rapt than terrified by the idea of apocalypse, though, having drawn that common comparison, it is necessary to qualify it. There's nothing of Martin's architectural shaping of natural grandeur here; very little, indeed, in the way of either large spatial shaping or detailed description of nature beyond the dotting in of 'yellow' and 'milky sweet' for the corn. The power of the images derives almost entirely from the thwarting of natural development – autumn in June, harvest before ripeness – and from a Miltonic insistence on negation that infects the very perspective: 'we with unregarding eyes / Saw'.

Part of the impact of the final couplet is that it brings the sharpness of individual perspective to a poem that has considerably disturbed any clarity of point of view: the ambiguous identity of the first-person plural, 'we ... self-cursers' in stanza one and the 'unregarding eyes' of a wasted generation in stanza two are, in the poem's closing couplet, replaced by a declarative first-person address. Where the speaker of the earlier poem was the mouthpiece of a degraded liberalism and a degraded poetry, this speaker rather recognises his condition impersonally as an extreme case of all humanity's degradation in the pursuit of power. Poem 127 keeps the possibility of lyric in view, but registers it as something from which this speaker is hopelessly, perhaps intransigently, excluded. Clearly audible behind the final couplet are famous lines of William Cowper, apt in so far as the experience of despair may be part of what one would want to be able to express 'sensibly' to God.

> But I, beneath a rougher sea,
> And whelm'd in deeper gulphs than he.[31]
>
> I doubly cursed on foriegn sod
> Fought neither for my home nor God

Unlike Cowper, Brontë strikes a consciously graceless closing note, rhyming 'God' on 'sod' – so that the final effect of the appeal to lyric is of lyric deliberately debased. Shelley had done it before, in his 'Sonnet to Byron' ('the worm beneath the sod / May lift itself in homage of the God'),[32] but the direction of his thought was emphatically upward – to emulate and vie with the god Byron – and one may reasonably assume that his identification of himself with the worm was not entirely serious. Brontë registers the longing for escape or return from a degraded condition, but gives no quarter to idealism.

It is tempting to see a critical development here: one that leads chronologically from the conservatively hierarchised, clearly Gondal-influenced narrative meditation on degradation that is Poem 126 (autumn 1846) through *Wuthering Heights*'s dual narrative/lyric modality, to the radically decentred lyricism of Poem 127 (1848), its speaker alienated from all humanity. *Wuthering Heights*, read in isolation, is an incorrigibly double literary form: a narrative of social degradation and revenge, ultimately dispensable/a lyric expression of love as at once the perfection of identity and an escape from identity. Framed by the two poems, the novel may instead appear as a stage in a longer development away from a narrative poetry dependent on character towards a fully impersonal lyricism, quite abstracted from scenes of human action. Lyric, as Poem 127 offers it, is what the I can say when the ego and its circumstances are finally left behind.[33] The problem with such a speculative account of the development of literary voice, as always with Emily Brontë, is that there is so small a textual base on which to anchor it. Had she written more, and for longer, it might be clearer whether the late drive towards abstraction of voice really is a drive, or just the impression given by a very short sequence of writings from a writer who was far from systematic and 'by nature' resistant to being known.

Notes

1. This chapter extends and complicates a line of thinking begun in my 'Introduction' to the World's Classics edition of *Wuthering Heights* (2009).
2. 'Desire and Metamorphosis', in Bersani, *A Future for Astyanax: Character and Desire in Literature* (London: Marion Boyars, 1978), pp. 189–219 (especially p. 214).
3. Bersani, 'Desire and Metamorphosis', p. 210.
4. Mary Ellen Snodgrass, ed. (New York: Infobase Publishing, 2006), p. 163.
5. 'Were I in your place,' Nelly tells him at the start of his sufferings under Hindley, 'I would frame high notions of my birth; and the thoughts of what I was should give me courage and dignity to support the oppressions of a little farmer!' But when Heathcliff does vindicate himself, it is not in this romantic vein: the man who returns from three years' absence is wealthy, 'dignified' with 'no marks of former degradation', but he is neither elegant nor cultured. Nelly perceives a 'half-civilized ferocity' in his face (p. 50; pp. 84–5).
6. Elizabeth Gaskell, *The Life of Charlotte Brontë*, Angus Easson, ed. World's Classics (Oxford: Oxford University Press, 1996), p. 177.
7. See Helen Small, 'Introduction' to *Wuthering Heights*, p. xii.
8. Mrs. Ellis H. Chadwick, *In the Footsteps of the Brontës* (London: Isaac Pitman, 1914), p. 113.
9. See Claire Harman, *Charlotte Brontë: A Life* (London: Viking, 2015), p. 124.
10. In Gaskell, *The Life of Charlotte Brontë*, p. 215.
11. Kate Soper, *What is Nature? Culture, Politics and the Non-Human* (Oxford: Blackwell, 2004), 27.
12. Amélie Oksenberg Rorty, 'A Literary Postscript: Characters, Persons, Selves, Individuals', in Rorty, ed., *The Identities of Persons* (Berkeley, CA: University of California Press, 1969), pp. 301–23, p. 304.
13. Juliet Barker, *The Brontës* (1994; London: Abacus, 2010), p. 344.
14. For an acknowledgement and direct rebuttal of this commonplace (though not often critically endorsed) position see Talia Schaffer, *Romance's Rival:*

Familiar Marriage in Victorian Fiction (New York: Oxford University Press, 2016), pp. 155–6.
15 Drawing on Giorgio Agamben, *Homo Sacer: Sovereign Power and Bare Life* (Stanford, CA: Stanford University Press, 1998).
16 Charlotte Brontë and Emily Brontë, *The Belgian Essays*, Sue Lonoff, ed. and trans. (New Haven: Yale University Press, 1996), pp. 224–31, pp. 228–330.
17 Charlotte Brontë, letter to Ellen Nussey, 3 March 1846 in *The Brontës: Their Lives, Friendships and Correspondence*, T. J. Wise and J. A. Symington, eds., 4 vols. (Oxford: Shakespeare Head, 1932), II, p. 84.
18 See Sue Lonoff, 'Brontë Scholarship: Retrieval, Criticism, Pedagogy', *Victorian Studies*, 39:1 (1995), 55–63, p. 61.
19 John Bayley's speculation, 'Kitchen Devil' [review of Frank, *Chainless Soul*], *London Review of Books*, 12:24 (20 December 1990), p. 16; perhaps drawing here on Edward Chitham, *A Life of Emily Brontë* (1987), p. 233.
20 Bayley, summarising Katherine Frank, *Emily Brontë: A Chainless Soul* (London: Hamish Hamilton, 1990), pp. 208–9.
21 Letter to Francis H. Grundy, 22 May 1842 in *The Brontës: Their Lives, Friendships and Correspondence*, T. J. Wise and J. A. Symington, eds., 4 vols. (Oxford: Shakespeare Head, 1932), I, pp. 263–4, p. 264.
22 Hannah Arendt, *Men in Dark Times* (London: Jonathan Cape, 1970), p. 4: 'The world lies between people, and this in-between…is today the object of the greatest concern'.
23 Bayley, 'Kitchen Devil', p. 16. See also Joyce Carol Oates's fuller articulation of the novel's pre-adolescent quality in 'The Magnanimity of *Wuthering Heights*', *Critical Inquiry*, 9:2 (1982), 435–49 (especially 440–1).
24 Edward Chitham, *The Birth of Wuthering Heights: Emily Brontë at Work* (Basingstoke: Palgrave, 1998), pp. 95–6 (and *passim*).
25 Ibid., at pp. 51–2, 146–57.
26 *The Poems of Emily Brontë*, Derek Roper and Edward Chitham, eds. (Oxford: Clarendon Press, 1995), pp. 184–93.
27 It is possible that 'The Prisoner' should also be included, but it is undated – sitting between 'Silent is the House' (9 October 1845) and 'No Coward Soul' (2 January 1846). It is likely that Brontë worked at revisions of other poems situated earlier in the transcription book during 1846 and 1847.
28 W. D. Paden, *An Investigation of Gondal* (New York: Bookman Associates). Excerpted in Roper and Chitham, eds., *Poems*, pp. 306–7.
29 *The Four Ages of Poetry*, in *The Works of Thomas Love Peacock*, H. F. B. Brett-Smith and Clifford Ernest Jones, eds., 10 vols. (London: Constable & Co. Ltd, 1924–1934), VIII, pp. 1–23 at pp. 1, 18.
30 Joseph R. Slaughter, *Human Rights, Inc: The World Novel, Narrative Form, and International Law* (New York: Fordham University Press, 2007), pp. 75–7.
31 'The Cast-Away', in *The Poems of William Cowper*, John D. Baird and Charles Ryskamp, eds., 3 vols. (Oxford: Clarendon Press, 1980–95), III, pp. 214–16 (p. 216).

32 *The Complete Poetical Works of Percy Bysshe Shelley*, Thomas Hutchinson, ed. (Oxford: Oxford University Press, 1905), p. 658: 13–14.
33 Mutlu Konuk Blasing offers a more generous formulation than Brontë's impersonal 'I' perhaps quite allows, though one glimpses something very close to it in Poem 127: 'a radically public language, [that] will not submit to treatment as a social document...because there is no "individual" in the lyric in any ordinary sense of the term.' *Lyric Poetry: The Pain and the Pleasure of Words* (Princeton: Princeton University Press, 2007), p. 4.

CHAPTER 8

'Angels ... Recognize Our Innocence'
On Theology and 'Human Rights' in the Fiction of the Brontës

Jan-Melissa Schramm

Upon the publication of *Jane Eyre* in October 1847, Charlotte Brontë's treatment of theological topics provoked distinctly different reactions from contemporary readers. Some decried its apparent challenge to orthodox social and religious values: as an anonymous reviewer observed in *The Christian Remembrancer*, its style is hard and coarse, '[t]he humour is frequently produced by a use of Scripture, at which one is rather sorry to have smiled' and '[e]very page burns with moral Jacobinism'.[1] Others interpreted virtually the same features of the novel as a necessary aspect of the proper exploration of the human passions in general and the female heart in particular: 'the style, though rude and uncultivated here and there, is resolute, straightforward, and to the purpose'.[2] In her comprehensive study of religion in the works of all the Brontë sisters, Marianne Thormählen suggests that this ambiguity was characteristic of the critical response to their fiction more generally: no authors quoted more liberally from the Bible, and with more evident appreciation of its tenets, than Charlotte and Anne, but at the same time, 'the fiction of [all] the Brontës accords very little space to the person of Christ'[3] and while their works reflect 'a thorough-going, if sometimes reluctant, respect for the Church of England, and a fundamental adherence to the values it attempted to uphold ... many of its representatives [are] described in ... unflattering terms'.[4] Their complex approach to the representation of Christian belief in fiction has, thus, seen them acclaimed, simultaneously, as almost blasphemous in their insistence on freedom of religious enquiry and overly ready to introduce arguably unnecessary scriptural allusion and quotation into their characters' dialogue. This apparent conflict between intellectual liberty (which sees self-development as a para-legal 'right', a necessary precondition to full enfranchisement) and creaturely submissiveness to God is the focus of this chapter. The Brontës' distinctive appreciation of human passions and, indeed, their recognisable rhetorical style, emerges from this conceptual battle between Romantic individualism and the inherited legacies of orthodox Christian thought.

Recent scholarship offers detailed accounts of the Brontës' Christian devotional practices: Thormählen's study of the family's beliefs is comprehensive and Heather Glen describes the extent to which Charlotte Brontë's fiction, in particular, registered the effect of Evangelical pedagogic theory and the proper instruction of young women.[5] Thormählen and Glen trace the sisters' inherent antagonism to Calvinist ideas of predestination in all its forms, their later enthusiasm for the writings of Samuel Taylor Coleridge and F. D. Maurice, and their suspicion of Tractarian or Anglo-Catholic practices which stressed the power of clerical hierarchies and potentially reduced the spontaneous devotional feelings of the heart to mere mechanised rituals. Thormählen locates the roots of the Brontës' belief system in an insistence on the work of conscience as 'the voice of God in man' – 'a direct intermediary between God and created beings ... an infallible prompter of action' whose 'commands must be obeyed'.[6] This call into direct relationship with the divine was inevitably gendered – 'Evangelicals championed the liberty of discernment and conscience for all believers, but also prized a model of marriage in which wives were spiritually subordinate to their husbands'.[7] In this chapter, then, I want to probe in more detail precisely how this liberty of enquiry and the concomitant entitlement of thinking subjects to plot their own path in life sits alongside Victorian Evangelical virtues of self-sacrifice and self-abnegation in the interests of the wider community. One of the reasons for the success of *Jane Eyre* – and indeed part of the appeal of *Agnes Grey* and *The Tenant of Wildfell Hall* – is the ability of Charlotte and Anne to reconcile a genuine and whole-hearted submission to the Protestant God of the Bible with an instinctive sense that self-determination can be understood as a *religious* (as opposed to merely an economic or intellectual) good.

The Submissive Self in Victorian Evangelical Theology

While the Evangelical revival which swept England in the late-eighteenth and early-nineteenth centuries stressed the effusive feelings of the Christian's heart towards God, it also emphasised the personal guilt which must follow a conviction of one's essential depravity: in Boyd Hilton's memorable phrase, the 'telos' of Evangelical belief and practice was not 'happiness, but justice, that is, punishment ... and this priority in turn led to an emphasis on sin'.[8] And 'sin' here was not only intentional acts of malevolence, but, far more destructively, inherited or 'original' sin, from which the only means of redemption was faith in the Atonement, the transaction in which Jesus Christ paid the penalty on humanity's behalf by offering himself up on the cross as a sinless sacrifice to satisfy a righteous God.[9] If man's inner essence was rebellious and incapable of autonomous acts of virtue, then self-improvement could only be understood as a part of the process of conversion by which Christ's goodness was made visible in the believer. The duties of the Christian were to stifle the sinful proclivities of the pre-conversion self, to suppress the bodily appetites, to renounce desire and to maintain an attitude of self-abnegation before God: in St Paul's words, 'Mortify therefore your members which are upon the earth; fornication, uncleanness, inordinate affection, evil concupiscence, and covetousness, which is idolatry' (Colossians 3: 5, King James Version). In this quest for heavenly wisdom, the believer was exhorted to acquire saintly qualities of 'goodness' far in excess of that shared communal minimum which the law demands of its citizens and subjects.[10] Despite initial Evangelical concerns about the role of the imaginative and narrative arts in religious devotion, the Victorian realist novel went on to play an important part in forming and instilling this 'culture of altruism',[11] publicising the value of extreme other-centred virtues like self-sacrifice in order to offer a much-needed antidote to the emphasis on competition promoted by laissez-faire economic policies. Alongside public moralists such as Thomas Carlyle and John Stuart Mill and Broad Churchmen such as F. D. Maurice and Benjamin Jowett, the novel sought to intervene in public debate and to take sides against the alleged exponents of unbridled capitalism and utilitarianism. In fact, so great was the literary suspicion of self-interest – so great the need to nurture corrective benevolence in an age of economic competitiveness – that authors from Charles Dickens to George Eliot and Elizabeth Gaskell felt compelled to advocate an ethics of self-sacrifice:[12] 'better to be Abel than Cain' in Dickens's terms;[13] 'more comfortable to be the calf than the butcher', in Eliot's.[14] The trope of self-sacrifice proved to be extraordinary supple, offering variations according to age, gender and sectarian conviction, thus extending to novelists a rich range of theological options for narrative closure.[15] In its more literal manifestation, the idea of sacrifice unto death – performed so valiantly by Dickens's Sydney Carton in *A Tale of Two Cities* (1859) and Eliot's Savonarola in *Romola* (1863) – provided a template for exploring the painful processes of scapegoating and martyrdom which sometimes occurred when the needs of the individual 'one' could not be accommodated alongside the demands of the many.

In its metaphorical Protestant formulation, the idea of self-sacrifice could inspire a life of charity lived mindfully for others – for example, in the cases of Dickens's Esther Summerson in *Bleak House* (1852–1853) and his eponymous heroine Amy in *Little Dorrit* (1857).

Such metaphorical self-sacrifice was often seen as a quintessentially female activity, thus putting pressure on the ways in which a noble life of renunciation could be reconciled with the teleological rewards of the courtship plot. Charlotte Brontë showed herself acutely aware of the extent to which marriage could be understood as both a labour of love for another and an act of self-gratification at the same time: as Jane explains to Rochester, when he suggests she is marrying him only out of a sense of duty,

> 'To be privileged to put my arms around what I value – to press my lips to what I love … Is that to make a sacrifice? If so, then certainly I delight in sacrifice'.[16]

As Mill observed, without the affective satisfaction wrought by such 'delight' and consent,

> [T]he wife is actual bondservant of her husband … No slave is a slave to the same lengths, and in so full a sense of the word, as a wife is … Marriage is the only actual bondage known to our law. There remain no legal slaves, except the mistress of every house.[17]

Because of its metaphorical slippage, such logic was controversial: while Mill's critique of the doctrine of coverture was admirable, the famous American abolitionist Frederick Douglass cautioned that 'the term slavery is sometimes abused by identifying it with that which it is not'.[18] The Brontës were alert to what Julia Sun-Joo Lee describes as the aesthetic and ethical potential of these moments of 'friction':[19] for each of the sisters, the idea of human selfhood emerges from this passionately experienced tension between self-abnegation and desire, repression and free will, ownership by another and the assertion of self-sovereignty. By claiming the right to speak freely of these matters, they prioritised the validity of first-person experience and brought pressing contemporary concerns before the expanding audience of public opinion – insisting on the equal value of passion and reason and faith and progress in an era of rapid social change.

Romantic Autobiography and the Language of Experience

The investment of political value in lived experience and first-person narration was crucial to Romantic art and literature. The initial impulse to such self-expression seems to have been Jean-Jacques Rousseau's

Confessions, published posthumously in two parts between 1782 and 1788, which deployed an unprecedented tenor of narrative sincerity to generate a distinctive radical sensibility: as Gregory Dart has noted, Rousseau 'implicitly represented the autobiographical subject as an anticipation, in individual form, of the transparency and virtue which would be the defining feature of the ideal political community of the future'.[20] If all men were perfectly transparent in their dealings with one another, the need for state regulation of human affairs would be minimised. The events of the Reign of Terror (1793–1794) in the course of the French Revolution revealed the tragic flaws of this utopian vision and reformers on both sides of the Channel felt duly disillusioned with the ideal of narrative transparency as the basis of a re-imagined political constitution. In this climate, William Wordsworth's epic study of the maturation of a poet's mind, *The Prelude* (1805, revised in 1850), must be read as bravely experimental because, as Gregory Dart observes, 'fully-wrought autobiography was, at least during the early years of the nineteenth century, a dangerously radical form in both England and France, not least because of its continuing potential to challenge existing notions of the relationship between private reflection and public politics, the individual personality and history'.[21] The suspect nature of the form was confirmed by Thomas De Quincey's *Confessions of an English Opium-Eater* (1827) which reminded its readership that autobiography recorded experience rather than innocence and was, thus, closely tethered to the confession of guilt as well as instruction by example.

In his seminal review of *Jane Eyre: An Autobiography* (1847) in *Fraser's Magazine* in December of the same year, George Henry Lewes identifies Brontë's work as the epitome of the virtues of this powerful first-person form:

> Reality – deep significant reality – is the great characteristic of the book. It is an autobiography – not, perhaps, in the naked facts and circumstances, but in the actual suffering and experience. The form may be changed, and here and there some incidents invented; but the spirit remains such as it was ... This gives the book its charm: it is soul speaking to soul; it is an utterance from the depths of a struggling, suffering, much enduring spirit: *suspiria de profundis!*[22]

Although he suspected that some of the more melodramatic and improbable elements of the plot were 'borrowed from the circulating libraries',[23] Lewes's enthusiastic reception of imaginatively translated autobiographical or 'real' experience was shared by the mid-Victorian readership: first-person narration remained Charlotte's narrative staple – only *Shirley*

(1849) is written in the third-person – and *Jane Eyre* was quickly followed by Anne's *Agnes Grey* (1847), Dickens's *David Copperfield* (1849–1850), Charlotte's *Villette* (1853), Esther's portion of the narrative in *Bleak House* (1852–1853), Charlotte's posthumously published *The Professor* (1857) and Dickens's *Great Expectations* (1861). If this narrative form was no longer seen as tainted by the implied revolutionary fervour of Rousseauvian sincerity, the women's novels (and Dickens's ventriloquism of Esther's plight) were nevertheless understood as literary contributions to political debate. From the 1790s onwards, self-representation had been idealised as the basis of participation in the public sphere: to be politically visible was to speak for oneself. In the words of the *Universal Declaration of Human Rights* (1948) written 150 years later, 'Everyone has the right to recognition everywhere as a person before the law'.[24] More recent work in the nascent interdisciplinary field of human rights and literature has argued that literature created the preconditions of such legal recognition, in effect offering paralegal acknowledgement of the full humanity of its speaking subjects in advance of that conferred by the law:[25] as Christine Krueger has observed in relation to the history of women's rights in the nineteenth century, 'literary advocacy … under specific circumstances, moved forward recognition for excluded groups'[26] – to prepare the way, in other words, for more formal proclamations of liberty, equality and fraternity. This is an exciting argument, but it is not without difficulties, as, although the novel often interrogates precisely this question of who is entitled to speak within a fictional community, it also condones and participates in various forms of exclusion: a novel, by virtue of its spatial, temporal and material limitations, can hardly aspire to wholly democratic conditions of representation[27] and first-person narration, as the Brontë sisters well knew, was not always a vehicle of transparent self-expression.

The Rhetoric of Rights

Lynn Hunt has argued that the philosophical framework of 'human rights' – 'the set of entitlements held to belong to every person as a condition of being human'[28] – first began to assume something like its modern shape with *The American Declaration of Independence* (1776) and *The Declaration of the Rights of Men and of Citizens, by the National Assembly of France* (1789).[29] In the latter, the Assembly's representatives, 'considering that ignorance, neglect, or contempt of human rights, are the sole causes of public misfortunes', set out the belief that men are born free and equal, and their 'natural, imprescriptible, and unalienable rights' are liberty, property, security and resistance of oppression.[30] Thomas Paine's *Rights of Man* (1791) endorsed the values which underpinned the early years of the French Revolution and urged equivalent reforms in England. But, to begin with, these rights were imagined as arising within national rather than global contexts and extending only to certain members of the community. As Mary Wollstonecraft pointed out in *A Vindication of the Rights of Woman* (1791) "'to see one half of the human race excluded by the other from all participation of government, was a political phaenomenon that, according to abstract principles, it was impossible to explain" (Talleyrand). If so, on what does our constitution rest? If the abstract rights of man will bear discussion and explanation, those of woman, by parity of reasoning, will not shrink from the same test'.[31] The rhetoric of 'rights' here references the nation-state, but contains a potentially expansive force that will flourish in increasingly global contexts as maturing cultural and political contexts permit.

Wollstonecraft's hopes that the unwritten, national constitution (and by implication, the idea of the fully 'human' and enfranchised) might become increasingly inclusive over time, were shared by many nineteenth-century novelists: as Joseph Slaughter has shown, the *Bildungsroman*, like nascent formulations of human rights frameworks, depends on the assumption that individuals should be free 'to pursue a story line, a life plot ... a right to narration' and nineteenth-century authors increasingly claimed this right for female protagonists as well as male:

> The novel genre and liberal human rights discourse are more than coincidentally, or casually, interconnected. Seen through the figure and formula of human personality development central to both the *Bildungsroman* and human rights, their shared assumptions and imbrications emerge to show clearly their historical, formal, and ideological interdependencies. They are mutually enabling fictions: each projects an image of the human personality that ratifies the other's idealistic visions of the proper relations between the individual and society and the normative career of free and full human personality development.[32]

As the protagonist narrates what Slaughter calls a 'petition for incorporation in the franchise of human rights',[33] the novel positions itself as an intellectual and social technology that mediates between the demands of 'one' and the many:

> Both the *Bildungsroman* and human rights law recognize and construct the individual as a social creature and the process of individuation as an incorporative process of socialization, without which individualism itself would be meaningless ... Both human rights and the idealist *Bildungsroman* posit

> the individual personality as an instance of a universal human personality, as the social expression of an abstract humanity that theoretically achieves its manifest destiny when the egocentric drives of the individual harmonize with the demands of social organization.[34]

Mid-Victorian authors were acutely aware of the compelling power of the first-person voice to 'make ... legible the inequities of an egalitarian imaginary'[35] and to articulate calls for greater equality, using sympathy – in Adam Smith's influential formulation, the imaginative capacity by which we 'change places in fancy with another' as we read – as an aid for the extension of our compassion.[36] As Charlotte's 'Preface' to the second edition of *Jane Eyre* makes clear, first-person narration enabled her to 'protest against bigotry' and the cause of her indignation is as much religious as legal, as she attacks the hypocrisy which masquerades as virtue in polite form.[37] Robert Ryan has argued that '[r]eligion was the crucial mediator between the cultural and the political-economic spheres' in early nineteenth-century England and authors directed themselves towards 'the spiritual and moral rehabilitation of their society, a renovation that presupposed an alteration in the national religious consciousness'.[38] If the first assertion of 'rights' in the late eighteenth century had been the product of secular forums in France, the rise of Evangelicalism problematised this vocabulary in England and ensured that assertions of female autonomy would have to be both uttered and received in a Protestant environment which privileged devotional feeling and filial submission to God the Father. Although the 'chains' of Catholic servitude had been thrown off at the Reformation (as Lucy Snowe notes several times in the course of *Villette*)[39] for Victorians, 'rights' could nevertheless only be understood in association with 'duties' and with the correct understanding of the Christian creeds which emphasised the importance of charity and benevolence in place of 'liberty, fraternity and equality'. Charlotte senses, then, that religion requires continual purification of its doctrines and self-scrutiny among its practitioners:

> Conventionality is not morality. Self-righteousness is not religion. To attack the first is not to assail the last. To pluck the mask from the face of the Pharisee, is not to lift an impious hand to the Crown of Thorns.

Charlotte seeks to invoke the precedent of William Thackeray – 'the first social regenerator of the day' – as the authority for her attempt to join 'that working corps who would restore to rectitude the warped system of things', but she is adamant that the way forward is Christian as well as socially inclusive: 'narrow human doctrines, that only tend to elate and magnify a few, should not be substituted for the world-redeeming creed of Christ'.[40]

Jane Eyre and the Extension of the Franchise of the Human

While Jane's voice has been lauded as emotionally compelling from the very first sentence of the novel, she must learn to temper her impetuosity and passion if she is to persuade the audience of her 'truthfulness'. And what she receives in the first chapters is nothing short of an evidentiary education in which she practises defending herself from accusations of deceitfulness and composing a persuasive account of her character and history. Jane's stature as a representative of the oppressed is established early – Mrs. Reed 'exclude[s her] from privileges' intended for richer children (*JE*, p. 13), and she imagines herself subordinate to all, as a 'rebel slave' (*JE*, p. 19) a phrase which alludes to the submission of women in Ancient Rome and in the seraglio (while gesturing obliquely to the slave trade in the West Indies which undergirds the action of the novel itself).[41] This is compelling but also controversial representative symbolism: as Julia Sun-Joo Lee asks, is this designed 'to disseminate antislavery sentiment? To illuminate alternate forms of injustice? Or simply to cannibalize the verbal and semantic forms of an exceedingly popular genre [the slave testimony]?'[42] The answer is hard to identify, but the emancipatory arc of the narrative is consistent with the plot structures of so-called 'Condition of England' novels like Gaskell's *Mary Barton* (1848) and Dickens's *Hard Times* (1854) in which the trial of the oppressed character's testimonial creditworthiness is forensic in nature and political in effect. Wrongfully accused of lying by Mrs. Reed and Mr. Brocklehurst at Lowood School, Miss Temple offers Jane the opportunity to clear her name and her para-legal acquittal serves, by extension, to justify the wider class for which she stands as the representative.[43] While Helen (the practitioner of saintly, supererogatory virtues) invites her to find comfort in the knowledge that 'angels see our tortures, [and] recognize our innocence' (*JE*, p. 82), Jane seeks earthly justification and welcomes the opportunity to 'prove' her integrity:

> 'Well now, Jane, you know, or at least I will tell you, that when a criminal is accused, he is always allowed to speak in his own defence. You have been charged with falsehood: defend yourself to me as well as you can. Say whatever your memory suggests as true; but add nothing, and exaggerate nothing'.

> I resolved in the depth of my heart that I would be most moderate: most correct; and, having reflected a few minutes in order to arrange coherently what I had to say, I told her all the story of my sad childhood. Exhausted by emotion, my language was more subdued than it generally was when it developed that sad theme; and mindful of Helen's warnings against the indulgence of resentment, I infused into the narrative far less of gall and of wormwood than ordinary. Thus restrained and simplified, it sounded more credible: I felt as I went on that Miss Temple fully believed me. (*JE*, p. 83)

Miss Temple seeks corroboration of Jane's tale as well – 'I know something of Mr. Lloyd; I shall write to him; and if his reply agrees with your statement, you shall be publicly cleared from every imputation: to me, Jane, you are clear now' (*JE*, p. 83). In the course of this testimonial transaction, Jane undertakes a rather typical Romantic journey – performed most famously by the eponymous protagonist of William Godwin's *Caleb Williams* (1794) – from the rhetoric of passion and 'wild supplication' (*JE*, p. 83) to the sincere language of plain 'facts' which eschewed ornamentation. By conforming to the evidentiary standards of the courtroom, Jane enfranchises herself within the public sphere and the socially marginal, orphaned charity-child is thus re-categorised as fully 'human' and a 'sister' to all mankind. Such narrative reliability is, of course, crucial to the success of a novel subtitled *An Autobiography* and this forensic template also reveals the indebtedness of the genre of Victorian 'realism' to standards of probability more generally.

And what Jane seeks to effect, in this journey towards full evidentiary efficacy, is nothing short of the rhetorical and conceptual union of human desire and religious devotion. As she studies the examples of others, she scrutinises her own heart, disciplining it before God and yet accepting the fulfilment of its legitimate demands as a sacred task. Alongside Jane's professional commitment to the Christian education of children, we see her interest in the rhetoric and the work of 'self-culture', defined by its most famous proponent in the period, the prominent American Unitarian William Ellery Channing, as 'the care which every man owes to himself, to the unfolding and perfecting of his nature'.[44] Channing's works were widely acclaimed in England after his travels in Europe in 1821-2 (not least because of his friendship with the Lake Poets), and while some of his beliefs may have been too liberal for Charlotte, they shared a rejection of predestination.[45] Channing observes that, as human potential lies latent within the soul 'like a living coal in the hand', with ongoing intellectual and moral labour, 'we can see in ourselves germs and promises of a growth to which no bounds can be set, to dart beyond what we have actually gained to the idea of Perfection as the end of our being':

> Self-culture is possible, not only because we can enter into and search ourselves; we have a still nobler power, that of acting on, determining, and forming ourselves. This is a fearful as well as glorious endowment, for it is the ground of human responsibility ... We can stay or change the current of [our] thought: we can concentrate the intellect on objects which we wish to comprehend: we can fix our eyes on perfection, and make almost everything speed us toward it. This indeed is a noble prerogative of our nature ... It transcends in importance all our power over outward nature. There is more of divinity in it, than in the force which impels the outward universe; and yet how little we comprehend it! How it slumbers in most men unsuspected, unused! This makes self-culture possible, and binds it on us as a solemn duty.[46]

For Channing (as opposed to say, Samuel Smiles, whose eminently successful *Self-Help* was published just over twenty years later in 1859) self-culture is religious rather than secular in its orientation and it is to be distinguished from other forms of ambition with more worldly goals:

> [I]t marks out a being destined for higher communion than with the visible universe. To develop this, is eminently to educate ourselves. The true idea of God, unfolded clearly and livingly within us, and moving us to adore and obey him, and to aspire after likeness to him, is the noblest growth in human and, I may add, in celestial natures. The religious principle and the moral are intimately connected, and grow together. The former is indeed the perfection and highest manifestation of the latter: they are both disinterested. It is the essence of true religion to recognise and adore in God the attributes of imperial justice and universal love, and to hear him commanding us in the conscience to become what we adore.[47]

A believer's work in life, then, is 'to fasten on this culture as our great end, to determine deliberately and solemnly, that we will make the most and the best of the powers which God has given us':[48] the specific choices made along the way will vary according the individual, for '[a]ll means do not equally suit us all. A man must unfold himself freely, and should respect the peculiar gifts or biases by which nature has distinguished him from others'.[49] Channing's sense that self-realisation rather than self-immolation is the task of the Christian ennobles mental labour, but also attempts to conscript ambition into the service of duty, a project that offered men far more freedom than women. Of the several books on the topic apparently owned by Charlotte, Glen notes that the volume *Thoughts on Self-Culture: Addressed to Women* (1850) by Emily Anne Shirreff and Maria Shirreff Grey was significantly more prescriptive than Channing's celebratory exhortations.[50]

John Beer notes that *Jane Eyre* was written at a time when an 'honest thinker' was bound to confront contemporary intellectual and scientific change – when an idea like 'providence ... was already a site of enquiry, even a battleground, no longer a rock of defence'.[51] Jane must adhere to a *via media* located between differing versions of Christian commitment, as she negotiates her way between Helen Burns's supererogatory saintliness and St. John Rivers's call to missionary duty which would place public service above affairs of the heart. Neither Helen's virtue nor St. John's missionary calling are wholly disparaged; their choices are respected as valid alternative templates of the *imitatio Christi* and Jane recognises this when she offers textual hospitality in the final paragraphs of the novel to St. John's vision of his own imminent demise in the service of the Christian empire abroad – 'a stranger's hand will write to me next, to say that the good and faithful servant has been called at length into the joy of his Lord' (*JE*, p. 502). All through the course of Jane's ethical education, Charlotte shows a sure-footed appreciation of the significance of contemporary debates regarding this definition of ethical action and 'the good life'. In her conversations with St. John, Jane tests current anxieties about whether charitable benevolence begins with domestic affections, or whether service to humanity in the abstract must take priority – there are distinctive affinities here with George Eliot's subsequent exploration of the same dilemma in *Romola* (1863) – and in her exchanges with Helen, Jane is compelled to scrutinise the idea of 'just deserts' as the foundation of justice.[52] Jane's belief that she return acts of goodness only to those who treat her well offends against Helen's articulation of the golden rule – 'to do as you would be done by', not 'to be done by as you did', in the terms of Charles Kingsley's *Water Babies* (1863).[53] Helen chastises Jane to the effect that her virtues are only proportional, directed by animal affections: 'There is not much merit in such goodness' and instead she calls upon Jane to 'Read the New Testament, and observe what Christ says, and how he acts – make his word your rule, and his conduct your example' (*JE*, p. 69), for 'the Bible bids us return good for evil' (*JE*, p. 66). Such moral instruction is predicated on the generous assumption that all men share the potential for goodness and this democratic trajectory of redemption is an important feature of both Charlotte and Anne's fiction.

Motivated by the Brontë sisters' rejection of Calvinist predestination, both Helen Burns and Helen Huntingdon in *The Tenant of Wildfell Hall* believe in the doctrine of universal salvation – the hope that all might be saved, albeit after a period of posthumous repentance, in something akin to the Catholic purgatory. This is clearly seen in Helen Huntingdon's dialogue with her aunt, immediately prior to her disastrous marriage to Arthur:

> '"If any man's work abide not the fire, he shall suffer loss, yet himself shall be saved, but so as by fire," and He that "is able to subdue all things to Himself, will have all men to be saved," and "will in the fullness of time, gather together in one all things in Christ Jesus, who tasted death for every man, and in whom God will reconcile all things to Himself, whether they be things in earth or things in Heaven."'
> 'Oh, Helen! Where did you learn all this?'
> 'In the bible, aunt. I have searched it through, and found nearly thirty passages, all tending to support the same theory.'
> 'And is that the use you make of your bible? And did you find no passages tending to prove the danger and falsity of such a belief?'
> 'No: I found indeed some passages that taken by themselves, might seem to contradict that opinion; but they will all bear a different construction to that which is commonly given, and in most, the only difficulty is in the word which we translate "everlasting" or "eternal". I don't know the Greek, but I believe it strictly means for ages, and might signify either endless or long-enduring.' (p. 150)

Helen Burns also insists on the democratic reach of Christ's creed, which extends hope to all and emphasises mercy rather than Calvinistic judgment: 'with this creed, I can so clearly distinguish between the criminal and his crime; I can so sincerely forgive the first while I abhor the last: with this creed revenge never worries my heart, degradation never too deeply disgusts me, injustice never crushes me too low: I live in calm, looking to the end' (*JE*, p. 70). Both these novels suggest the influence of Maurice – and Maurice would later be dismissed from his Chair of Theology at King's College London (in October 1853) for preaching precisely the ambiguous definition of 'eternity' which Helen articulates to her aunt.[54] Anne and Charlotte express the passionately held hope that all lives are ends in themselves: they are not created as somehow surplus to the narrative of (economic) progress and (fiscal) salvation popularised by the school of political economists – initiated by Thomas Malthus's *Essay on Population* (1798) in which the poor were positioned as largely expendable to the national (material and theological) interests. In their reflections on the education of children, expressed with some consistency in *Jane Eyre*, *Villette*, *Agnes Grey* and *The Tenant of Wildfell Hall*, Charlotte and Anne both reject the idea of original sin, stressing that all children are capable of moral cultivation, provided they are subject to effective parental discipline and thoughtful scholarly instruction: without such moral direction, Agnes's charge Matilda is full of 'animal ... vigour'

(*AG*, p. 59) and Tom displays a terrible 'propensity to persecute the lower creation' (*AG*, p. 42), but neither are predestined to spiritual damnation.

Yet, despite this impulse towards an inclusive constitution based on mutual appreciation and strength of fraternal feeling, the Brontë protagonists are not governed wholly by sympathy to the detriment of doctrinal conviction. Jane's and Agnes's plain speech is accompanied by unshakeable confidence in the religious righteousness of their own judgments. Convinced that conscience calls her to leave Rochester when his bigamous plan is revealed, she asserts the value of a humble woman's self-worth:

> 'Do you think I am an automaton? – a machine without feelings? and can bear to have my morsel of bread snatched from my lips, and my drop of living water dashed from my cup? Do you think, because I am poor, obscure, plain, and little, I am soulless and heartless? You think wrong! – I have as much soul as you – and full as much heart! And if God had gifted me with some beauty and much wealth, I should have made it as hard for you to leave me, as it is now for me to leave you. I am not talking to you now through the medium of custom, conventionalities, nor even of mortal flesh: it is my spirit that addresses your spirit; just as if both had passed through the grave, and we stood at God's feet, equal – as we are!' (*JE*, p. 350)

The radical implications of such an assertion are clear: as Patsy Stoneman observes, *Jane Eyre* was staged many times in the course of the nineteenth century (from 1848 onwards) and six of the extant versions, which make otherwise very significant revisions to the rest of the novel, include this speech virtually word for word as it appears in Brontë's text[55] – it demonstrates all the spirited defiance of social oppression on which melodrama depended for its success. When Rochester insists that she remain at Thornfield and care for him because her life is by implication less valuable, Jane counters,

> 'I care for myself. The more solitary, the more friendless, the more unsustained I am, the more I will respect myself. I will keep the law given by God; sanctioned by man. I will hold to the principles received by me when I was sane, and not mad – as I am now. Laws and principles are not for the times when there are no temptations: they are for such moments as this, when body and soul rise in mutiny against their rigour; stringent are they; inviolate they shall be. If at my individual convenience I might break them, what would be their worth? They have a worth – so I have always believed; and if I cannot believe it now, it is because I am insane – quite insane, with my veins running fire, and my heart beating faster than I can count its throbs. Preconceived opinions, foregone determinations are all I have at this hour to stand by: there I plant my foot.' (*JE*, p. 356)

Charlotte demonstrates that the moral maxims of the Ten Commandments ('Thou shalt not commit adultery') trump her own passionate inclinations here: in contradistinction to Eliot's narrator's assertion in *The Mill on the Floss* (1859–1860) that moral maxims must always be subordinate to the complexities of 'the individual lot',[56] it is the abstract rule of Christian law which Jane must adhere to when feelings become suspect in their intensity or their aim. After various trials and tests of this commitment to the law of conscience, she enjoys a marriage which duly reconciles affect and duty: again, in Channing's terms, her social self-culture has been perfected by the trials of the plot:

> [S]elf-culture is social, as one of its great offices is to unfold and purify the affections which spring up instinctively in the human breast; which bind together husband and wife, parent and child, brother and sister; which bind a man to friends and neighbours, to his country, and to the suffering who fall under his eye, wherever they belong. The culture of these is an important part of our work; and it consists in converting them from instincts into principles, from natural into spiritual attachments; in giving them a rational, moral, and holy character.[57]

In its Victorian formulation, then, moral perfectionism was never solely about self-realisation: as Andrew Miller has observed, narrative trajectories of maturation also included the promotion of cooperation and social union as 'second-person relationships such as friendship and marriage' were often required to model the reconciliation of wider competing claims based on intimacy and distance, affection and contract.[58] If Emily Brontë's *Wuthering Heights* (1847) offers something of a case study in which the protagonists' transition from adolescence to adulthood, and the formation of effective second-person intimacies, is not successful (at least among the members of the older generation at Wuthering Heights) then *Jane Eyre*, *Shirley*, *Agnes Grey* and *The Tenant of Wildfell Hall* all suggest that the individual can both pursue self-perfection and find companionship along the way. In Slaughter's analysis, these novels are examples of the providential or 'comic' *Bildungsroman*, which 'normalize ... the story of enfranchisement', affording, for example, moral recognition to the orphaned Jane, and confirming Agnes as a suitable person to undertake the instruction of children. For Charlotte and Anne, this 'incorporative process of socialization' partly manifests itself in equality of men and women before God (as Jane proclaimed so powerfully in her dialogue with Rochester) and partly in equality of opportunities to perform works of social utility, as Shirley and Caroline acknowledge:

> 'Shirley, men and women are so different: they are in such a different position. Women have so few things to think about – men so many: you may have a friendship for a man, whilst he is almost indifferent to you. Much

of what cheers your life may be dependent on him, while not a feeling or interest of moment in his eyes may have reference to you ...'

'Caroline', demanded Miss Keeldar, abruptly, 'don't you wish you had a profession – a trade?'

'I wish it fifty times a day. As it is, I often wonder what I came into the world for. I long to have something absorbing and compulsory to fill my head and hands, and to occupy my thoughts.'

'Can labour alone make a human being happy?'

'No, but it can give varieties of pain, and prevent us from breaking our hearts with a single tyrant master-torture. Besides, successful labour has its recompense; a vacant, weary, lonely, hopeless life has none'. (*Shirley*, pp. 192–3)

The Tenant of Wildfell Hall is also attentive to the limitations on women's economic autonomy and financial independence, but neither Charlotte nor Anne deploy the rhetoric of 'rights' to advance their protagonists' cases. Indeed, one of the criticisms of Channing's influential doctrine of 'self-culture' was that it encouraged the individual to turn away from rights-based arguments in favour of maintaining the political status quo, focusing instead on inwards improvement and an acceptance of hierarchies and forces external to the individual. As Ryan has argued, and as I noted previously, the renovation of religious belief was regarded as a necessary precondition for the establishment of a newly inclusive constitution, and the novel provided a generous and capacious forum for such ideological experimentation and refinement to occur.

And just as Channing was seen by some as an at best lukewarm supporter of abolitionism in America,[59] any attempt to conceptualise *Jane Eyre* as a powerful contribution to the paralegal advocacy of women's rights runs aground on the representation of Rochester's incarcerated first wife, Bertha Antoinetta Mason. On seeing her for the first time, Jane fails to perceive her shared humanity: 'What creature was it, that, masked in an ordinary woman's face and shape, uttered the voice, now of a mocking demon, and anon of a carrion-seeking bird of prey?' (*JE*, p. 327). A line is drawn, then, between the young woman who can be enfranchised and empowered by the ever more polite rehearsal of her tale, and the entrapped young woman in the attic whose humanity the text refuses to recognise and whose voice the reader is never allowed to hear. Slaughter acknowledges that the cultural work of recognition undertaken by the *Bildungsroman* is gradual and constrained by conditions in the public sphere:

> [A]s a human rights claim, [the *Bildungsroman*] is a narrative instrument for historically marginalized people to assert their right to be included in

the franchise of the public sphere and to participate in the deliberative systems that shape social normativity itself by setting the limits between the enfranchised and the disenfranchised. In this sense, the democratic social work of the *Bildungsroman* is to demarginalize the historically marginal individual – to make the socially unrepresentative figure representative. Thus, the exemplarity of the affirmative *Bildungsroman's* protagonist lies in its relatively modest marginality, a model marginality that is capable of being transcended or transformed by proper training in the socially acceptable forms of speech and the social norms of the bourgeois reading public. Thematically and sociologically, the viability of a candidate for affirmative *Bildungsroman's* incorporative work appears largely pegged to the historical conditions of the public sphere itself – to the actual or imminent expansions and contractions of the franchise, which the novels both reflect and catalyze.[60]

That Jane need only refine her tale of grievance, expunging bitterness and exaggeration and acquiring a tone of impartiality is in effect evidence of precisely this 'model marginality that is capable of being transcended or transformed by proper training in the socially acceptable forms of speech': Bertha, on the other hand, is inarticulate, expressing herself only with growls and mutterings. While Slaughter places self-determination and the telling of one's own tale at the heart of human rights discourse, Bertha's plight reminds us that children, the elderly and the seriously ill are left without protection when 'being human' is equated exclusively with a sophisticated usage of language. Yet if Charlotte seems blind to Bertha's predicament in *Jane Eyre*, she manifests a more nuanced awareness of what it means for women to be less confident of their narrative prowess in *Villette* when Lucy Snowe's perpetual disappointments manifest themselves in personal reticence, but above all it is Jean Rhys's *Wide Sargasso Sea* (1960) which emphasises the capacity of the first-person perspective to afford new political insights into the power relations which Charlotte had taken for granted. In *Wide Sargasso Sea*, Rhys empowers Antoinette (Bertha) to narrate her own version of events, enabling her to articulate her fears of inheriting her mother's catastrophic mental instability, her lonely experiences in childhood and at school (which mirror Jane's own) and finally her abandonment by Rochester. When Rochester accuses Antoinette of madness, she suggests he 'listen to the other side ... There is always the other side, always'.[61] While Charlotte used first-person narrative to incorporate working women into the wider public sphere, twentieth-century literature could revisit the form's potential, placing ever more socially marginal figures centre stage – thus extending the limits of the 'fully human' even further.

The Brontë sisters died before agitation for female suffrage gained strength and intensity. They rarely used the language of 'rights': in an era in which Christian belief remained the foundation of the British polis, the equality of all before God was something which had to be proclaimed aloud and securely established in the public imagination before the law of 'rights' could follow. The Brontës' representation of 'the human' thus serves as a case study of the extent to which modern ideas of autonomy and self-development might be successfully accommodated alongside – or grafted onto – Christian ideas of humility and self-abnegation before God. As Thormählen observes, '[t]he courage with which all the Brontë heroines oppose attempts to shatter their integrity derives from a sense that they are not defending their own personal interests so much as upholding a core of moral values with which they regard their selves as affiliated'.[62] As they learn from the examples of others, the Brontës require their heroines to discriminate carefully between acts of reckless selfishness and admirable attempts to be true to the embryonic potential of selfhood which God has implanted in all men and women alike. The realist novel was a powerful vehicle for the interrogation of these contrasting ethical systems of value: it registers the tensions (and simultaneously effects an aesthetic reconciliation) between the discourse of rights, emancipation and enfranchisement and the economy of 'creatureliness' and of submission to a loving God. One of the great achievements of the mid-Victorian realist novel, then, was to create a discursive space in which the fusion of these different philosophical traditions, each of such value to the English public sphere, might take place in receptive readerly imaginations. In the works of Anne and Charlotte Brontë, Biblical quotation underpins trajectories of female empowerment. But at the same time, as the sisters selected, parsed and glossed those allusions from the New Testament which seemed most fully to support human autonomy, they (paradoxically) helped to create a more liberal sense of what the Bible contained once edited for modern readers.[63] As Joss Marsh has argued,

> [T]he Bible entered the nineteenth century as the sacred repository of historical truth and exited it a literary masterpiece, instated at the head of a different 'canon', while Literature in counterbalance, took on the role ... of a secular 'Scripture', the subject of serious university study and the cornerstone of mass educational strategy.[64]

Jane Eyre, Shirley, Agnes Grey, Wuthering Heights, Villette and *The Tenant of Wildfell Hall* all play an important part in the complex process by which Biblical injunctions are simultaneously amplified, illustrated and diluted. The fiction of the Brontë sisters responded with precision to contemporary theological debates about the nature of 'the human' and, in turn, transformed the inheritance they received: although the manifold richness of a novel always exceeds any attempt to reduce it to a slogan, Jane's intuition that she stands with Rochester, 'at God's feet, equal – as we are!' was to be the rallying cry for women in the second half of the nineteenth century.

Notes

1. Anon., 'Review' in *The Christian Remembrancer*, (15 April 1848), 396–409, reproduced in Miriam Allott, ed., *The Brontës: The Critical Heritage* (London: Routledge & Kegan Paul, 1974), pp. 88–92, pp. 89–90.
2. Anon. [Albany Fonblanque], 'Review' in *The Examiner*, 27 November 1847, 756–7, reproduced in Allott, ed., pp. 76–8, p. 76.
3. Marianne Thormählen, *The Brontës and Religion* (Cambridge: Cambridge University Press, 1999), p. 64.
4. Ibid., p. 173.
5. See Thormählen, cited in note 3, and Heather Glen, *Charlotte Brontë: The Imagination in History* (Oxford: Oxford University Press, 2002), pp. 65–143.
6. Thormählen, p. 165.
7. Maria Lamonica, 'Jane's Crown of Thorns: Feminism and Christianity in *Jane Eyre*', *Studies in the Novel*, 34:3 (2002), 245–63, p. 247.
8. Boyd Hilton, *The Age of Atonement: The Influence of Evangelicalism on Social and Economic Thought 1785–1865* (Oxford: Clarendon Press, 1988), p. 21.
9. See the discussion in Hilton, pp. 3–26, and Jan-Melissa Schramm, *Atonement and Self-Sacrifice in Nineteenth-Century Narrative* (Cambridge: Cambridge University Press, 2012), pp. 1–37.
10. John Rawls, *A Theory of Justice* (Cambridge, MA: Harvard University Press, 1971, rev. edn. 1999), p. 155, pp. 298–99.
11. Stefan Collini, *Public Moralists: Political Thought and Intellectual Life in Britain, 1850–1930* (Oxford: Clarendon Press, 1991), pp. 62–3.
12. For discussion of this theme in more detail, see Schramm, *Atonement and Self-Sacrifice*, pp. 1–32.
13. Charles Dickens, *Our Mutual Friend* (1864–1865) (Harmondsworth: Penguin, 1985), p. 770.
14. George Eliot, *Daniel Deronda* (1876) (Harmondsworth: Penguin, 1986), p. 218.
15. On habits of renunciation appropriate to age and gender, see Glen, pp. 69–95.
16. Charlotte Brontë, *Jane Eyre* (1847) (Harmondsworth: Penguin, 1996), p. 225. Other references to this edition are included within the body of the text with the abbreviation *JE*. On marital sacrifice, see Ilana Blumberg, *Victorian Sacrifice: Ethics and Economics in Mid-Century Novels* (Columbus: Ohio State University Press, 2013), pp. 99–138.

17 John Stuart Mill, 'On the Subjection of Women' (1869) in Stefan Collini, ed., *On Liberty and Other Writings* (Cambridge: Cambridge University Press, 1989), pp. 147–8.
18 Cited in Julia Sun-Joo Lee, *The American Slave Narrative and the Victorian Novel* (Oxford: Oxford University Press, 2010), pp. 20–1.
19 Lee, p. 21.
20 Gregory Dart, *Rousseau, Robespierre, and English Romanticism* (Cambridge: Cambridge University Press, 1999), p. 9.
21 Ibid., at p. 180.
22 George H. Lewes, 'Recent Novels: French and English', *Fraser's Magazine*, 36 (December 1847), 686–95.
23 Ibid., at p. 689.
24 *Universal Declaration of Human Rights* (adopted by the UN General Assembly, 10 December 1948), Article 6.
25 See, for example, Lynn Hunt, *Inventing Human Rights: A History* (New York: Norton, 2007), and Joseph Slaughter, *Human Rights, Inc.: The World Novel, Narrative Form, and International Law* (New York: Fordham University Press, 2007).
26 Christine Krueger, *Reading for the Law: British Literary History and Gender Advocacy* (Charlottesville: University of Virginia Press, 2010), p. 3.
27 For extensive discussion of this point, see Alex Woloch, *The One vs. the Many: Minor Characters and the Space of the Protagonist in the Novel* (Princeton: Princeton University Press, 2003).
28 'Human Rights' in 'Human' adj. and n. in *Oxford English Dictionary* online, accessed 25 June 2017. www.oed.com.
29 See Hunt cited in note 25. Some historians and literary critics object to this attempt to trace lines of continuity between the eighteenth-century and twentieth-century language of 'rights': see for example, Peter de Bolla, *The Architecture of Concepts: The Historical Formation of Human Rights* (New York: Fordham University Press, 2013), and Samuel Moyn, *Human Rights and the Uses of History* (London: Verso, 2014).
30 *The Declaration of the Rights of Men and of Citizens, by the National Assembly of France* (1789), Article 1.
31 Mary Wollstonecraft, *A Vindication of the Rights of Woman* (1791), Sylvana Tomaselli, ed., (Cambridge: Cambridge University Press, 1995), pp. 68–9.
32 Slaughter, p. 4.
33 Ibid., at p. 157.
34 Ibid., at pp. 19–20.
35 Ibid., at p. 39.
36 Adam Smith, *The Theory of Moral Sentiments* (1759, rev. 1790), Knud Haakonssen, ed. (Cambridge: Cambridge University Press, 2002), pp. 11–12.
37 Charlotte Brontë, 'Preface' to *JE*, p. 5.
38 Robert Ryan, *The Romantic Reformation: Religious Politics in English Literature 1789–1824* (Cambridge: Cambridge University Press, 1997), p. 4.

39 See for example, Charlotte Brontë, *Villette* (1853) (Harmondsworth: Penguin, 1986), Chapters 14 and 26.
40 Brontë, 'Preface' to *JE*, p. 6.
41 Slavery was abolished in England in 1807 and the slave trade was abolished in the British Empire in 1833, but it remained active in the United States up until the Civil War (1861–1865). Critics disagree as to the extent to which *Jane Eyre* deals with the contemporary slave trade: see Sue Thomas, *Imperialism, Reform, and the Making of Englishness in Jane Eyre* (Basingstoke: Palgrave Macmillan, 2008), pp. 8–30, and Lee, pp. 25–52.
42 Lee, p. 20.
43 See Jan-Melissa Schramm, 'Towards a Poetics of (Wrongful) Accusation: Innocence and Working-Class Voice in Mid-Victorian Fiction', in Yota Batsaki, Subha Mukherji, and Schramm, eds., *Fictions of Knowledge: Fact, Evidence, Doubt* (Basingstoke: Macmillan, 2011), pp. 193–212.
44 William Ellery Channing, *Self-Culture* (London: 1838), p. 6. Channing had first visited England in 1821–1822, and by 1840, when *Self-Culture* was lauded in London periodicals, he was regarded as among the most important public moralists in the United States – his influence on the Transcendentalists was significant.
45 On the relationship between Channing, Wordsworth and Coleridge, see John Beer, *Providence and Love: Studies in Wordsworth, Channing, Myers, George Eliot, and Ruskin* (Oxford: Clarendon Press, 1998), pp. 98–115.
46 Channing, *Self-Culture*, p. 7.
47 Ibid., at p. 10.
48 Ibid., at p. 17.
49 Ibid., at p. 22.
50 Glen argues that the volume called '"Self-Culture" in a "List of Books from Smith & Elder 18 March 1850" found in Charlotte Brontë's desk was probably Emily Shirreff and Maria Grey, *Thoughts on Self-Culture: Addressed to Women* (London: Moxon, 1850)': Glen, pp. 84–5.
51 Beer, p. 14.
52 For the Victorian definition of 'just deserts', see John S. Mill, 'Utilitarianism' (1861) in *Essays on Ethics, Religion, and Society*, J. M. Robson, ed. (Toronto: University of Toronto Press, 1969), p. 255.
53 On the tension between mercy and judgment in Victorian literature, see Schramm, *Atonement and Self-Sacrifice*, pp. 25–33 and pp. 216–36: see also Charles Kingsley, *The Water Babies* (1863) (Harmondsworth: Penguin, 2008), pp. 111–13.
54 See Thormählen, pp. 88–9.
55 See, for example, John Courtney, *Jane Eyre, or the Secrets of Thornfield Manor* (1848), Act II sc. Iii, John Brougham, *Jane Eyre: A Drama in Five Acts* (1849), Act III, sc. Iii, Charlotte Birch-Pfeiffer, *Jane Eyre, or the Orphan of Lowood* (1870), Act III, sc. Vii, Madame von Heringen Hering's *Jane Eyre* (1877), Act IV, sc. Viii: all discussed in Patsy Stoneman,

'Introduction' to *Jane Eyre on Stage, 1848–1898: An Illustrated Edition of Eight Plays with Contextual Notes* (Aldershot: Ashgate, 2007), pp. 7–16.
56 George Eliot, *The Mill on the Floss* (1860) (Harmondsworth: Penguin, 1985), p. 528.
57 Channing, *Self-Culture*, p. 12.
58 Andrew Miller, *The Burdens of Perfection: On Ethics and Reading in Nineteenth-Century British Literature* (Ithaca: Cornell University Press, 2008), p. 25.
59 See for example, William Ellery Channing, *Slavery* (London: Hunter, 1836).
60 Slaughter, pp. 156–7.
61 Jean Rhys, *Wide Sargasso Sea* (1966) (Harmondsworth: Penguin, 2011), p. 99.
62 Thormählen, p. 156.
63 On the selective sanitation of the Bible, especially the Old Testament, in the nineteenth century, see Yvonne Sherwood, *Biblical Blaspheming: The Trials of the Sacred in a Secular Age* (Cambridge: Cambridge University Press, 2012).
64 Joss Marsh, *Word Crimes: Blasphemy, Culture, and Literature in Nineteenth-Century England* (Chicago: University of Chicago Press, 1998), p. 169.

CHAPTER 9

'A Strange Change Approaching'
Ontology, Reconciliation, and Eschatology *in* Wuthering Heights

Simon Marsden

> The ledge, where I placed my candle, had a few mildewed books piled up in one corner; and it was covered with writing scratched on the paint. This writing, however, was nothing but a name repeated in all kinds of characters, large and small – *Catherine Earnshaw*, here and there varied to *Catherine Heathcliff*, and then again to *Catherine Linton*. (*WH*, p. 15)

Lockwood's reading of the names etched into the window ledge of Cathy's childhood bedroom at Wuthering Heights is followed later that night by the appearance at the window of a ghostly child who calls herself Catherine Linton.[1] Lockwood himself acknowledges the strangeness of the name used by the ghost: 'why did I think of *Linton*?' he wonders; 'I had read *Earnshaw* twenty times for Linton' (p. 20). The sequence of names records the conflicted desires of the young Cathy, compelled to choose between bonds of emotional and imaginative sympathy with Heathcliff and more conventional erotic attraction to – and the promise of economic security and social status with – Edgar Linton. The use of the married woman's name by the ghostly child seeking entrance into her childhood bedroom signifies the fragmentation of identity that begins with Cathy's choice of Edgar as husband. For many readers, this fragmentation has been a core truth of *Wuthering Heights*: Brontë's novel tells the story of an original, symbiotic union divided by external and/or internal pressures and seeking to regain the lost union and wholeness in death. The specific forces of division may vary in accordance with the theoretical perspective of the critic, but the narrative of union, division and (perhaps) reunion remains constant across a wide range of interpretations.

Yet, this familiar narrative is not without difficulties. Lynne Pearce observes that critical readings of *Wuthering Heights* have tended to conceal 'their own susceptibility to the discourse of romantic love: that is to say, the way in which the apparently overwhelming force of Catherine and Heathcliff's love for one another becomes the text's baseline "truth" and *raison d'être*.[2] In Pearce's view, readings of the novel across a wide range of theoretical perspectives have rested upon a common – and frequently unargued – assumption that the novel celebrates the Heathcliff–Cathy relationship and that this celebration is the starting point for critical interpretation. Of course, there are exceptions to this view of the novel, but these have often been exceptions that prove the rule: they perpetuate the romance paradigm by shifting the focus of celebration to Hareton and the second Catherine. The second-generation couple is seen as an improved version of their predecessors, repeating many of their trials but avoiding their failures. As Janet Gezari observes, readers 'agree to differ over whether the love affair of the younger pair represents a diminished alternative to the more authentic love affair of Heathcliff and Catherine, or a more fruitful version of it, modulated and humanized so as to give some promise of social happiness'.[3] Whichever couple one favours, these critical narratives persist in asking us to choose between two models of romance as the foundation for our interpretation of the novel.

The often-silent privileging of the romance plot in readings of *Wuthering Heights* has played a significant role in shaping critical reflection on the novel's engagement with religion. Indeed, discussions of religion have often reproduced, implicitly or overtly, the paradigm of romance as the novel's hermeneutic key. In J. Hillis Miller's groundbreaking study *The Disappearance of God* (1963), the love of Heathcliff and Cathy is nothing less than a theological and spiritual revolution that casts off the angry, transcendent God of Joseph's Calvinism and replaces it with the benevolent presence of an immanent divine spirit. By the close of the narrative, Miller argues,

> God has been transformed from the transcendent deity of extreme Protestantism, enforcing in wrath his irrevocable laws, to an immanent God, pervading everything, like the soft wind blowing over the heath. This new God is an amiable power who can, through human love, be possessed here and now ... The love of Heathcliff and Cathy has served as a new mediator between heaven and earth, and has made any other mediator for the time being superfluous. Their love has brought 'the new heaven and the new earth' into this fallen world as a present reality.[4]

For Miller, then, the overwhelming force of the Heathcliff–Cathy relationship is sufficient to refigure the relationship between heaven and earth. Miller's is a theologically nuanced reading, alert to early-Victorian tensions between dogmatic strains of Evangelical Protestantism and the immanentist theologies favoured by Romanticism.⁵ Yet it remains vulnerable to Pearce's charge that the narrative logic of the romance plot has become an unacknowledged assumption for readers of *Wuthering Heights*. If a high view of the Heathcliff–Catherine relationship – and it is difficult to imagine a higher view of it than Miller's – can be shown to be flawed, the theological edifice built upon it is in danger of collapse.

Where Miller reads the romance itself in theological terms, other critics have viewed it – and, by extension, the novel – as profoundly anti-Christian. Sandra Gilbert and Susan Gubar's influential feminist study *The Madwoman in the Attic* (1979) reads Christianity as one of the patriarchal forces against which Heathcliff and Cathy must contend, a view echoed in subsequent studies by Margaret Homans, Irene Tayler and Stevie Davies.⁶ In these readings, Heathcliff and Cathy are seen as Romantic rebels against a Christian orthodoxy that is implicated in structures of patriarchal power. Marianne Thormählen offers an alternative reading in *The Brontës and Religion* (1999), the most detailed account yet produced of the religious contexts of the Brontës' novels. For Thormählen, *Wuthering Heights* is a 'nineteenth-century Revenger's Tragedy in which the avenger is never reconciled – Heathcliff seeks no forgiveness and grants none – but ultimately disarmed by the one force [i.e., love] that is stronger than his hatred'.⁷ On this reading, the novel is more readily compatible with at least some components of nineteenth-century Protestant doctrine than has been recognised by critics who regard Christianity as one of the patriarchal powers contested by Brontë's writing. Indeed, much of the critical discussion of religion in *Wuthering Heights* has seemed to present readers with two options: Heathcliff and Cathy are either heroic rebels against an oppressive, dogmatic Christianity, or they are a cautionary tale that can be understood in terms of (sometimes unorthodox) Christian doctrine.⁸

The purpose of this chapter is not to choose between these competing ways of reading *Wuthering Heights*, but rather to move beyond the increasingly obsolete terms in which much of the debate about the novel's engagement with religion has been framed. The tendency of critics to identify Christianity either with institutional and hierarchical power structures or with particular versions of doctrinal orthodoxy has neglected the extent to which Christian theology concerns itself with the ontological status of the human person, the nature of human flourishing and the relationship between language and meaning. Because criticism – even when sympathetic to religious concerns – has largely reflected a secular view of religion as a discrete category of discourse, the ways in which theology might shape and illuminate the novel's depictions of human personhood, community and language have remained relatively unexplored.[9] This essay seeks a way forward in this debate by reading *Wuthering Heights* in relation to more recent theological accounts of human ontology and semiotics developed by Catherine Pickstock and John Milbank. It begins by examining representations of social fragmentation in *Wuthering Heights* and situating these in relation to theological ideas of human ontology explored elsewhere in Brontë's writing. It proposes an eschatological reading of the names written on the window ledge as signs open to the possibility of radical transformation and reinterpretation. Drawing on Pickstock's account of repetition, the essay concludes by reading the novel's second-generation characters as non-identical repetitions, or redemptive returns, of their first-generation predecessors.

Refusing the Stranger

To cross the threshold of Wuthering Heights is to enter a world in which conventions of social and linguistic exchange break down. This is a world in which an invitation to 'walk in' is 'uttered with closed teeth and [expresses] the sentiment, "Go to the Deuce!"' (p. 1). 'We speak because we are in search of recognition', Rowan Williams argues, and because 'we want to have opened for us the possibility of new kinds of shared action, ways of "going on" in the company of others'.[10] If this is correct, then *Wuthering Heights* illustrates the extent to which our speech might work, paradoxically, to resist or disavow the social participation upon which the very act of speaking is predicated. This is not simply because the world of the novel is set apart from the language and behaviour of 'civilised' or 'polite' society, as Charlotte Brontë suggested in her preface to the 1850 edition.[11] On the contrary, Emily Brontë is alert to the ways in which violence, inequality and exclusion are embedded within the linguistic and behavioural conventions of 'civilised' society itself. Thus Heathcliff's initial exclusion by the Earnshaw household is repeated by the more socially refined Lintons, who reject him as 'quite unfit for a decent house'; a classification prompted as much by his ambiguous ethnicity and socio-economic status – they call him 'a gipsy', 'a little Lascar, or an American or Spanish castaway' and a 'wicked boy' who should be hanged 'before he shows his nature in acts, as well as features' – as by his behaviour (pp. 43–4). Significantly, these remarks are addressed not to Heathcliff, but to other members of the Linton household. In this instance, the social function of speech is deliberately restricted: the speakers seek recognition and shared endeavour only with other members of their own bounded community. Heathcliff's exclusion from the Lintons' conversation is as real and literal as his exclusion from their home.

At several points in the narrative, this depiction of social breakdown and linguistic corruption takes on specific theological reference. The most overt instance occurs in Lockwood's dream of Jabes Branderham's sermon, in which the preacher's parsing of the text of Matthew 18: 21–22 yields a ludicrous taxonomy of transgression: Christ's command to forgive 'Until seventy times seven' times (in other words, constantly) is transformed into a list of 490 discrete sins that must be forgiven before vengeance can ensue.[12] This detailed accounting of transgressions produces the opposite of the moral order that it seeks. Branderham's command to enact vengeance upon the sinner guilty of the 'First of the Seventy-First' yields a scene of violence in which '[e]very man's hand was against his neighbour' (p. 20). Brontë's nineteenth-century Christian readers would have recognised Branderham's misreading of his biblical text which, in its original context, places upon Christians an obligation to recognise their own status as forgiven sinners and to respond by forgiving freely the sins of others.[13] Branderham and his congregation corrupt a text calling for forgiveness and the restoration of relationship by reading it as a justification of vengeance and division.

Other points in the novel show a similar distortion of Christian social ethics. The Christian duty of hospitality is refused on multiple occasions: the rejections of Heathcliff by the Earnshaws and Lintons; Heathcliff's imprisonment of and abusive behaviour towards the second Catherine; Heathcliff's cruel reception of his son, a tragic echo of his own arrival at Wuthering Heights; and Heathcliff's hostility towards Lockwood. Indeed, Lockwood's own attempt to exclude Cathy's ghost in the second part of his dream is emblematic of the refusal to welcome the stranger that characterises social behaviour throughout the novel. In the context of a narrative in which hospitality is so often denied, Edgar Linton's use of a biblical allusion on the occasion of Heathcliff's return from self-imposed exile is intriguing. 'Catherine', he says, 'try to be glad, without being absurd! The whole household need not witness the sight of your welcoming a runaway servant as a brother' (p. 84). The allusion is to St. Paul's letter to Philemon, in which Philemon is urged to receive his runaway slave, Onesimus, as a brother in Christ:

> For perhaps he therefore departed for a season, that thou shouldst receive him forever; Not now as a servant, but above a servant, a brother beloved, specially to me, but how much more unto thee, both in the flesh, and in the Lord?[14]

Edgar's allusion to the letter to Philemon ironically refuses its social ethic. The reception of the 'runaway servant as a brother', urged as a Christian duty in the letter, becomes for Edgar an action to be concealed from the 'whole household': including, of course, its domestic staff. Brontë's use of this biblical allusion signals a gap between the subversive social ethic of the New Testament text and the implicit refusal of this ethic by a Christian social elite determined to preserve hierarchical socio-economic divisions.

A similar critique of a Christian social and political elite that violates Christian ethics emerges elsewhere in Brontë's writing. In her final poem – 'Why ask to know the date – the clime?' – Brontë imagines a civil war in which public worship is performed by perpetrators of violence:[15]

> Why ask to know the date – the clime?
> More than mere words they cannot be:
> Men knelt to God and worshipped crime,
> And crushed the helpless even as we –[16]

Brontë deploys a critique that has a long history within Christianity and, indeed, in pre-Christian biblical literature. It has a clear precedent in the book of Isaiah, which similarly contrasts public acts of religious worship with the injustices perpetrated by the worshippers. Repeatedly in Isaiah, the prophet voices God's condemnation of religious ritual performed in the absence of social justice:

> Is not this the fast that I have chosen? to loose the bands of wickedness, to undo the heavy burdens, and to let the oppressed go free, and that ye break every yoke?

> Is it not to deal thy bread to the hungry, and that thou bring the poor that are cast out to thy house, when thou seest the naked, that thou cover him; and that thou hide not thyself from thine own flesh?[17]

Like Isaiah, Brontë condemns religious observance unaccompanied by the pursuit of justice for the oppressed. She extends this critique by framing it in terms of the New Testament's emphasis on mercy. The speaker of her poem is a soldier guilty of the gratuitous torture of a prisoner. Tormented by the memory of his crimes, the soldier recalls his prisoner 'Pleading in mortal agony/To mercy's Source but not to me' (188: 184–5). In a passage that recalls various New Testament texts – including the text of Branderham's sermon in Matthew 18 – that link the gift of divine mercy to demonstration of its human equivalent, the soldier anticipates God's judgment of his actions:

> I know that Justice holds in store,
> Reprisals for those days of gore –
> Not for the blood, but for the sin
> Of stifling mercy's voice within. (187: 161–4)

The refusal to show mercy to others is simultaneously a refusal of God's mercy for oneself. The same theology appears elsewhere in Brontë's poetry. In 'Shed no tears o'er that tomb', a dead man is imagined as facing divine judgment for a life of cruelty:

> The time of grace is past
> And mercy scorned and tried
> Forsakes to utter wrath at last
> The soul so steeled by pride
>
> That wrath will never spare
> Will never pity know
> Will mock its victim's maddened prayer
> Will triumph in his woe (109: 21–8)

These poems suggest a specific theological basis for the violent results of Branderham's sermon in *Wuthering Heights*. William A. Madden is surely correct when he observes that 'Emily Brontë makes it clear that for her the unforgivable sin consists in judging the human offenses of others as unforgivable'.[18] It is ironically appropriate that Branderham's call to 'execute upon [Lockwood] the judgment written' (p. 19) produces a brawl. The scene implicitly acknowledges the fact that the only 'judgment' written in Branderham's source text is upon the unmerciful servant in Christ's parable who has a debt forgiven by his master, but offers no similar forgiveness to his own debtor. Informed of the servant's actions, the master withdraws his mercy and condemns the servant to the full punishment of the law. In their zeal to punish the guilty, it is ironically necessary that the members of Branderham's congregation must punish each other.

Brontë's essay 'Filial Love', written as an exercise in French composition at the Pensionnat Heger in Brussels in 1842, offers a glimpse into the theological aspects of her view of human ontology and relationship. The essay grounds its view of the love between parent and child in a theological understanding of creation:

> Parents love their children; this is a principle of nature. The doe does not fear the dogs when her little one is in danger; the bird dies on its nest. This

instinct is a particle of the divine spirit we share with every animal that exists. Has God not put a similar feeling into the heart of the child? Truly there is something of the kind, yet still the voice of thunder cries out, "Honor your parents or you will die!" Now, that commandment is not given, that threat is not added, for nothing. There may be men who scorn their happiness, their duty, and their God to such a point that the spark of heavenly fire dies out in their breast, leaving them a moral chaos without light and without order, a hideous transfiguration of the image in which they were created.[19]

Brontë's choices of theological language and imagery in this passage are intriguing. She appears to make no distinction between parental love as a 'principle of nature' and as a 'particle of the divine spirit': both terms are deployed as descriptions of reality. Unlike many of the natural theologians of the eighteenth and early-nineteenth centuries, Brontë does not here think of nature as an objective fact from which the existence of a divine creator might be deduced. Instead, she allows the concepts of the natural and the created to stand as synonymous.[20] Developing this idea, she depicts the child who denies this instinctive love for the parent as a distortion of the divine image in which the human was created. The divine spark fades within them, light gives way to darkness and the order of creation reverts to chaos. The theology of the essay is (broadly) Augustinian: it identifies 'nature' with the creative activity of God and evil as a violation or corruption of the good creation. As the modern Augustinian theologian John Milbank observes, 'evil for the Christian tradition was radically without cause – indeed it was not even self-caused, but was rather the (impossible) refusal of cause'.[21] In Brontë's essay, this 'refusal of cause' is the denial of one's own nature given by the creative God and its instinctive orientation towards familial love. This denial is presented less as conscious choice and more as inexplicable failure of the will to embody and perform one's own full humanity: those who do not show filial love 'have never given a thought to what their parents have done for them', while 'the memory of their youth has never recalled to them the hopes and affection of the father they disobey'.[22]

Perhaps the most compelling instance in *Wuthering Heights* of the privation of natural familial love occurs when Linton Heathcliff arrives at the Heights.[23] Heathcliff describes his son as 'it' – a tragic echo of his own hostile reception by the Earnshaws – and claims him as 'property' rather than family (p. 182). Linton is justifiably uncertain that 'the grim, sneering stranger [is] his father'; nothing in Heathcliff's welcome conveys paternal affection. The only value that Heathcliff attaches to his son is economic. Heathcliff sees in Linton an opportunity to reverse his own socio-economic deprivation and to punish those whom he holds responsible: 'I want the triumph of seeing *my* descendent fairly lord of their estates; my child hiring their children, to till their fathers' lands for wages – That is the sole consideration which can make me endure the whelp – I despise him for himself, and hate him for the memories he revives!' (p. 184). The failure to recognise the inherent worth of others is central to the Christian understanding of evil. Keith Ward points out that '[s]in is a turning away from the call of love towards egoistic desire. We fall into hating and fearing others, because we are not prepared to value them for what they are, and seek their good as much as we seek our own'.[24] If Branderham's sermon and its violent aftermath are emblematic of social fragmentation in the world of the novel, Heathcliff's reception of Linton demonstrates the extent to which this fragmentation pervades all kinds of human relationship.

Approaching Change: Repetition and Difference

Like the sequence of names written by the young Cathy on her window ledge, Linton Heathcliff's name is a signifier of seemingly irreconcilable division. His name is chosen by his mother as a defiant insult to his father; it is an assertion both of Isabella's possession of their son and of the Linton family name that Heathcliff seeks to ruin. Isabella's naming of her son is tragically ironic: she seeks to protect Linton from Heathcliff, yet she gives him a name that is itself a provocation. Linton's name symbolises the legacy of conflict that he – like Hareton and Catherine – inherit from the previous generation. In *Wuthering Heights*, children are born into a world that is always already fallen; to use a biblical image, the sins of the fathers are visited upon the children.[25] Whether bestowed upon children or inscribed upon the building, names function as signifiers of the history within which the lives of the second-generation characters take shape. Yet names are also signs available to be reread and reappropriated by the new generation, as they are when Catherine and Linton search for balls with which to entertain themselves at the Heights:

> We found two in a cupboard, among a heap of old toys; tops, and hoops, and battledoors, and shuttlecocks. One was marked C., and the other H.; I wished to have the C., because that stood for Catherine, and the H. might be for Heathcliff, his name; but the bran came out of H., and Linton didn't like it. (p. 219)

This brief scene illustrates both the persistence of names as linguistic signifiers of the previous generation's legacy and the availability of those names for reinterpretation. It shows the ambiguity of the signifiers themselves: the ball marked 'H.' is at least as likely to have belonged to Hindley as to Heathcliff, despite Catherine's association of the inscription with Linton's surname. Crucially, this reinterpretation takes place in the social context of Catherine's attempts to build a friendship with Linton, an act of love towards the unlovely boy that stands in sharp contrast to his treatment by the inhabitants of the Heights. If names in *Wuthering Heights* record a history of division and conflict, the availability of those names for reinterpretation signifies the possibility that the legacies of the past might be transformed.

I am here taking a different view of names and symbols in *Wuthering Heights* to that of J. Hillis Miller in his book *Fiction and Repetition* (1982). In Miller's influential post-structuralist account, the repetitions in Brontë's narrative disclose the impossibility of locating a hermeneutic centre or key to the novel and, thus, testify to the incompleteness of each act of reading. Miller points to the multiple images of Cathy as signifiers of absence; signs without content:

> Each thing stands not for the presence of Catherine as the substance behind the coin, the standard guaranteeing its value, the thing both outside the money system and dispersed everywhere in delegated form within it. In this case, each thing stands rather for the absence of Catherine. All things are memoranda, written or inscribed memorials, like a note I write myself to remind me of something. They are memoranda that she did exist and that Heathcliff has lost her, that she is dead, vanished from the face of the earth. Everything in the world is a sign indicating Catherine, but also indicating, by its existence, his failure to possess her and the fact that she is dead. Each sign is both an avenue to the desired unity with her and also the double barrier standing in the way of it.[26]

Heathcliff's inability to 'possess' Cathy becomes a metaphor for the reader's inability to 'possess' the truth of the novel. Like Heathcliff, Miller argues, we seek to push beyond the signs, to grasp the substance of what they signify. Yet each time we do so, we find that there is no final truth to be grasped. The novel's words and images are signifiers of absence instead of presence. Each reading of the novel is already the beginning of the next; our interpretations are haunted by the alternative meanings that they exclude.[27]

Is it necessarily the case, however, that *Wuthering Heights* holds its reader within the cycle of rereading that Miller proposes? It is striking that despite his emphasis on repetition, Miller pays little attention to time. In his account, we seem to encounter the novel's repetitions not in temporal sequence, but rather as simultaneously occurring hermeneutic *loci*: that is to say, whichever symbol one chooses as emblematic of the novel's meaning, one's reading is always disrupted by the presence of other symbols that make equal claim to emblematic status. At this point, we might note Pickstock's claim that for postmodern theory 'the systematic exaltation of writing over speech has ensured within Western history the spatial obliteration of time, which in seeking to secure an absolutely immune subjectivity, has instead denied any life to the human subject whatsoever'.[28] What Miller calls 'repetitions' turn out not to be repetitions in the temporal sense at all: they are a collection of similar or identical symbols scattered throughout the space of the text, like the images of Cathy that Heathcliff sees in every part of the physical space of Wuthering Heights. For Pickstock, in contrast, repetition in time is constitutive of being:

> For a thing to exist at all or to be observable by us, it must hold beyond the instant; it must have some continuity. It must paradoxically exceed its own identity in order to occupy a terrain or persist in being. So every thing that exists must be something non-identically repeated.[29]

If Pickstock's analysis is correct, then Miller's view of repetition as exposing the absence of the signified rests on his (silent) spacialisation of the novel's symbols. Because Miller does not consider repetition as occurring in time, he does not attend to the ways in which repetition might work to refigure, transform or – to use a theological idiom – redeem the words that for his reading must remain symbols of absence.

An approach that understands repetition as occurring within time enables us to see the names written on the window ledge at Wuthering Heights – '*Catherine Earnshaw*, here and there varied to *Catherine Heathcliff*, and then again to *Catherine Linton*' (p. 15) – as available for reinterpretation with the passage of narrative time. When Lockwood reads the names, they invoke a history: his reading (literally) raises the spectre of the girl who once occupied the bed on which Lockwood now lies. Yet, in bringing this history into view, the words invite us to imagine the situation of their writer, for whom they represented not the past, but the present and the future. Attempting to navigate her conflicted desires for Heathcliff and Edgar, the young Cathy imagined multiple versions of her adult identity: the words that Lockwood reads as traces of the past were once anticipations of the future, or of multiple possible futures. The history of the words reminds us that Lockwood's (and our) reading of them occurs *in media res*. The words were not always symbols of Cathy's absence; that they are when Lockwood reads them shows only that his reading is historically contingent, not that absence must be the signified of all signifiers that Miller takes it to be.[30] I am not arguing here for a naive or pre-critical relationship between word and object. I am suggesting, rather, that because the repetitions of names and symbols in *Wuthering Heights* occur within narrative time, it is appropriate to pay attention to the ways in which they are reinterpreted and transformed within the flow of this time. In other words, it is necessary to understand the ways in which particular meanings are replaced – or superseded – by new interpretations and new iterations.

The names on Cathy's window ledge demonstrate the openness of signs to new meanings and redemptive transformation. When Cathy writes the names, they are signifiers of possible futures and of her own conflicted desires; when Lockwood reads them, they are signifiers both of Cathy's physical absence and of the enduring legacy of history at the Heights. As Nelly's narrative unfolds, however, the inscriptions are given new meaning. The names belong not only to Cathy, but to her daughter, the only character in the novel to embody the entire sequence. The three names are held legally by the daughter, not the mother (whose surname is never Heathcliff). They point not only to the physical absence of the first Cathy, but also to a new kind of embodiment in the second generation. In the terms of Pickstock's philosophy, this narrative reinterpretation of the symbols can be considered as a redemptive return or resurrection. Pickstock writes:

> Above all, there is only story because of the resurrection. Resurrection is the process at work in non-identical repetition by which that which is repeated is not unmediably different, but analogously the same. This redemptive return is what allows a person to tell a story, since for there to be a story, there must be "analogous" subjects and objects, persisting as same-yet-different.[31]

The second Catherine is a non-identical repetition of her mother, compelled by circumstance to repeat some of her mother's story yet also able to transform this story into something new.

As we have seen, critics have often understood the relationship between the first and second generations in *Wuthering Heights* in terms of the romances between Cathy-Heathcliff and Catherine-Hareton, with the latter seen either as a diminished version of the passion of the former, or as a stable and sustainable alternative to the former's destructive intensity. I have attempted to demonstrate that the critical focus on romance has tended to obscure the novel's focus on a wider network of societal and familial relationships and, indeed, on their fragmentation and collapse. The remainder of this section argues for a view of the second generation not simply as offering an alternative model of romance, but as a rewriting of the familial, social and religious failures of its predecessors. On my reading, the second generation constitutes (in Pickstock's terms) a resurrection or redemptive return for the community. This social redemption occurs in two ways: through demonstration of agapeic love and through the refusal of Joseph's theology of vengeance.

The tendency to read the second generation as defined primarily by the Catherine–Hareton romance is misleading not least because, as the names on the window ledge remind us, the second Catherine marries twice. Catherine's marriage to Linton Heathcliff is ignored in most critical readings (repeating, ironically, the neglect endured by Linton in the narrative itself). This lack of critical attention underrates the moral and philosophical significance of the relationship. During Linton's residency at Wuthering Heights, Catherine alone demonstrates self-giving, compassionate love for the unlovely boy. She rejects the premise of Heathcliff's belief that Linton's character will poison her love for him:

> 'I know he has a bad nature,' said Catherine; 'he's your son. But I'm glad I've a better, to forgive it; and I know he loves me and for that reason I love him. Mr. Heathcliff, *you* have *nobody* to love you; and, however miserable you make us, we shall still have the revenge of thinking that your cruelty arises from your greater misery! You *are* miserable, are you not? Lonely, like the devil, and envious like him? *Nobody* loves you – *nobody* will cry for you, when you die! I wouldn't be you!' (p. 254)

Catherine's love for Linton endures because it rests not on erotic attraction or intensity of feeling, but upon her own decision to love. She gives herself to Linton as a gift that is also a gesture of hope that she might receive the reciprocal gift of himself, however imperfect – and imperfectly given – this return might be. This is not the love of romance, but of solidarity and community against the cruelty of Heathcliff, whose hatred of all except the dead Cathy ensures that, as Catherine correctly observes, he lives in isolation from all true community. Catherine's words are borne out by her actions. She is consistent in her sympathy for Linton, shielding him when she is able from the worst of Heathcliff's abuses and nursing him patiently through his final illness.

If Catherine's marriage to Linton undermines Heathcliff's view of love as predicated upon the worthiness of its object, her relationship with Hareton is based upon her ability to find a way through the impasse encountered by her mother. For the first Cathy, Heathcliff and Edgar represented a choice of emotional and sympathetic attachment with the former or a life of social privilege with the latter. In her confession to

Nelly Dean, Cathy acknowledges the social and economic basis of her decision: 'I've no more business to marry Edgar Linton than I have to be in heaven; and if the wicked man in there had not brought Heathcliff so low, I shouldn't have thought of it. It would degrade me to marry Heathcliff, now' (p. 71). Yet when Catherine is presented with a repetition of her mother's circumstances – Hareton has been 'brought ... low' by Heathcliff as Heathcliff was by Hindley – she responds with love that undoes the effects of Heathcliff's degradation of Hareton. Under Catherine's influence, Hareton is changed from the 'clown' and 'boor' (p. 10) of Lockwood's first visit to 'a young man, respectably dressed, and seated at a table, having a book before him' (p. 273). Where Cathy chose marriage to a respectable man, leaving Heathcliff to effect his own social transformation, Catherine helps Hareton to become a gentleman. In this specific sense, Catherine's love is redemptive: it reverses much of the damage caused by Heathcliff's attempts to corrupt Hareton and, thus, rejects the destructive legacy of the previous generation to its children.

Yet there remains one point on which Catherine is wrong in her assessment of Heathcliff. Contrary to Catherine's expectations, there is someone to cry for Heathcliff when he dies. Nelly recalls that:

> poor Hareton, the most wronged, was the only one that really suffered much. He sat by the corpse all night, weeping in bitter earnest. He pressed its hand, and kissed the sarcastic, savage face that every one else shrank from contemplating; and bemoaned him with that strong grief which springs naturally from a generous heart, though it be tough as tempered steel. (pp. 298–9)

Hareton's grief at Heathcliff's death is consistent with his refusal to seek vengeance for his degradation at Heathcliff's hands. Heathcliff has attempted to corrupt Hareton, by depriving him of education and encouraging him in vice, as an act of revenge upon Hindley for his own similar degradation of Heathcliff. Hareton's love for Heathcliff breaks the cycle of vengeance. In one of the final confrontations between Heathcliff and Catherine, each antagonist attempts to claim Hareton's support: the former to turn Catherine out of the room, and the latter to strike back against Heathcliff. Yet Hareton refuses to take either side: he intervenes to protect Catherine from Heathcliff's assault, but does so without striking Heathcliff.

This scene marks the end of Heathcliff's campaign of vengeance; in his own words, it is 'an absurd termination to my violent exertions':

> My old enemies have not beaten me – now would be the precise time to revenge myself on their representatives – I could do it; and none could hinder me – But where is the use? I don't care for striking, I can't take the

trouble to raise my hand! That sounds as if I had been labouring the whole time, only to exhibit a fine trait of magnanimity. It is far from being the case – I have lost the faculty of enjoying their destruction, and I am too idle to destroy for nothing. (p. 287)

One might say that Heathcliff has committed an error of critical interpretation. Reading Hareton and Catherine as the 'representatives' – or representations – of his enemies, Heathcliff has failed to see the ways in which they have rewritten their inheritance. Heathcliff's revenge is undermined by a new generation that refuses, finally, to continue the conflicts of the past. With this refusal, their signification changes: no longer the representatives of Heathcliff's enemies, they become instead symbols of his own frustrated past and of his desire for Cathy. When he tells Hareton to go to Catherine because 'he wondered how I could want the company of any body else' (p. 291), he reveals the extent to which the second-generation lovers have become for him symbols of the reunion with Cathy for which he longs. As Heathcliff himself observes, there is 'a strange change approaching' (p. 287). Catherine and Hareton have not simply defeated Heathcliff's campaign of vengeance. They have rendered it obsolete.

The agapeic love and forgiveness demonstrated by the second generation undoes the vindictive theology that has held sway at Wuthering Heights since the childhoods of Heathcliff and Cathy. Branderham's sermon is emblematic of the judgmental, moralistic religion that has corrupted human relationships under Joseph's influence at the Heights (Joseph, of course, is a member of Branderham's congregation in Lockwood's dream). I have suggested elsewhere that Joseph's influence is more pervasive and destructive than has usually been recognised by critics who have seen him merely as a caricature of the extreme Calvinism that Brontë despised.[32] Joseph's theology poisons the relationship between the Earnshaw children and their father, as Nelly recalls of Cathy:

> 'Nay, Cathy,' the old man would say, 'I cannot love thee; thou'rt worse than thy brother. Go, say thy prayers, child, and ask God's pardon. I doubt thy mother and I must rue that we ever reared thee!'
> That made her cry, at first; and then, being repulsed continually hardened her, and she laughed if I told her to say she was sorry for her faults, and beg to be forgiven. (p. 37)

Told to ask for forgiveness that is always denied, Cathy becomes 'hardened' by her father's rejection of her. When Heathcliff begins his campaign of vengeance, he does so in imitation of the God of wrath worshipped by the household religion: 'God won't have the satisfaction that I shall', he says, when Nelly tells him that it is 'for God to punish wicked people' (p. 53).

As adults, Cathy and Heathcliff do not practice Joseph's religion, but their lives remain tainted by it; Heathcliff, in particular, enacts its vindictive tenets unrestrained by Joseph's physical and moral cowardice.

Hareton and Catherine, however, turn away from Joseph's influence, symbolised in their uprooting of Joseph's 'black currant trees' in order to plant flowers (p. 282). When they marry, they will move to Thrushcross Grange, leaving Joseph behind at the Heights to live in the kitchen while 'the rest will be shut up' (p. 300). Symbolically uprooted and then left alone in an abandoned house, Joseph's influence ends with the rise of the second generation. Milbank observes that 'evil can only pass away, be forgiven and forgotten, if not only the past can be revised, but also what is deficient in the past can be revised out of existence'.[33] The second generation in *Wuthering Heights* constitutes a redemptive return, a non-identical repetition of the past. Catherine's embodiment of the sequence of names written by her mother is a metaphor for the ways in which the flawed legacies of the past might be appropriated and transformed. The words written on the window ledge of Wuthering Heights are symbols that remain open eschatologically to new meanings to come to them in the future.[34] These new meanings are also the manifestation of the new social possibilities represented by the second generation's repetition with difference of a failed social inheritance.

Notes

1. In the interests of clarity, I use 'Cathy' to refer to the elder character and 'Catherine' for her daughter.
2. Lynne Pearce, *Romance Writing* (Cambridge: Polity, 2007), pp. 92–3. I note that Pearce – quite rightly – includes an early version of my own work among those which conceal this susceptibility to the romance plot.
3. Janet Gezari, *Last Things: Emily Brontë's Poems* (Oxford: Oxford University Press, 2007), p. 113.
4. J. Hillis Miller, *The Disappearance of God: Five Nineteenth-Century Writers*, 3rd edn. (Urbana and Chicago: University of Illinois Press, 2000), p. 211.
5. See Bernard M. G. Reardon, *Religion in the Age of Romanticism* (Cambridge: Cambridge University Press, 1985).
6. Sandra Gilbert and Susan Gubar, *The Madwoman in the Attic: The Woman Writer and the Nineteenth-Century Literary Imagination* (New Haven: Yale University Press, 1979); Margaret Homans, *Women Writers and Poetic Identity: Dorothy Wordsworth, Emily Brontë, and Emily Dickinson* (Princeton, NJ: Princeton University Press, 1980); Irene Tayler, *Holy Ghosts: The Male Muses of Emily and Charlotte Brontë* (New York: Columbia University Press, 1990); Stevie Davies, *Emily Brontë: Heretic* (London: Women's Press, 1994).

7 Marianne Thormählen, *The Brontës and Religion* (Cambridge: Cambridge University Press, 1999), p. 119.
8 For a detailed discussion of the reception history of Brontë's writing and religion, see Micael M. Clarke, 'Emily Brontë's "No Coward Soul" and the Need for a Religious Literary Criticism', *Victorians Institute Journal*, 37 (2009), 195–223; Simon Marsden, *Emily Brontë and the Religious Imagination* (London: Bloomsbury, 2014), pp. 1–28.
9 Exceptions include Francis Fike, 'Bitter Herbs and Wholesome Medicines: Love as Theological Affirmation in *Wuthering Heights*', *Nineteenth-Century Fiction*, 23 (1968), 127–49; Emma Mason, 'The Key to the Brontës? Methodism and *Wuthering Heights*', in Mark Knight and Thomas Woodman, eds., *Biblical Religion and the Novel, 1700–2000* (Aldershot: Ashgate, 2006), pp. 69–77. These essays read the novel not simply as rooted in specific Christian doctrines, but as exploring theological understanding of (respectively) love and religious enthusiasm.
10 Rowan Williams, *The Edge of Words: God and the Habits of Language* (London: Bloomsbury, 2014), p. 91.
11 See Brontë, *Wuthering Heights*, pp. 307–10.
12 For a fuller discussion of Branderham's sermon, see Marsden, *Emily Brontë and the Religious Imagination*, pp. 72–86.
13 Marianne Thormählen helpfully contextualises the significance of forgiveness for nineteenth-century Christian ethics. See *The Brontës and Religion*, pp. 119–43.
14 Philemon 15–16, King James Version.
15 Stevie Davies is correct, I think, to read this poem as engaging with revolutionary politics and violence in Britain and Europe in the late 1840s. See Davies, *Emily Brontë: Heretic*, pp. 240–7.
16 Emily Brontë, 'Why ask to know the date – the clime?', in *Emily Brontë: The Complete Poems*, Janet Gezari, ed. (London: Penguin, 1992), 183–90: 1–4. A revised version of the poem remained incomplete at the time of Brontë's death. All subsequent quotations from Emily Brontë's poetry in this chapter are from this edition.
17 Isaiah 58: 6–7.
18 William A. Madden, '*Wuthering Heights*: The Binding of Passion', *Nineteenth-Century Fiction*, 27:2 (September 1972), 127–54, p. 131.
19 Emily Brontë, 'Filial Love', in Charlotte Brontë and Emily Brontë, *The Belgian Essays*, Sue Lonoff, ed., and trans. (New Haven and London: Yale University Press, 1996), pp. 156–9, p. 156.
20 On the history of natural theology, see John Hedley Brooke, *Science and Religion: Some Historical Perspectives* (Cambridge: Cambridge University Press, 1991).
21 John Milbank, *Being Reconciled: Ontology and Pardon* (London: Routledge, 2001), p. 17.
22 Brontë, 'Filial Love', p. 156.

23 The concept of privation is a key component of Augustinian theology, which understands evil as the distortion or absence of the good, rather than as an ontological reality. See G. R. Evans, *Augustine on Evil* (Cambridge: Cambridge University Press, 1982); Charles Mathewes, *Evil and the Augustinian Tradition* (Port Chester, NY: Cambridge University Press, 2001).
24 Keith Ward, *Christianity: A Short Introduction* (Oxford: Oneworld, 2000), p. 37.
25 Exodus 20: 5.
26 J. Hillis Miller, *Fiction and Repetition: Seven English Novels* (Oxford: Basil Blackwell, 1982), pp. 64–5.
27 Miller, pp. 42–72.
28 Catherine Pickstock, *After Writing: On the Liturgical Consummation of Philosophy* (Oxford: Blackwell, 1998), p. 118.
29 Catherine Pickstock, *Repetition and Identity* (Oxford: Oxford University Press, 2013), p. 43.
30 For a detailed conceptual response to Miller's approach, see Valentine Cunningham, *In the Reading Gaol: Postmodernity, Texts, and History* (Oxford: Blackwell, 1994). Cunningham argues that language represents not the absence claimed by postmodern theory, but rather a drama of both presence and absence, or emptiness and fullness.
31 Pickstock, *After Writing*, p. 265.
32 Marsden, *Emily Brontë and the Religious Imagination*, pp. 106–12.
33 Milbank, *Being Reconciled*, p. 54.
34 I am deliberately echoing Paul S. Fiddes' claim that '[a]ll texts are eschatological, both in being open to the new meaning which is to come to them in the future, and also in being "seriously" open to the horizon which death brings to life'; *The Promised End: Eschatology in Theology and Literature* (Oxford: Blackwell, 2000), p. 6.

CHAPTER 10

'Surely Some Oracle Has Been with Me'
Women's Prophecy and Ethical Rebuke in Poems by Charlotte, Emily, and Anne Brontë

Rebecca Styler

As Joseph Bristow has shown, there was a lively debate in the early Victorian era about how the poet's otherworldly vision should relate to the real world. For some, the poet's role was to elevate the world-weary and oppressed to a higher spiritual sphere, while for others it was ultimately political, to transform and not transcend troubled social reality.[1] In the culturally dominant formations of the poet shaped by Romanticism, the vatic capacity was assumed to be exclusively male, yet countering this commonplace was a tradition of women's prophecy spanning the Romantic and Victorian eras which originated in religious dissent. In the spirit of women prophet-preachers, female writers of poetry and fiction claimed direct divine sanction to correct the faults of the public sphere, from the viewpoint of a superior 'feminine' moral sensibility. This female prophetic literary tradition eschewed escapist visionary flight, always having as its focus the world of human affairs and the advancement of humanity.

This article considers the poetry of Charlotte, Emily and Anne Brontë within this tradition, an approach that has been fruitfully applied to a number of Romantic and Victorian women poets, including Charlotte Smith, Anna Barbauld, Elizabeth Barrett Browning, Christina Rossetti and George Eliot. That such an approach has not been taken with the Brontës' poetry is perhaps surprising, given the critical attention that has been given to prophecy in Charlotte Brontë's fiction and to the complex interactions between poetic vision and the gendering of the poetic voice in Emily Brontë's verse. In this chapter, I consider a selection of Charlotte, Emily and Anne Brontë's poems which present a female outsider exerting spiritual authority over a patriarchal establishment and its associated 'masculine' cultural values to promote an alternative 'feminine' ethic, applied in different ways and to various aspects of culture. Particular attention will be given to 'Pilate's Wife's Dream' and 'Gilbert' (by Charlotte), 'The Prisoner' and 'The night was dark yet winter breathed' (by Emily) and 'A Word to the Calvinists' (by Anne). I show how each poet combines the inheritance of biblical and dissenting models of the prophet with Romantic ideals of the charismatically inspired visionary, as well as the more Gothic symbol of the avenging ghost. Hence, I will identify some of the ethical and political concerns of the Brontës' poetic works and connect their rhetorical strategies with a tradition in women's poetry that spanned the Romantic and Victorian eras.

Prophecy and Women's Poetic Tradition

In the vatic ideal of the prophet that prevailed in the Romantic and early Victorian eras, poetry was identified as the special capacity to perceive spiritual mysteries beyond the material surfaces of the real. To Thomas Carlyle the poet was a 'genius' who felt more at home in 'the true Unseen World' than on earth and who 'leads us to the edge of the Infinite', eschewing the world's 'formulas and hearsays' to find a more authentic truth in 'the great Deep of Nature'.[2] Carlyle summed up a common Romantic understanding when he claimed this prophetic capacity solely for the *man* of letters, a view at home in a culture where spiritual authority and literary genius were gendered male, women being barred from the former by orthodox church teaching and assumed incapable of the latter through intellectual inferiority. Women's poetry was typically framed in terms of the figure of the 'poetess', who celebrated matters of the heart and home, posing no challenge to 'separate spheres' gender ideology. As Cynthia Scheinberg states, the poetess was constructed as a signally 'non-prophetic' figure, 'a local, spontaneous and sympathetic voice, rather than the philosophical and theological voice of "truth"'.[3] In masculine Romantic discourse, women poets lacked the cultural authority or breadth of vision to address the public domain.

However, scholars have shown an alternative tradition of women's prophecy, which drew its inspiration more from religious models than from Romanticism. This evoked the model of the biblical prophet who speaks as an outsider to the establishment, claiming direct divine inspiration to rebuke the nation for falling into unrighteous ways. The prophet was 'critic, rather than sanctifier, of the status quo', denouncing religious as well as secular authorities and conventions for betraying the ways of God and bowing to worldly moral expediencies.[4] The Hebrew scripture offered examples of women prophet-poets, such as Deborah and Miriam, and injunctions for both 'sons and daughters' to prophesy (Joel 2:28), examples which Methodist and Quaker women preachers cited to authorise their vocation despite orthodox denials of female spiritual authority.[5] As scholars including Ann Mellor and Christine Krueger have shown, this tradition of women prophet-preachers had a legacy in literature by women: Mellor writes of a group of women Romantic poets who refused the confining model of the 'poetess' to embrace an alternative model of the 'female poet' who 'self-consciously and insistently occupies the public sphere' and boldly judges its failings from the viewpoint of 'a more refined virtue than the average man could attain'. Laying claim to a superior female virtue, they aimed to advance humanity by feminising the public sphere, preaching 'an ethic of love, compassion, peace, and liberty'.[6]

While the same ethos has been widely recognised in relation to Victorian fiction, energised also by the women prophet-messiahs and female divine symbols of radical millenarian movements, it is only fairly recently that Victorian women's poetry has been considered in this prophetic mode.[7] Elizabeth Barrett Browning, Christina Rossetti and George Eliot have been shown to employ a prophetic poetic voice to sanction women's literary and spiritual authority and to preach transformation of the cultural landscape through the wider application of the sympathetic 'feminine' values that were celebrated in the discourse of 'woman's mission'.[8] While Mellor separates the prophetic 'female poet' from the 'poetess' on account of their contrasting spheres of influence, Charles LaPorte emphasises these models' interconnection for poets who based their prophetic authority on possession of the affective values celebrated by the poetess.[9] A striking review of Barrett Browning's 1856 *Aurora Leigh* praised the author's union of 'masculine vigour, breadth and culture' with 'feminine subtlety of perception, feminine quickness of sensibility, and feminine tenderness'. 'Mrs Browning has shown herself all the greater poet', the reviewer declared, 'because she is intensely a poetess'.[10] This combination of authoritative public voice, and identification with a superior morality, is a feature of some of the Brontës' poems.

It is salutary to bear in mind Linda Shires's warning against segregating Victorian traditions of poetry too firmly according to gender. For, as Shires argues, women poets engaged in a 'double-pronged dialogue' with both male and female poetic traditions, to adopt, adapt and 'disidentify' with their rhetorical strategies and subject positions.[11] Early Victorian women poets responded to the male poetic tradition in which they were immersed, as well as biblical texts and the context of women's public voices in dissent and radicalism. That Victorian Bible criticism itself increasingly espoused a 'Romantic-charismatic' view of the prophet, whose authority derived from independent creative inspiration rather than obedience to the external Mosaic law,[12] further complicates the separation of these different inheritances which poets combined in order to address ethical and social concerns of their own times.

This combined inheritance characterises the Brontës, steeped as they were in the poetry of Wordsworth, Byron and Shelley and the Romantic model of the poet 'who lived in the inward sphere of things, in the true, Divine, and Eternal, which exists always, unseen to most, under the Temporary, Trivial'.[13] But they were also conscious of dissenting women preachers and their self-vindications, for, as Marianne Thormählen shows, despite their upbringing as daughters of an Anglican cleric, the Brontës were also exposed to Methodist practices and beliefs and negotiated their ways independently through multiple religious influences.[14] The Brontë children witnessed their father's Methodist connections and inherited their aunt's run of the *Methodist Magazine* (1798–1812) which, as K. Cubie Henck reminds us, featured items by and about women preachers. Most notable among these was the renowned Mary Bosanquet Fletcher with whom Patrick and Maria Brontë were acquainted and whose autobiography (which went through twenty editions between 1817 and 1850) kept alive the idea of women's public voice and spiritual vocation despite denominational rulings against these. It has been argued that the model of Bosanquet Fletcher informs the writing of *Jane Eyre*.[15] My purpose is not to demonstrate specific influences, but to draw attention to the Brontës' employment of an ethos, or structure, in which the female outsider, aligned with divine authority, denounces the prevailing social order whose faults are connected to male dominion and the prevalence of culturally 'masculine' values. All three Brontë poets engage, at times, with this rhetorical strategy to give their poetry a public dimension and to challenge and educate their readers.

Charlotte Brontë: Prophecy as Feminist Critique/Revenge

Of the three writers, Charlotte Brontë makes the greatest use of the female prophetic voice, as is shown in a number of discussions of her juvenilia and fiction.[16] These rich studies show Brontë asserting women's spiritual right to interpret the scriptures for themselves and deconstruct androcentric orthodoxies, as well as her creation of heterodox female divine symbols. In Brontë's poetry the female outsider manifests as both victim and critic of the wrongs of the masculine order. In 'Pilate's Wife's Dream', which opens the 1846 *Poems*, Brontë (like many other Victorian women poets) uses the dramatic monologue to imaginatively inhabit a female character from history or myth, and by voicing her subjectivity offers a counter-perspective to the dominant cultural narrative. As Dorothy Mermin notes, women's dramatic monologues typically invited a high level of sympathy and identification with the speaker, being a vehicle for the author's own political concerns.[17] Brontë's poem also shares strategies with the popular genre of scripture biography, in which Victorian women writers filled out sparse biblical accounts with interpretation to make scriptural women into role models.[18] Pilate's wife features in only one Bible verse, in the context of Jesus' trial: 'When he [Pilate] was set down on the judgment seat, his wife sent unto him, saying, "Have thou nothing to do with that just man: for I have suffered many things this day in a dream because of him"' (Matthew 27:19). Brontë seizes on the image of the outspoken, righteous woman dreamer and embellishes the account with details from apocryphal texts and legend, including Pilate's wife's conversion to Christianity, her veneration as St Claudia Procula and critical accounts by Jewish historians (such as Philo) of Pilate as a cruel, colonial oppressor.

Brontë magnifies Pilate's wife from bearer of a single omen into a sybilline critic who indignantly condemns Pilate's entire value system and foretells his merited demise, much in the spirit of Old Testament prophets. Although 'forced to sit by his side and see his deeds' as an onlooker, not agent, of history, the woman is identified with a night-time dreamworld that permits alternative insights to the wisdom of the hegemonic social order.[19] She is intensely conscious of her prophetic capacity:

> Surely some oracle has been with me,
> The gods have chosen me to reveal their plan,
> To warn an unjust judge of destiny:
> I, slumbering, heard and saw; awake I know,
> Christ's coming death and Pilate's life of woe.[20]

In her presentation of Pilate's wife as one who identifies with the abject – the innocent victim Christ – Brontë credits her protagonist with 'the hermeneutical privilege of the oppressed', who achieves critical insight into the operations of power from the viewpoint of those oppressed by them.[21]

Far exceeding the role allotted her by scripture or legend, Pilate's wife launches an excoriating denunciation of Pilate's character as governor – her primary concern is public morality. She regards his condemnation of Jesus as symptomatic of a tendency to govern by expediency rather than principle, motivated by personal ambition in 'A triple lust of gold, and blood, and power' (p. 3). Pilate lacks paternalistic concern for his people's wellbeing, becoming a caricature of cruelty 'whose cold and crushing sway / No prayer can soften, no appeal can move; / Who tramples hearts as others trample clay' (p. 3). Ignoring the gospel texts' details of the governor's reluctance to convict Jesus, Brontë has her protagonist emphasise Pilate's collusion in imperial oppression as 'Rome's servile slave' and 'Judah's tyrant scourge' who brutally subjugates the local populace with 'his slaughtering, hacking, sacrilegious sword' (pp. 3–4). The sibyl spares no verbal blows in her denunciation of the man's cynical self-serving pragmatism and casual violence.

In his discussion of Charlotte Brontë's anti-colonial poems that are more obviously directed at contemporaneous British imperialism, Carl Plasa shows how Brontë identifies a construction of masculinity in which a colonialist mentality extends to sexual relations also through a common ethos of objectifying, subjugating and conquering the 'other' to create a self-glorifying myth of dominance.[22] So, in 'Passion' the wife is desired in ways that replicate – and reinforce – the soldier's aggressive colonial enterprise: 'If, hot from war, I seek thy love, / Darest though turn aside? ... No, my will shall yet control / Thy will ... I'll read my triumph in thine eyes' (*Poems*, p. 114). While 'Pilate's Wife's Dream' is less overt, a parallel between the colonial and conjugal, between public and domestic morality, is suggested in the wife's description of Pilate's advances:

> Has he not sought my presence, dyed in blood –
> Innocent, righteous blood, shed shamelessly?
> And have I not his red salute withstood? (p. 4)

Her boast of resistance presents her more as judge than victim, the subaltern female refusing to be conscripted into her oppression, or the oppression of innocent others, by submitting to Pilate's sexual advances which are tainted with the same greedy aggression that characterises his rule. For her sake, but far more for her people's, Pilate's wife foretells Pilate's solitary, grisly death in the wilderness, with some satisfaction.

However, in this poem female prophecy does not survive the dawn of Christianity, but becomes subsumed in veneration of a male messiah, whose apotheosis entails the woman's loss of divine authority. Janet Larson argues that Brontë's evocation of biblical male prophetic models could compromise her religious feminism, for while they provided grounds for opposition to the majority's beliefs and practices that could be adapted for feminist purposes, they were imbued with 'problematic androcentrism' that could undermine the author's feminist message. Only in Brontë's later fiction, argues Larson, does she abandon deference to external religious authorities and create a 'heretic logos' from within a woman's own experience.[23] 'Pilate's Wife's Dream' evidences this gender contradiction as the sybil conventionally celebrates Christianity's superiority to the 'rotten' and 'defiled' rites of Roman religion and abandons her own oracular power to the truer 'oracle of God' (pp. 6–7). She replaces her own symbol of the moon with the 'Glorious Sun' and seeks inspiration now 'from on high' and not from within her dreams (p. 7). It is true that, like many Victorian Christian women's writings, the poem foretells the slow triumph of a religion in tune with womanly values – Jesus' 'wise and mild' ethos (p. 6) – over brutal patriarchal politics, but it also presents the paradox that in this feminised Christ-religion, women's autonomous spiritual power is superseded by the male divine.[24] The narrative voice is not self-evidently ironic – Brontë perhaps unwittingly presents the paradox that Christianity's dawn signalled the twilight of female spiritual autonomy, even while it exalted 'feminine' social values.

Female prophecy takes a Gothic turn in the poem 'Gilbert' where Brontë intensifies the association of abjected womanhood with an otherworldly power source through the figure of a ghost. Brontë evades the gender complications involved in biblical models by avoiding religious reference almost altogether, substituting this with Romantic and Gothic images of the otherworldly. The ghost is identified, like Pilate's wife, with the night-time dream-world and also with nature which operates as an autonomous spiritual force, manifest in moon, tree and sea. As Gail Turley Houston shows, the image of maternal nature – mythologised by many Romantic poets as a passive muse to the male poet's genius – could be appropriated by women writers as a feminist power source.[25] The poem evokes a neo-platonic logic to suggest an authentic, eternal feminine whose moral ethos includes gender justice versus a 'prison house' of culture (to use Wordsworth's phrase from the 'Intimations' ode) which is patriarchal. The repeated phrase 'Above the city hung the moon' is metonymically suggestive of feminised divine nature standing in judgment over patriarchal culture's deformations (pp. 60, 73).

The ghost in this poem is that of a woman, Elinor, who returns from eternity to haunt the man who abused her in life with an impunity endowed by codes of masculine privilege. 'Gilbert' tells a classic fallen woman narrative in which a young girl is beguiled into a sexual relationship with a lover who promises fidelity, but subsequently casts her off for a more materially advantageous match, following which she flees in shame and meets an untimely death at sea. Diana Wallace writes of the female ghost's potency for feminist writers because it symbolises 'the ghosting of women within patriarchy' in their own lifetimes: women's invisibility and status of non-being in the symbolic order.[26] The ghost story generally dramatises 'the return of the repressed' and in a number of gothic stories the abjected woman returns from the world of the dead with vengeful agency to punish her tyrants in a variant on the 'female gothic' sub-genre wherein a woman victim acquires agency.[27] For example, in the 1839 story 'A Chapter in the History of a Tyrone Family' – published in the *Dublin Review* and almost certainly an influence on the writing of *Jane Eyre* – the ghost of a cast-off wife whom the law did nothing to protect haunts her Bluebeard husband and finally drives him through terror to suicide.[28] This rough justice was diluted in Brontë's novel, but is certainly present in 'Gilbert' where the boundary between divine justice and vindictiveness becomes rather blurred.

Brontë departs from the common depiction of the fallen woman's exploiter as an aristocrat with exceptional social power and immunity (e.g., Gaskell's *Ruth* or Eliot's *Adam Bede*), instead presenting her protagonist as a fairly typical man of the middle class. Male privilege and female self-abnegation are, therefore, analysed as bourgeois social norms and they are internalised as much by Gilbert's lover as by himself. It is the power of being idolised that Gilbert fondly recalls in his illicit romance, for 'truly it was sweet / To see so fair a woman kneel, / In bondage, at [his] feet', recognising his 'despot-might / Her destiny to wield' (pp. 62–3). Her subservient devotion becomes his greatest weapon against her, enabling him to embark on sexual relations confident in his immunity from social consequence, because 'I knew her blinded constancy / Would ne'er my deeds betray, / And, calm in conscience, whole in heart, / I went my tranquil way' (pp. 63-4). Brontë presents an astute analysis of the mentality of entitlement and others' collusion in their own oppression as Elinor joins the conspiracy of silence that protects the man from any consequences. What happened to Elinor he takes no trouble to inquire, suppressing occasional stirrings of conscience for fear of cost to his reputation and worldly success should he attempt some sort of redress. Gilbert is no exceptional brute, but merely 'a wise and worldly man' who knows his vested interests are served by a code of privilege that he half-recognises is unjust (p. 75).

I suggest that the ghost's role is prophetic rather than merely vengeful, because her initial manifestations are intended to re-educate her oppressor. On two occasions she appears uncannily within domestic contexts, harnessing nature's power to disturb his complacency and provoke his conscience. She first provokes the young Gilbert who experiences some 'inward trouble' when remembering his descent 'like a god' to the maid he ruined, recalling the pattern enacted in many classical myths, which often resulted in the maid being turned into a shrub to spare her violation (pp. 63–4). Nature responds ironically to this pattern of sexual power relations, animating the tree on which he leans in his garden and instilling momentary terror, although Gilbert dismisses the incident. Her second visitation comes closer to home, when as a more visible 'sullen shade' she invades the home and hearth of his more mature years, surrounded by his 'blooming wife' and 'smiling children', and overpowers him with a sensation of being engulfed by the sea (pp. 66–7). Nature – which receives Elinor in her 'isle-conceiving womb' (p. 70) – is not just an avenging mother here, but attempts to teach him empathy as he is brought to experience something like the drowning that was the fate of his cast-off lover. Gilbert again represses the incident and its truth-bearing potential, not so much through rationalist scepticism, but through a commitment to his own position within the status quo. Drily the narrator reports:

> Gilbert has reasoned with his mind –
> He says 'twas all a dream;
> He strives his inward sight to blind
> Against truth's inward beam.
> He pitied not that shadowy thing,
> When it was flesh and blood;
> Nor now can pity's balmy spring
> Refresh his arid mood ...
> 'Conceal her, then, deep, silent sea,
> Give her a secret grave!
> She sleeps in peace, and I am free,
> No longer terror's slave.
> And homage still, from all the world,
> Shall greet my spotless name,
> Since surges break and waves are curled
> Above its threatened shame.' (p. 72)

Brontë's Wordsworthian allusions to 'inward sight' and 'pity's balmy spring' suggest Gilbert's wilful suppression of innate feelings of humanity, leaving him no excuse. It is clear that the moral framework of the poem demands not just remorse, but redress in a public admission of wrong and a willingness to lose the worldly benefits bought with his lover's 'ghosting' in life and through death.

Because of Gilbert's refusal to go through this mortification, the prophetic message of the poem is ultimately one of retribution, like Old Testament prophets who preach doom to the unrepentant. The ghost's final visitation is as a menacing figure who parodically greets Gilbert at his own front door in dripping dress and who, like the ghost-wife in the 'Tyrone' story, terrifies her oppressor into suicide. It is ironic, given the widespread association of Jesus with 'feminine' pacific virtues, that the only explicitly Christian reference in the poem is a quoted New Testament text which authorises punishment, not forgiveness: 'The measure thou to her didst mete, / To thee shall measured be!' declares the ghost before sending the man to his grotesque and ignominious end (p. 75; see Matthew 7:2). This emphasis on retribution is at odds with the emphasis of many women writers' 'feminine' social gospel of mercy; Brontë makes clear that forgiveness of bourgeois male oppressors is conditional on not only repentance but reform, and Gilbert's refusal to enact this brings down divine wrath. The destiny of Gilbert's soul is not a matter that needs addressing in the logic of the gothic ballad: its fantastic mode allows Brontë to symbolically punish male privilege and tyranny within the temporal realm, empowered by an otherworldly feminine who exacts justice for offended womanhood.

Emily Brontë: Woman Vates, Feminine Ecology

Emily Brontë was deeply invested in the notion of the poet as prophetic visionary of the kind imagined by Carlyle, more at home in the 'true Unseen world' than on earth.[29] Brontë drew from Romantic poets like Wordsworth and Shelley ways to express her own sense of life as 'a process of estrangement from transcendent origins, a story of lost glory and dream'.[30] Many of her poems depict the spirit's quest to escape corporeal and temporal bounds and its achievement of epiphanic communion with the eternal spirit in nature that 'Pervades and broods above, / Changes, sustains, dissolves, creates and rears', with a conviction so intense that some have called Brontë a mystic.[31] As Simon Marsden argues, Brontë's poetry expresses the longing to experience divine immanence in the world; prophecy enables her to achieve this momentarily and points to its greater fulfilment after death.[32] In contrast, Brontë views reality disdainfully as an alienating 'dark world' whose love, the poet insists, she 'will not, will not share', showing what Stevie Davies calls 'breath-taking cynicism' towards earthly life and a general tone of 'exclusive superiority to the herd.'[33] Brontë's world-rejection has been interpreted by some as a feminist protest against the conditions of patriarchy, but these claims rely on an archetypal reading of Brontë's ideal world as symbolically maternal and a too-easy equation of such symbolism with feminist politics.[34]

Overwhelmingly Brontë's *contemptus mundi* theme seems an existential, rather than political, protest against the terms of reality, and prophetic vision is sought for escape and consolation rather than as a mandate for reform.

But, on occasion, Brontë uses the trope of the prophetic female outsider in ways that are more specific to women's social experience and the values they were stereotypically associated with. In 'The Prisoner', Brontë celebrates a female visionary to vindicate female vatic authority itself, presenting a reposte to culturally powerful Romantic models of authorship that denied women such originary capacity. In this poem, a female captive confounds her male captors, who constrain and underestimate her, through a demonstration of her imagination's power. Her 'tyrants' are presented in terms of a masculine stereotype of arrogance and aggression, 'scoff[ing]' and taunting her fetters (15:33, 14:10). Seeing her sleeping, her chief captor describes her unworldliness in condescending terms, her face 'soft and mild / As sculptured saint, or slumbering unwean'd child' (14: 13–14). This description is suggestive of objectification, sentimentality and inarticulacy – traits which resonate with conventional estimates of women's role in poetry as silent muse, not creator. This diminutive impression is immediately belied by the woman's invocation of the sublime as she wakes to proclaim her nightly visitations by her muse, the West Wind (evocative of Shelley's 'Ode to the West Wind'):

> But first, a hush of peace – a soundless calm descends;
> The struggle of distress, and fierce impatience ends;
> Mute music soothes my breast – unuttered harmony,
> That I could never dream, till Earth was lost to me.
>
> Then dawns the Invisible, the Unseen its truth reveals;
> My outward sense is gone, my inward essence feels:
> Its wings are almost free – its home, its harbour found,
> Measuring the gulph, it stoops and dares the final bound. (15: 45–52)

Although her muse is given a male pronoun, as Irene Taylor suggests he is not inevitably patriarchal, but mediates between poet and the lost prenatal paradise: 'Though the muse was male, he was derived from the mother-world and was thus an angel and a comforter'.[35] The woman's capacity to transcend the real and experience the Ideal establishes her literary-spiritual authority on equal terms with male peers, in accordance with the Carlylean model. The 'agony' that accompanies the prisoner's return to reality is not (contra Emma Mason) proof of the delusive nature of the prisoner's imaginings, but is rather evidence of the captive's belonging in the Unseen, and her consequently feeling 'homeless on earth'.[36]

The changes which Brontë made when she adapted 'The Prisoner' from a much longer poem for the 1846 *Poems* intensify the gendered schema and its symbolic potential. The longer poem 'Julian M. and A. G. Rochelle' specifies the captor and prisoner as figures in the Gondal saga who, once childhood friends, become enemies in the civil war that besets the realm. Following the prison scene extracted for the 1846 poem, the original goes on to erode the oppositional stances of prisoner and captive, female and male: the two characters recall their childhood affection, Julian frees the prisoner and over time wins her love and the captive's world-rejection quickly fades to find that 'Earth's hope was not so dead' and 'heaven's home was not so dear' (180: 111). 'The Prisoner' is a new poem altogether, gaining universality through the removal of Gondal references and from the starkly oppositional schema. Vitally, too, a new final stanza is added in which the chastened male speaker admits the woman's moral triumph:

> She ceased to speak, and we, unanswering, turned to go –
> We had no further power to work the captive woe:
> Her cheek, her gleaming eye, declared that man had given
> A sentence, unapproved, and overruled by Heaven. (16: 61–4)

The once-loquacious men who disparaged the female visionary are confounded and silenced, forced to recognise that she has within her the divine spark. Although Tayler puts these textual changes down to Charlotte's interference, Derek Roper argues a detailed case for seeing 'The Prisoner' as Emily's work entirely (with which Gezari concurs).[37] Either way, the new poem becomes the lynchpin for the Brontës' first entry into print – and placed immediately after 'Gilbert' continues its schema of female judge of male oppressors – to make an indisputable case for women's literary-spiritual authority despite their male detractors.

There is certainly no prevailing myth of superior female sensibility in Brontë's poetic oeuvre – Gondal's queen is embroiled in her world's heroic violence, and in one of Brontë's most powerful narrative poems – 'Why ask to know the date - the clime?' – the means of grace to the soldier's conversion is a display of paternal love (189–90: 221–59). However, in another Gondal poem, 'The Night was dark yet winter breathed', the otherworldly female is identified with a 'feminine' ethical code, according to cultural conventions, in terms of the human relationship with the natural world. As Charlotte Brontë does in 'Gilbert', Emily

Brontë uses a ghostly nature-spirit to symbolise what is denied in the symbolic order, and anthropocentrism is her target. A soldier, exhausted and lost after fleeing a battle in one of Gondal's civil wars, finds himself wandering through a forest and lets his horse, 'his worn companion', free to wander at will while seeking rest for himself (92: 10). The environment is charged with the numinous, as winter 'breathed / With softened sighs', the wind 'repining' and clouds 'sullen' (92: 1–3, 93: 13). Night brings an encounter with an earth-mother figure, a 'shadowy spirit' in the forest, who explains her benign role in nurturing and care for all creatures, human and animals alike: she describes how she guides a deceased shepherd's spirit to his 'native glen', helps sailors to safety and protects timid sheep (93: 24, 44). But this is no passively comforting feminine: the spirit who at first bends (alluringly, tenderly) over the soldier's couch becomes an avenging earth mother who punishes the soldier for his mistreatment of his animal companion.[38] She tells how she rescued the horse whom the soldier had 'left to die', reframing the soldier's act of liberation (as he saw it) as one of carelessness and indirect cruelty (94: 54). Castigating by implication his exploitation of the horse as a vehicle of war, to be cast off once exhausted, she assumes a model of fellowship – rather than hierarchy – in the human–animal relationship, describing the horse as 'friendless' (94: 53). Like the ghost in 'Gilbert', the nature divinity delivers retribution, prescribing years of exile and 'toilsome pain' through which the solder will experience manifold what the horse suffered (94: 62). Her ultimate aim is 'mercy' and the soldier's return home after a long education of the soul, but the punishment is extreme, reminiscent of the disproportionate suffering meted out to the bewildered Ancient Mariner after his shooting of the albatross. Coleridge's moral framework in *The Rime of the Ancient Mariner* is deeply ambiguous, but Brontë's punitive earth mother acts with a consistent, if demanding, anti-anthropocentric moral code.[39]

There is something of the 'eco-feminine' in Brontë's use of a female earth spirit to reveal to the soldier his sins of omission. While eco-femin*ism* calls for an ethos of care, kinship and non-violence to replace the 'logic of domination' that operates in patriarchal social relations and likewise in human exploitation of the natural world, the term 'eco-femin*ine*' has been used to describe the essentialist identification of that sympathetic ethos with women's natures.[40] For some, this archetypal 'feminine' principle finds symbolism in the figure of the ancient earth goddess, which has been taken to signify the sacred life immanent in all of nature that harmonises all living beings in a spirit of harmony and mutuality.[41] This ethos of sympathy and affinity is expressed in Brontë's wood spirit, correcting the logic of domination and conflict between creatures that is implied in the soldier's exploitation of his horse in the context of warfare. Brontë's image of divine maternal nature judging and educating a man into seeing animals as his fellows directs female prophecy towards a new concern beyond the social, which can be contextualised in the nascent environmental consciousness present in other early nineteenth-century poetry.[42]

Anne Brontë: Dissenting from the 'Prophetic Sublime'

It is perhaps commonplace to observe that Anne Brontë was more wary of Romantic ideals than were her sisters and identified with writers of an earlier tradition of rational moralists who placed more faith in reason than feeling. Many of her poems present a subject who longs for, but fails to achieve, communion with the Unseen and, thus, turns to more mundane sources of comfort to render the fallen world tolerable, including faith, practical virtue and emotional self-management.[43] Anne's poems engage in a critical dialogue with Emily's, especially in the 1846 *Poems* where, through juxtaposition, Emily's celebrations of disdainful ecstasy are undercut by Anne's images of thwarted longing (e.g., 'If this be all' follows 'The Prisoner', 'The Doubter's Prayer' follows 'Imagination'). In some of her later dialogue poems, Anne voices overt scepticism towards what Larson calls 'the prophetic sublime',[44] criticising the individual's claim to transcendence for its implicit arrogance and disdain for the common lot (e.g., 'The Three Guides', stanzas 10–18). While Anne Brontë clearly responded to poets like Wordsworth and Coleridge, she shares many women Romantic writers' commitment to 'quotidian values' and their investment in reason and sensibility, rather than imaginative flight, as the sources of virtue.[45] She read works by Hannah More, who praised sensibility as 'the sympathy divine' flowing through all and uniting humanity in a community of empathetic feeling.[46] On occasion, Brontë abandons her lyrical focus to adopt the voice of the dissenting woman preacher, speaking as judge and teacher to condemn mentalities that deny common humanity.

This strategy is evident in 'A Word to the Calvinists' in which the poet deconstructs the doctrine of election (influential in early Victorian orthodoxy) from the basis of an ethical commitment to compassion and inclusiveness. Countering the spiritual privilege assumed by the 'elect', the poem moves from an Armenian argument (that all are free to choose salvation) to a Universalist declaration (that all will ultimately be saved).

Brontë's poem joins other voices within Anglicanism who in the 1840s were beginning to question orthodox theology on the grounds that it offended humane sensibilities and painted God as a tyrant.[47] Brontë's speaker likewise assumes that divine nature must be in accordance with the most advanced moral sensibilities and denounces Calvinists for their lack of compassion:

> You may rejoice to think *yourselves* secure,
> You may be grateful for the gift divine ...
> But is it sweet to look around and view
> Thousands excluded from that happiness,
> Which they deserve at least as much as you? (1846 *Poems*, pp. 104–5)

Brontë criticises the elect's conviction of exclusive intimacy with God, basing her case on humanist moral logic and humanitarian feeling. The ironic rhetorical questions, delivered in stately iambic pentameter, imply incredulity at the cruel logic implicit in their dogma: 'Is yours the God of justice and of love / And are your bosoms warm with charity? / Say does your heart expand to all mankind?' (105). In her proud dissent – 'May God withhold such cruel joy from me!' (105) – Brontë adopts the position of the sage writer who 'unveils the real spiritual marginality of his readers from his own location at the centre of Reality'.[48]

Brontë does not explicitly gender either of the theological positions set out in this poem, although her critique rests on values culturally associated with the feminine (in a similar protest, Theodore Parker wrote that 'there is nothing feminine in the popular image of God').[49] Likewise, it is in a 'feminine' lyrical voice that Brontë takes the logic of her anti-prophetic stance to its full conclusion, undercutting her own speaker's self-righteousness as she adopts a tentative, personal tone to declare her Universalist faith (developed more boldly in her novel *The Tenant of Wildfell Hall*). The form changes to shorter lines and opinion is expressed as a heartfelt 'hope' rather than convinced statement, although Brontë does support her view with a number of biblical allusions to suggest that her case has scriptural support. She cites Paul's promise that 'In Christ shall all men live' (cf. 1 Corinthians 15:22) and the prophecy that '[t]he metal' will one day be 'purified' (cf. Isaiah 1:25). In the 'radical protestant' spirit which Ruth Jenkins sees in Charlotte Brontë's fiction, Anne Brontë exercises her spiritual right to interpret scripture for herself, to challenge its prevailing interpretations and reveal the God of conservative religious authorities to be a false idol created in the image of those who enjoy domination, hierarchy and privilege. The change of mode from didactic sermon to tentative lyric, in a sense, embodies her point – that doctrines based on hierarchies and oppositions fall short of the divine ideal. Brontë abandons the voice of the indignant preacher because it is imbued with the self-inflation and antagonism that she wants to dispel, and adopts a form and tone that take a far less confrontational stance towards the reader. This poem ultimately exhausts the prophetic sublime, because it lays bare the paradox of using the aggressive prophetic voice as a vehicle to preach compassion, humility and inclusion.

Of the three writers, Charlotte most closely adopts the tradition of female prophecy, while Anne and Emily use it very selectively, wary of some of the model's implications. Charlotte and Emily use the voice of the otherworldly woman outsider to condemn the suppression and silencing of women in the literary and social worlds. Their women prophets identify with tropes of feminised nature as a potent source of moral authority for women to defy cultural norms. While Charlotte always directs her poetic prophecy towards gender politics, denouncing the sins of male privilege and domination, Emily and Anne evoke it to promote sympathetic values, to counter oppressive hierarchies that separate human beings from one another and from their fellow creatures. While Emily and Charlotte show confidence in imaginative inspiration as an ethical source, Anne reveals some of the ethical contradictions inherent to prophecy, whose self-righteous and even vindictive spirit could be at odds with the values it claims to speak for, as her sisters' poems demonstrate. That most of the poems discussed here appear in the 1846 *Poems* (Emily's 'The night was dark' being the exception) shows the Brontës' first publication entering the early Victorian debate on poetry's relation to the social sphere, for while these poets often wrote of the inner life, they all used the inherited figure of the female prophet, in nuanced and varying ways, to give their poetry a public dimension and to claim a place for women's literary and moral authority in matters to do with gender justice, theology and environmental ethics.

Notes

1 Joseph Bristow, 'Reforming Victorian Poetry: Poetics after 1832', in Joseph Bristow, ed., *The Cambridge Companion to Victorian Poetry* (Cambridge University Press, 2000), pp. 1–24.
2 Thomas Carlyle, *On Heroes, Hero-Worship and the Heroic in History* (London: Chapel & Hall Ltd, 1897), pp. 43, 96, 83, 54, 62.
3 Cynthia Scheinberg, '"Measure to yourself a prophet's place": Biblical Heroines, Jewish Difference and Women's Poetry', in Isobel Armstrong and

Virginia Blain, eds., *Women's Poetry, Late Romantic to Late Victorian: Gender and Genre 1830-1900* (Basingstoke: Macmillan, 1999), pp. 236–91, pp. 264–5.

4 Rosemary Radford Ruether, 'Feminist Interpretation: A Method of Correlation', in Letty M. Russell, ed., *Feminist Interpretation of the Bible* (Oxford: Blackwell, 1985), pp. 111–34, p. 117.

5 See Christine L. Krueger, *The Reader's Repentance: Women Preachers, Women Writers and Nineteenth-Century Social Discourse* (London: University of Chicago Press, 1992).

6 Anne K. Mellor, *Mothers of the Nation: Women's Political Writing in England 1780–1830* (Bloomington: Indiana University Press, 2002), pp. 71–2, 77.

7 For discussions of women's social fiction as an extension of this prophetic tradition, see Krueger; Ruth Y. Jenkins, *Reclaiming Myths of Power: Women Writers and the Victorian Spiritual Crisis* (Lewisburg: Bucknell University Press, 1995); and Gail Turley Houston, *Victorian Women Writers, Radical Grandmothers and the Gendering of God* (Columbus: Ohio State University, 2013).

8 See Amanda Benckhuysen, 'The Prophetic Voice of Christina Rossetti', in Christiana de Groot and Marian Ann Taylor, eds., *Recovering Nineteenth-Century Women Interpreters of the Bible* (Atlanta: Society of Biblical Literature, 2007), pp. 165–80; Charles LaPorte, 'George Eliot, The Poetess as Prophet', *Victorian Literature and Culture*, 31:1 (2003), 159–79; Houston, Chapter 4 (on Barrett Browning); Scheinberg discusses Felicia Hemans and Christina Rossetti in this light, contrasted with Grace Aguilar.

9 LaPorte, pp. 159–60.

10 Quoted in LaPorte, pp. 159–60.

11 Linda Shires, 'Victorian Women's Poetry', *Victorian Literature and Culture*, 27:2 (1999), 601–9, p. 606.

12 Janet L. Larson, '"Who Is Speaking?" Charlotte Brontë's Voices of Prophecy', in Thais E. Morgan, ed., *Victorian Sages and Cultural Discourse: Renegotiating Gender and Power* (New Brunswick: Rutgers University Press, 1990), pp. 66–86, pp. 67–8.

13 Carlyle, p. 155.

14 Marianne Thormählen, *The Brontës and Religion* (Cambridge University Press, 1999), pp. 13–17.

15 K. Cubie Henck, '"That Peculiar Voice": Jane Eyre and Mary Bosanquet Fletcher, an Early Wesleyan Female Preacher', *Brontë Studies*, 35:1 (2010), 7–22, pp. 7–9, 16.

16 See Henck, Jenkins, Houston, and Larson.

17 Quoted in Glennis Byron, *Dramatic Monologue* (London: Routledge, 2003), pp. 57–8.

18 See Rebecca Styler, *Literary Theology by Women Writers of the Nineteenth Century* (Farnham: Ashgate, 2010), pp. 69–87.

19 Houston observes that the moon is frequently 'harbinger of the visionary' in Brontë's fiction, p. 36.

20 *Poems by Currer, Ellis, and Acton Bell* (1846; London: Smith, Elder and Co., 1848), pp. 1–7, 3. All poems are quoted from this edition unless otherwise indicated, and are cited only by page number because the edition has no line numbers. Note that all quotations from Emily's poems are taken from Janet Gezari, ed., *The Complete Poems* (London: Penguin, 1992).
21 Lee Cormie, 'The Hermeneutical Privilege of the Oppressed: Liberation Theologies, Biblical Faith, and Marxist Sociology of Knowledge', *Proceedings of the Annual Convention of the Catholic Theological Society of America*, 32 (1978), 155–81.
22 Carl Plasa, *Critical Issues: Charlotte Brontë* (Basingstoke: Palgrave Macmillan, 2004), pp. 40–4.
23 Larson, pp. 69, 80–3.
24 It is interesting to compare this with H. D.'s mid-twentieth-century prose retelling 'Pilate's Wife', in which Jesus' centrality is displaced for the recognition of divine potential within all men and women. By removing emphasis from Jesus (who is taken down alive from the cross and given a human, not divine, identity) H. D. locates the divine within the 'poetry' not the 'poet', a logic by which Pilate's wife retains the divine with herself. The name H. D. gives her, 'Veronica' (vera ikon) signals that she is herself still the 'true image' of God. H. D., *Pilate's Wife*, Joan A. Burke, ed. (New York: New Directions, 2000), pp. 106, 134, 86, 121.
25 Houston, p. 13.
26 Diana Wallace, 'Uncanny Stories: The Ghost Story as Female Gothic', *Gothic Studies*, 6:1 (2004), 57–68, pp. 60, 57.
27 See for example Alison Milbank, *Daughters of the House: Modes of the Gothic in Victorian Fiction* (London: Macmillan, 1992), pp. 10–13.
28 *The Oxford Book of Gothic Tales*, ed. Chris Baldick (Oxford University Press, 1992), pp. 102–32. See also the anonymous 'The Ruins of the Abbey of Fitz-Martin' (1801) in the same collection, pp.31–50. There is a not dissimilar theme in Wordsworth's poem 'Goody Blake and Harry Gill', where a poor, living woman's curse against her aristocratic male oppressor gains effect.
29 Carlyle, p. 96.
30 Irene Tayler, *Holy Ghosts: The Male Muses of Emily and Charlotte Brontë* (New York: Columbia University Press, 1990), p. 18.
31 'No coward soul is mine', in Emily Jane Brontë, *The Complete Poems*, Janet Gezari, ed. (London: Penguin, 1992), 182:19–20. All quotations from Emily's poems are taken from this edition. On Brontë as a mystic, see Caroline F. E. Spurgeon, *Mysticism in English Literature* (Cambridge University Press, 1913), pp. 80–4.
32 Simon Marsden, *Emily Brontë and The Religious Imagination* (London: Bloomsbury, 2014), pp. 29–30, 62–4.
33 'How clear she shines', *The Complete Poems*, p. 20; Davies, p. 43.
34 Stevie Davies, *Emily Brontë: Heretic* (London: The Women's Press, 1994), p. 18, and Christine Gallant, 'The Archetypal Feminine in Emily Brontë's

Poetry', in Tess Cosslett, ed., *Victorian Women Poets* (London: Longman Ltd, 1996), pp. 44–54. Tayler also reads Brontë's poetry archetypally as a quest for the lost mother.
35 Tayler, p. 27.
36 Emma Mason, '"Some god of wild enthusiast's dreams": Emily Brontë's Religious Enthusiasm', *Victorian Literature and Culture*, 31:1 (2003), 263–77, pp. 268, 273–4; Carlyle, p. 89.
37 Tayler, 67–8; Derek Roper, 'The Revision of Emily Brontë's Poems of 1846', *The Library*, 6th series 6 (1984), 153–67, pp. 154–61; Gezari, 'Introduction', *The Complete Poems*, pp. xix–x.
38 The change is reminiscent of John Keats's 'La Belle Dame Sans Merci'.
39 A moral preference for animals over human beings is evident in Brontë's essays 'The Cat', 'Filial Love' and 'The Butterfly', in *Five Essays Written in French*, Lorine White Nagel, trans., Fannie E. Ratchford, ed. (Austin: University of Texas Press, 1948).
40 Victoria Davion, 'Is Ecofeminism Feminist?', in Karen J. Warren, ed., *Ecological Feminism* (London: Routledge, 1994), pp. 8–29, pp. 8, 17, 23–4.
41 Davion, pp. 22–4.
42 See Jonathan Bate, *Romantic Ecology: Wordsworth and the Environmental Tradition* (London: Routledge, 1991), pp. 8–9.
43 See discussion in Styler, pp. 43–68.
44 Larson, p. 67.
45 Stuart Curran, 'Romantic Poetry: The I Altered', in Anne K. Mellor, ed., *Romanticism and Feminism* (Bloomington: Indiana University Press, 1988), pp. 185–203, pp. 189–90, 195–8; Anne K. Mellor, *Romanticism and Gender* (London: Routledge, 1993), p. 33.
46 More, *Poems* (1816), introduced by Caroline Franklin (London: Routledge/Thoemmes Press, 1996), p. 179.
47 Josef Altholz, 'The Warfare of Conscience with Theology', in *The Mind and Art of Victorian England*, Josef Altholz, ed. (Minneapolis: University of Minnesota Press, 1976), pp. 58–77.
48 Larson, p. 75.
49 Theodore Parker, 'The Public Function of Women', in *Collected Works*, 12 vols., ed. Frances Power Cobbe (London: Trubner and Co., 1863), VIII, p. 103.

CHAPTER 11

Jane Eyre, *A Teaching Experiment*

Isobel Armstrong

This chapter is about a teaching experiment in which *Jane Eyre* was the focal text. The students were working on a summer MA course at the Middlebury Bread Loaf School of English in Vermont.[1] For six weeks they worked on interpreting *Jane Eyre* through the creation of artworks that responded to the text. First through visual art, then specifically photography (*the* Victorian art form), then through sound and finally movement. After a close reading of the novel, the students launched into group projects on visual art.[2] This was one of the most exhilarating teaching experiences of my life, but also one of the most vertiginous, as the rhythms of this course were entirely different from a planned teaching schedule. I have never taught a class that plunged me into despair after one session and the next day produced a series of epiphanies that collectively inspired us. I certainly experienced such vicissitudes and I think the students went through this too.

My purpose in writing of this experiment here is twofold. First, our discussions were directly relevant to the theme of this collection of essays on Charlotte Brontë: through their work, the students repeatedly raised the question of Bertha Rochester's species being and her exclusion from the category of the fully human. What that category might be is a question throughout the novel. Secondly, I believed (and still believe) that this form of interpretation is as rigorous as any customary form of critical analysis. It is not an alternative to conventional ways of teaching, but simply adds to the many ways we can ask students to think about literature. It expands what we can think and feel about a novel and deepens an understanding of the language of the text. One of the happiest moments for myself and the students was to see their copies of the novel, by the end of the course, thumbed and underlined to the point of the physical disintegration of the novel's pages. *Jane Eyre* is almost always the first Brontë novel that people read. Readings of *Jane Eyre* can easily become fixed, but the students opened up this novel. We came to see it as a profoundly experimental, carefully shaped work. It is fiercely comic and just as fiercely erotic, working through imagery that has the force of analysis.

Because it seemed alien to academic practice there were many principled objections to the course. Some of my colleagues worried about the intellectual standing of such a course in an MA programme. Some thought of this as an escape course for less-able students who could not cope with textual criticism. A student who emailed a friend with a picture of her work on a panorama of Jane's artworks, on which she spent twelve hours, was exasperated to receive an instant reply – 'O, I see, Summer Camp' – implying that she was engaged on a form of superficial busy work. Equally problematic were those who thought that the students were creating a mere *transcription* of the novel into visual form and that this form of representation was naïve.

But this experiment prompted the closest of close readings and actually entailed a double reading of the text. To derive an artwork from the novel demanded, in the first place, a careful, detailed and open-ended interrogation of it. Secondly, once a decision had been made about the form of the artwork, the relation of the made object had to be continually checked against the text. The objective of the course was not to produce a high aesthetic work as a culmination of the course so much as to produce a work that was an *interpretation* of the text. It was interpretation, not translation. I regarded the construction of artworks, whether as visual representation, sound, or movement, as the equivalent of the research done in preparation for class, naturally taking place outside it, as occurs in conventional classes. Therefore, because the artworks were not made in class, protocols for our critical activity *in* class had to be developed. The groups made and constructed their work outside class according to a formal time table, only bringing it to the classroom for critical inspection when it was completed. The result was that the protocols developed for class always entailed an intensive reading of the text that took many forms. I will mention just one of these protocols among many before proceeding to the thematics of Bertha explored by the students.

The students were divided into four groups of between four and five students and each group worked on producing an interpretative artwork. In one session, I evolved the following exercise. Groups 1 and 2, *without describing their practical projects*, were to read out to the class the passages their project derived from and which inspired it. Withholding the nature of the artwork meant that we had to concentrate on these passages *as passages*. Similarly, Groups 3 and 4 repeated this exercise on the following day. Each of the listening students were asked to respond in turn to the passages by saying what images and colours arose for them as they listened to the passages and studied them in their own copies of the novel. In addition, they were asked to produce an extemporised four-line, free-verse poem prompted by the passage that attracted them most. This quite formal response fed the presenting groups with ideas and images for their as yet unfinished or nascent project and enabled them to respond in turn to their listeners without closing down on their plans. What they would do in their practical work was kept open.

For example, a group working on a panorama of Jane's experience at Thornfield quoted a passage from Chapter 26 when the failed 'marriage' brought Jane to despair: 'A Christmas frost had come at midsummer; a white December storm had whirled over June; ice glazed the ripe apples … My hopes were all dead'.[3] One student responded with this poem:

> Back on stool, not 'Liar' this time
> But Fool.
> Off with the wedding gown, on with the brown stuff frock.
> The chill has returned, the ice frozen in the pitcher again.

The student has recapitulated the pre-Thornfield ice, cold and humiliation of Lowood in response to the icy passage, remembering the description of the ice in the pitchers that had to be broken every morning before the girls could wash. Details in one part of the text were answered by details in another that indicated how closely the text had been read.

Indeed, I found that the stricter and more specific the protocols of the class work the fuller, richer and more focused an understanding of the text. When, the following week the almost-finished works were presented in class, each student was required to ask a question, and one question only, about a single detail of the work and its relation to the text. When we were preparing to think about photographs, the students were asked to turn to their neighbour and work with them to find three passages that had some relation to the chiaroscuro of the photograph. So, though the class began with open-ended discussion of the text as a preliminary to the creation of artworks, while the students were actually working on interpretations the exercises were much tighter. Consequently, the text was read both through broad critical interrogation and with great precision and detail.

The core of my discussion is Bertha, because this is what fascinated the class most, but I want to start with two readings that contextualise, though perhaps indirectly, her presence in the text.

Human Need and Rochester's 'Family'

I begin with two small instances of the insights – and the *process* by which we arrived at insights – about *Jane Eyre*.

During our initial close reading we were so intensely text-based that we almost seemed to move away from the creative project of the class. Therefore, before we had completed the reading of the novel, I asked the students to bring to the first class after the weekend an artwork, construction, or installation relating to the first ten chapters and specifically either to food, clothes or beds as they appeared in these chapters – all overdetermined elements in this part of the novel. The students arrived for Monday's class in an exuberant mood: they produced a range of artefacts; one made a miniature beaver hat and found out just how expensively the Brocklehurst women were dressed; another constructed an erotic bed from the Red Room episode, insouciantly associating death and sex; another ascetic bed referenced the miserable beds of Lowood; another made an oil painting of the bedroom scene where Jane plots to leave Lowood, while her room-mate snores: another made a collage incorporating seeds, the seeds of Miss Temple's seed cake and the generative seeds of the gardens the pupils were allowed to create even in the desolation of Lowood; another made a puppet who was able to eat the words of the psalms (lifting torn fragments of the print to his mouth) – those parts of the Bible Jane tells Brocklehurst she dislikes. (Brocklehurst's arrantly hypocritical riposte is the story of an obedient boy child who, though preferring a rote reading of the psalms to a gingernut is nevertheless rewarded with one for his performance.) One student, who had been a chef earlier in his life, brought in a tray that contained both found and made objects. The found object was a nest discovered in the grounds (a response to the images of nesting in this part of the novel) and the made objects were miniature replications of the delicate pastry on the bird of paradise plate that Jane refuses in her enervated psychic exhaustion after the red room episode. There were enough for the class to eat and we did not refuse them! It was not simply a festival. To understand how to make these objects it was necessary to read the text very precisely. These presentations were always followed up by a detailed discussion of those parts of the text to which they related.

Figure 11.1   The first tableau: Mrs. Fairfax, Jane and Adèle. Image supplied by Dana Olsen at Middlebury Bread Loaf.

Food, clothing, sleep: these are the primal human needs insistently reiterated in this part of the novel. Beyond that, yes, the comfort of love, the erotic, beauty. But these, though shaping needs, do not have the coerciveness of basic necessity as the first three elements do. Food, clothing and sleep define what it is to be human. If they are denied, or distorted so that a person becomes through their lack radically severed from intersocial life, that person's species being is in question. Such persons don't just fall out of the social world, they are treated as if they are not fully human. How far Bertha is denied all of these – food, clothing, sleep – is an open question.[4]

My second example comes from later in the novel. We created, what on the face of it, may seem like two absurd images.

A woman sitting upright in a newspaper corset, another knitting with two biro pens and the fringes of a scarf, another seated at her feet in a turban and skirt of paper flounces (Figure 11.1). Why are they staring fixedly in an unlovely way?

Figure 11.2 The second tableau: Rochester and Blanche. Image supplied by Dana Olsen at Middlebury Bread Loaf.

A woman standing beside a man sitting on a chair (Figure 11.2). She is dressed in a sheet that is actually spread over two chairs to obtain width.

The students were improvising the stylised and conventional images of *cartes de visite* photographs from tableau scenes described in the novel. *Cartes de Visite*, the earliest form of social media, were small photographs, about two by four inches, on pasteboard, exchanged between family and friends. *Jane Eyre* is a little early for the craze for such photographs, but the static groupings the novel frequently describes justify this slight anachronism and indeed anticipate the *cartes*.

In the first tableau, Mrs. Fairfax is sitting knitting with Jane at her side and Adèle at her feet. It is a domestic scene, taking place just after Jane has returned from visiting Mrs. Reed, that pleases Rochester as he glances voyeuristically into the room and remarks on the scene (p. 284). In the second image, deliberately alluding to bourgeois marriage as Rochester does so mockingly many times during the Blanche visit to Thornfield, 'Rochester' is seated and 'Blanche' stands just behind him.

What did we learn from pastiche-ing this early form of social media? First, the couple: it was customary for *cartes* to show married couples like this, the male seated and the woman standing behind him, generally touching his shoulder in token of the bond of matrimony. Strangely enough, female subordination is marked by the *standing* woman. This must have been partly to accommodate voluminous skirts, but had the strange effect of reversing the expected power relation of a standing man and subordinate, seated woman, so that other means of signing domestic power, such as the touch on the shoulder, were devised by the *cartes* conventions. In the novel, Blanche's ringlets almost touch Rochester's shoulder (p. 215), but the students decided not to portray this: this couple is *not* married, but is posing almost as if they were, exposing the dangerous edge of Rochester's carnivalesque travesty during the house party at Thornfield. The whole Thornfield section is about masquerade, parodying bourgeois conventions, from marriage to the arboreal family. It is dreamwork: in Chapters 11–27 the unconscious speaks. Class, gender and sexuality enter the realm of travesty and upheaval.

The 'family' group comprising Mrs. Fairfax, Jane and Adèle has slung a bed cover behind it to allude to the ubiquitous presence of the draped curtain in the *cartes* and in the text. A curtain protects, muffles, covers, conceals. Here it is a residual synecdoche for the domestic furnishing that is perhaps the most ambiguous of all elements – revealing and exposing by the very fact of its capacity for hiding something, as, for instance, in one of the very first episodes of the novel, when Jane hides behind the window curtain to escape from the Reed family and John in particular. The curtain is akin to a veil and with a 'family' relationship to the wedding veil that features so powerfully in the text. This tableau reveals the artefactual nature of domesticity. Rochester himself points this up by referring to Jane as the 'adopted daughter' of Mrs. Fairfax and to Adèle as the child of Jane, her 'maman' (p. 284). But no member of this 'family' group is actually incorporated into the institution of the family, with the consequence that family is destabilised and becomes a form of artifice. Mrs. Fairfax is a disenfranchised 'distant' relation of Rochester and *feels* so remote that she does not press the 'connection', she tells Jane (p. 119). Jane is a penniless orphan of eighteen, Adèle is an illegitimate child, questionably the biological daughter of Rochester – and 'French'. As one of the students remarked, Jane is proud of her French, but industriously tries to eliminate 'Frenchness' from Adèle. Brontë's interest is in the outsider – in creating an atypical social paradigm – and in the artifice of what constitutes belonging. All three women are in a state of existential homelessness that is denied in the fragile bonding of the group. We later noted that the relations Jane discovers in the latter part of the novel mean that she ends up adjacent to two sisters and a brother at either end of the novel, each representing the 'bad' and 'good' family group: but the latter includes another persecutor as the first does – and each of the persecutors is named John (John Reed, St John Rivers). The students' *carte* tableau suggests that this group of three females *is* a family of sorts, bonded as it is, but simultaneously a family against the odds and a dispossessed group. And yet, this group is drawn together in companionship and comfort. Rochester's nod of satisfaction, however, cannot be in good faith. The legal family partner he has contracted in marriage is concealed in the upper realms of the house, banned from the group.

These two experiments in posing also revealed the experimental structure of the novel – a block of more documentary material at each end of the text (I hesitate to name it realist) and a segment of dreamwork at the centre.[5]

Jane Eyre as Non-Subject

There is an important episode in the novel when Jane Eyre is condemned to lose her status as fully human and become perforce a beggar after leaving Thornfield. Her extremity is recognised with extraordinary acuity by Brontë. The rapid destruction not only of status but of selfhood that occurs in the three days she spends wandering on the moors almost destroys her and reduces her to the state of what Giorgio Agamben calls 'bare life'.[6] She starves. Sleep becomes progressively more disrupted, finally destroyed by the 'intruders' into the rainy wood where she tries to sleep on her final night in the open. She clings on to the marks of a lady – her scarf and gloves – and would have relinquished them for a bread roll had the woman in the bread shop accepted them as barter. Food, sleep and clothes (clothes as signifiers of standing), those primal entities, are withdrawn from her.

Brontë portrays this episode as agonistic and ontologically destructive. The full force of the novel's compassion is behind Jane here. I shall move to this episode in a moment, but as it plays out one cannot but become uneasy about the state of Bertha Rochester (*not* Bertha Mason as she is often designated), imprisoned, destitute, her fall from humanness given none of the scrupulous analytic attention that Jane's state receives. We must compare them because of the constant calibration of Bertha and Jane in the text: Eyre/heir; Bertha/birth. How do both women enter the structures organised by a patriarchal society? The white woman who becomes an heir by chance, the Creole whose birth determines her fate? One is granted a surname, the other seemingly not, though it is actually 'Rochester'. The question being asked is the extent to which birth determines, not only how you are regarded, but also your material fate. (Jane is treated as an outcast by the servants at Gateshead – servants sustain the class system in this house.) The students, surprising me, saw Bertha as a victim, a traumatised victim who moreover established a *liaison* with Jane, not an enmity. The opposition implied in their names – air and [b]earth – actually allies the two women in a shared cosmos. Bertha is not mad, they argued. Nor, incidentally, does she inhabit the attic of Thornfield, but a room on the third floor. It's from this floor that her laugh resounds.

We looked at the crucial Chapter 28, the point at which, escaping from Thornfield, Jane becomes a vagrant – a beggar virtually outside the category of the human.[7] 'Not a tie holds me to human society' (p. 371) Jane Eyre remarks as the coach she can afford no longer departs, leaving her alone on the isolated heath. The startling innovation and narrative shock of Chapter 28 was to place the middle-class woman in the lived social space occupied by working-class vagrants, beggars and the unemployed, as the starving Jane wanders into an isolated Yorkshire parish like the internal immigrants among England's poverty-stricken dispossessed. The importunateness of the starving body and the bondage of starvation is brought right into middle-class experience. Jane's unthinking response to beggary prior to this is exemplified in a moment that becomes retrospectively ironic – rejoicing in the thought that she will be married to Rochester, she gives 'all the money I happened to have in my purse' to a beggar woman and her son at the gates of Thornfield (p. 297). Random charity is the product of chance moods and irrational, fitful empathy. Her middle-class status means that Jane's destitution cannot be described in the Malthusian and ideological terms that blame the poor for their dispossession – the improvident actions of a 'surplus' population. Brontë offers the historical essentials of destitution and implicitly calls into question the commonplace explanations of it. Because the dating of the action is never precise, it is not clear whether the novel is set before or after the new bureaucratised Poor Law of 1834, with its punitive workhouses and attack on women as the source of poverty. But Brontë understood privation. 'There are great moors behind and on each hand of me … I see no passengers on these roads … Not a tie holds me to human society' (p. 371). Jane sees the moor as an encounter with a redeeming transcendental solitude in the presence of God, but her first thought is that she must disappear into the heath before she encounters a 'sportsman or poacher' (p. 372) to whom she would have to explain her presence. The unsaid is that the moor is owned: the sportsman – on a moor he could only be shooting game nurtured for the purpose – is the guardian of this space, while the poacher attempts to encroach on it; the poacher and sportsman pair as a necessary consequence of the power of exclusion. Jane is a trespasser.

Jane's account of destitution is in the present tense, rarely used in the novel:

> Two days are passed. It is a summer evening; the coachman has set me down at a place called Whitcross; he could take me no farther for the sum I had given, and I was not possessed of another shilling in the world. The coach is a mile off by this time; I am alone. At this moment I discover that I forgot to take my parcel out of the pocket of the coach, where I had placed it for safety; there it remains, there it must remain; and now, I am absolutely destitute. (p. 371)

'[A]bsolutely destitute': the present tense drives the narrative on through the stark economics of travel and the almost casual, religious reference to the cross in the place name. The chapter charts three days of destitution before Jane finds the Rivers family. The present tense forces real-time experience upon a narrative – it unfolds experience as it is happening. For readers the peculiar cognitive affect of narrative that describes events as yet to be over is intensified as, caught in the present, with an indeterminate future, the experience has not become part of the past. It *is* not *was*. This is a generic overturning after the dreamwork narrative of Thornfield, where so many crises occur at the time of sleep – the fire in Rochester's bedroom, the attack on Mason, the bridal veil episode. The plainest and starkest of description replaces the erotics of Thornfield with naturalism – though ironically this naturalises the self as lack.

'Not a tie holds me to human society'. This statement tends to be ignored. Chapter 28 is generally read through Biblical allusion to *Exodus* and the plight of the Israelites in the wilderness as well as through Bunyan's *Pilgrims Progress*. These allusions are powerfully present, but they direct us to social as well as religious experience. The chapter can be too easily elided as the bridge between one episode and another, the move from Thornfield to Marsh House, from Rochester to St John, from financial and sexual exploitation of the West Indies to missionary exploitation in East India. But Chapter 28 is crucial, not only in its portrayal of the destitution and dispossession of the social outcast, its material and existential condition but because it makes a sustained critique of conditions where some subjects can be defined as deficit beings, not fully human. It asks ontological and political questions about the ways we guarantee personhood to someone. Who is deemed worthy of empathy, who is not?

I shall return to the particulars of Jane's experience. For the moment I want to emphasise the empty space of the heath and its recall of exposure in another empty space, the heath of *King Lear*. Finally rescued, Jane says to Diana: 'If I were a masterless and stray dog, I know that you would not turn me from your hearth tonight'. She has indeed characterised herself as like a 'lost and starving dog' in the time of starvation. 'Mine enemy's dog, though he had bit me, should have stood / That night against my fire' (*King Lear*, 4.6. 30–1). This is Cordelia on the exposure of Lear in the storm scenes and through subtle verbal convergences – 'heath', 'pelt' (recalling *Lear's* 'pelting villages'), for example – *King Lear* is present in this chapter. It is perhaps strange that the text evokes the archaic Elizabethan Poor Laws behind *King Lear*, in which beggars and workless outcasts could be punished by violence and even put to death. But it seems that Brontë moves instinctively to *Lear* precisely because the play has problems with what makes us fully human; problems that expose the questions at work beneath the manifest text of the novel. King Lear offers a deficit model of the human. 'Thou art the thing itself': Edgar as 'unaccommodated man' falls out of the category of the human. Edgar is a 'thing', an ostensive definition of the living being in a state of lack (3. 4. 100–3) – remember that Jane Eyre thinks of herself as a 'thing' to the Reed family in the first chapter of the novel. Like the 'poor naked wretches' Lear speaks of earlier, Edgar is a 'poor bare forked' *animal* (1. 11. 98). For such beings, their essence is lack. Forced back into nature, not endowed with human status, they can belong to human culture only by fiat of the entitled. They invite arbitrary treatment, but since lack is their essence their condition of depletion can never be assuaged except by being co-opted into humanity by the sovereign.

Jane drops out of human personhood on her 'heath': as unaccommodated woman she loses any right to a recognition of humanness, with its concomitant entitlements, and is on the way to becoming a 'thing' and indeed an animal. A vagrant, a prostitute dressed as a 'lady', or, as Davies suggests, a woman fleeing the punitive poor law of 1834, Jane could mean any form of derelict to those who encounter her: but for all she falls out of subjecthood. When Jane approaches the parsonage of the hamlet into which she wanders she believes that the clergyman's duty of aid gives her 'something like a right to seek counsel here' (p. 377). The phrasing, juridical in form though in a religious context of charity and Christian loving kindness, raises the secular question of natural and political rights and her eligibility for inclusion in them. The discourse of rights was well-advanced at this point in the nineteenth century and this phrase cannot but recall them. Are there, perhaps, trans-historical and core human rights to which all have access without the need for justification? This question implicitly valorises Jane's status as always already human. The text's analysis gives full force to an underlying egalitarian reading that refuses the deficit subject of lack. Jane's beggar status as deficit subject or bare life is created by common consent of the inhabitants of the hamlet she finds herself in. It is not intrinsic.

'Bare life' directs us for a moment to the biopolitics of Agamben who underwrites the deficit subject requiring a supplement of humanity. In an interesting moment, the only time that gender features in his book, Agamben describes the Roman law, the *ius patrium* that gives the father the right of magistracy to kill his sons, as the model of sovereignty: yet there is a large 'but' here; *the father's power should not be confused with the power to kill, which lies within the competence of the father or the husband who catches his wife or daughter in the act of adultery.*[8] It seems that in this purely domestic context, women belong to a sphere of disposability that does not even endow them with the 'sacred' aura of bare life. The killing of a wife or daughter is an unimportant matter of personal choice. There seem to be two kinds of irreducible biopolitics, one for men and one for women, a gender distinction that Agamben seems unaware he has made. In this reading, in which the subject of bare life is not human until proved otherwise, women are doubly deprived of humanity since they even fall outside the status of bare life, which is itself outside the human. Since it is the very power of sovereignty to enable the supposition of bare life to pass over into fact by fiat in Agamben's paradigm, this gender issue does not trouble him. But what he presents as an analysis of the modern condition often emerges as an acceptance of it. The scandal of Bertha, a kind of unconscious of Brontë's white woman's destitution, is pointed up by Agamben's impercipience. It will never be clear how Brontë intended Bertha to be read. All that is possible is to understand the text's problems.

Chapter 28 charts Jane's descent from situated space on the moors, open to the cosmos, to the limits of the body in pain, from transcendental, religious space to abjection, from the standing of a lady to the sub-human state of beggary. Quite soon the importunateness of need attacks her: 'But I was a human being, and had a human being's wants' (p. 374). She moves from the transcendental to the intensely material needs of the body. But thereafter her 'human being's wants' fail to be recognised. Driven by these wants, her movements are restlessly from, and to, the human and social, which is the rhythm of the chapter; the psychic and spatial rhythm of the outcast. Approaching a village, she enters a bread shop but dare not beg, wanders to houses and asks without success for work – 'the white door closed' (p.376): 'I rambled round the hamlet, going sometimes to a little distance and retuning again … I … sat down under a hedge … I drew near houses; I left them, and came back again ... instinct kept me roaming round abodes where there was a chance of food ... I thus wandered like a lost and starving dog' (p. 377). She is approaching the condition of Lear's 'animal'. Eventually, desperation overcoming shame, she tries to exchange her scarf and gloves – the mark of a lady – for bread from the bread shop woman, and fails. Compulsively leaving and returning to the village, longing for and yet dreading human contact – because it evokes 'suspicion', 'distrust' and the 'shame' of rejection (p. 378), space closes down on her. After the second night in the woods soaked by rain and disturbed by passers-by whom she significantly terms 'intruders' (p. 379) as if to claim the wood as her own space, she begs a little girl for the cold porridge being thrown to a pig trough – 'Mother! ... there is a woman wants me to give her these porridge ... Give it her if she's a beggar. 'T pig doesn't want it'. Reduced to the sub-human she is handed it without a word, exiled from language, beyond the pale even of animals. 'I devoured it ravenously' (p. 379). The 'stiffened mould' of the cold porridge is a poignant comment on the earlier Lowood school episode, when even at their hungriest the girls could not eat burned porridge. She is only just saved from death by starvation and exposure.

Bertha

But what of Bertha's dehumanisation, so little recognised in comparison with Jane's? From the start the class was fascinated by Bertha. But they did not take the view that feminists and anti-racists have taken. Most feminists approach Bertha as it were with garlic round their necks, uneasily convinced of Charlotte Brontë's uninterrogated racism, of her assent to colonial violence and of her denial to Bertha of the women's rights that are accorded to Jane however problematically. That Bertha is regarded as a creature of less-than-human status is an assumption of feminist criticism. That Bertha is in some way also the dark double of Jane is also an assumption, taking back Bertha's humanity with one hand and cancelling it with the other.[9] The students insisted that with continuing imprisonment in Rochester's Bastille she has become progressively more degraded. Her degradation is humanly made, the result of a crushing diminution of her physical and mental world. Bertha's bare life is produced by Rochester's male prerogative.

Parallels between Jane and Bertha have always been made but the students were particularly specific here, from our very first close reading of Chapter 1. Jane is dragged upstairs to the Red room (just as Bertha belongs to the upper part of the house, the third floor) with fulminations on her animal-like behaviour ('I was borne upstairs', p. 14). Jane is grappled by Bessie and Abbott and almost tied to a stool. We noted the comedy here; the stout Abbot is asked to give up her garters to tie Jane down – Bessie's are too small. Bertha is grappled by Rochester and tied down to a chair – successfully and without the comedy. Animalism is attributed to both women: the same feral language appears: 'Rat! rat!' (Jane, p. 14), 'some strange wild animal' (Bertha, p. 338). 'Shaking my hair from my eyes, I lifted my head and tried to look boldly round the dark room' (Jane, p. 20). 'She parted her shaggy locks from her visage, and gazed wildly at her visitors' (Bertha, p. 338): Bertha has been incarcerated for nearly ten years, we hear. The incarceration dates not far from the year that Jane experiences the trauma of the red-room. She was 'about ten' when she went to Lowood she tells Rochester and is eighteen when she begins her Thornfield service. So, Jane and Bertha begin their lives of enclosure in parallel.

The exactitude of the novel's timing emerges – something we associate with *Wuthering Heights* but not with *Jane Eyre*: for instance, it is exactly a year from the false 'marriage' that Jane hears Rochester's voice while she is almost being inveigled into another false marriage by St John.

To reclaim Bertha's humanity, one of the groups decided to imagine what her artwork would be like. Jane is granted access to the aesthetic through her painting and drawing. Throughout the text her practice as an artist is intensely significant and grants her both autonomy and status. Through her art she is an interpreter, handling the real and symbolic. She can draw from life – self-portrait and portraiture are one of her skills – and she can draw allegorically and symbolically, as the surreal land and seascapes she shows Rochester indicate – extravaganzas of existential desolation derived from the re-imagined landscapes in Bewick's book of birds introduced at the beginning of the novel (Chapter 13). Indeed, these water colours, with their dispersed body parts, severed heads, darkness and intimations of a violated feminine sexuality, could just as easily reference Bertha's damaged body proleptically as they reference Jane's bruising history. Think of the first picture's gold bracelet, torn from a dead arm by a cormorant. This image could be a stand-in for Rochester's colonialist stripping of the assets of a Creole woman, her violation. What would happen if we saw Bertha's art? To imagine Bertha's art meant attributing interiority to her, attributing a psyche, attributing her own, not Rochester's history.

Some students imagined the flower and nature drawings that Bertha's training as a genteel lady early in life would have demanded of her – inflected with the exotic and paralleling the no less exotic imagery of

Jane's own art. One student made tatted lacework incorporating red stripes – one for every year of captivity – and tore it up as if in rage and frustration. There were two self-portraits, setting up a damaged face, both highly lyrical. One made use of broken eggshells to indicate the crushing of a psyche. Collectively the students created types of graffiti, including flames of red, and in their final presentation of the scene of painting scrawled 'No ... No ...' across the wall. I understood the violence and anger, but when I questioned the lyricism the students insisted that this was legitimate. Bertha's face, they argued, is always mediated and distorted by darkness, either by darkness itself or the mirror, a 'dark oblong'. The racialised distortion of her features that conventionally suggest negritude, were, they argued, a confusion of Jane's perception. The blackened, distorted features associated with Bertha suggested damage from smoke and light deprivation as much as racialised distortion. It is dark when Jane sees Bertha for the second time. Not everyone assumes that Bertha is black (Jean Rhys in *Wide Sargasso Sea* does not assume this, for instance) and the students certainly thought her racial blackness was questionable, a way of diverting interrogation from her real abjection. And certainly, when one looks closely at the description of Bertha's face, it is purple that dominates the swollen features when Bertha is viewed in her windowless room (p. 338). In the wedding veil scene, it is 'red' and 'bloodshot eyes' that dominate (p. 327). The eyebrows are 'black' and the features 'blackened' as if charred. The students also insisted, as the use of broken eggshells was meant to suggest, that there was a tragedy of fertility in Bertha's case. She had been married for four years without bearing a child (another aspect of the pun on birth in her first name): what happened? Miscarriage, abortion? Her anger is a justifiable anger. Ten years in a windowless room attended by an alcoholic. This depends on unpacking the term 'Gothic' and imagining what such experience would be like as lived experience. The rage of these hypothetical artworks brought us back to the mourning and melancholy of the paintings Jane presents to Rochester, with their severed limbs and drowning bodies. Masochistic mourning and rage are equal and opposite states, pairing the two women.

Moreover, until Chapter 25 and Bertha's invasion of Jane's bedroom, there is no mention of the trauma of the red-room. That episode and its violence is repressed and surfaces only after the abortive wedding just before Jane leaves Thornfield – 'I dreamt I lay in the red-room' (p. 367) – the delayed action of trauma. Bertha and Jane are both the traumatised subjects of violence – displaying rage on the one hand and masochism on the other. Jane's exasperating self-deprecation and her insistence on addressing Rochester as 'master', comes into focus as an act of self-abnegation with its own pleasures. '*I* care for myself' (p. 365), Jane affirms in response to Rochester's manipulation of her feelings after the marriage debacle. But, in this dreamwork Thornfield section of the novel, self-preservation is the other side of masochism.

There were two further aesthetic engagements with Bertha. One was the crisis of the tearing of the wedding veil and the second a response to her sounds – the preternatural laugh.

It is Rochester who speaks of 'the spiteful tearing of the veil' (p. 328), but Jane who speaks of 'that unfortunate lady' (p. 347). The students insisted that Bertha Rochester was issuing a warning to Jane when she held the candle over her at the end of the pre-wedding visit to Jane's room, and not a threat: 'the fiery eyes glared upon me – she thrust up her candle close to my face, and extinguished it under my eyes' (p. 327). Bertha is symbolically extinguishing the flames of the erotic here, the warning 'glare' of her face a kind of anti-sexual signal. I was worried by the incipient sentimentality of this view. I pointed out that a burning candle was left outside Jane's room on the night of Rochester's bed burning by Bertha earlier (Chapter 15). But they argued that Bertha had not attempted to kill Jane on either of her descents from the third floor. She *had* attempted to kill Rochester, setting his bed alight on the very evening of his long self-justifying account of the love affair with Celine, a tale that is a mask for his earlier dealings with Bertha. The tale of Celine recalls Bertha for him and it is clear that this West Indian story almost displaces the French story on which he is embarked: 'Lifting his eye to its [Thornfield's] battlements he cast over them a glare such as I never saw before or since. Pain, shame, ire …' (p. 167) The students argued that the burning bed is a retribution for Rochester's dissimulating language and failure to understand or recognise Bertha's own. His initial obliviousness to the fire – he sleeps while '*Tongues* of flame darted round the bed' (p. 174, emphasis added) – betokens his deafness to her feminine language.

On the night of the fatal tearing of the veil (Chapter 25), Jane sees Bertha for the first time: 'it removed my veil from its gaunt head, rent it in two parts, and flinging both on the floor, trampled on them' (p. 327). A group of students made a photocollage of Chapter 25 and they took the same view as the group who had seen the candle as a warning rather than a threat. They pointed out that this chapter was full of halves and doubles: the 'cloven halves' of the chestnut tree, they argued, intimated, not the punitive disaster of the illicit marriage, but Bertha and Jane, the two 'wives' or 'comrades' as these severed pieces are designated, of

Figure 11.3 Collage: Chapter 25 (Bertha and Jane). Image supplied by Dana Olsen at Middlebury Bread Loaf.

Rochester. The daemonic language of 'cloven' suggests the doubleness of the devil, a consciously fabricated plot rather than a God-created sign. Doubleness runs through the chapter: the dividing of the ripe and unripe apples, the dividing moment of midnight, when Jane tells her story, the two dreams – there is not one, but two dreams of ruin. And finally, the veil. This too is doubled – Jane sees the veil in the morning, 'the veil, torn from top to bottom, in two halves' (p. 328). The veil intimates Bertha, who has a right to it, and Jane, two halves of Rochester's marriage plot. The tearing of the veil carries powerful religious and Biblical language. The veil of the temple was rent in two at the moment of Christ's death as the secrets of the tabernacle were exposed.

The collage that emerged from these discussions was divided by a fluttering piece of lace. With a background of violent reds and blacks, a cloven tree, two apples, two beds (the place of dreams), and two mouths, the doubling of lips deliberately exaggerated, intended to indicate the *lack* of speech on the two occasions Bertha visits Jane's room (once outside, once inside it), were placed on the collage surface in overdetermined repetition, creating a drama of images (Figure 11.3).

To turn now to Bertha's sounds. The students worked on the formation of a soundscape for her, bearing in mind that Bertha is associated with sound and never with speech. They created a hypothetical history of sound for her, but in the process we looked carefully at every reference to Bertha's sounds in the text. Is there a semiotics to this sound? Bertha's sounds are non-referential, but ask for interpretation. I track them here with our discussions. Bertha's sounds occur in Chapters 11, 12, 15, 20 and 27 – she makes no sound in Chapter 25, the moment of the veil visit.

Chapter 11: The first sound: 'a laugh struck my ear. It was a curious laugh – distinct, formal, mirthless ... The sound ceased, only for an instant. It began again, louder – for at first, though distinct, it was very low. It passed off in a clamorous peal that seemed to wake an echo in every lonely chamber ... The laugh was repeated in its low syllabic tone, and terminated in an odd murmur'. The laugh 'was as tragic, as preternatural a laugh as any I ever heard; and, but that it was high noon ... I should have been superstitiously afraid' (pp. 126–7). The laugh, then, is drawn out as if syllable by syllable: it is 'formal', not a continuous inchoate noise. Though the note says that this is inchoate noise, the syllabic laugh is organised sound. Jane calls it a 'cachination' – immoderate laughter.

The aspect of organised sound should alert us to the epithet, 'tragic'. This relates Bertha to a long, perhaps the longest, tradition of human creation in culture – tragedy as an aesthetic form and an enquiry into the human experience of loss and violence. She is credited with the capacity for tragedy and incorporated into aesthetic tradition. Laughing to oneself implies dialogue, a dialogue with the self – some form of recognition. What has Bertha recognised? Surely the sounds of Jane and the housekeeper exploring the third floor and the leads of the roof. The advent of another woman, a possible rival, is a significant event.

This in turn made us reflect on the animality attributed to Bertha. Animals do not laugh. Animals do not commit suicide as Bertha does, deliberately hurling herself from the burning roof of Thornfield. The aural experience of uninterpretable 'murmur' coming to Jane's ear is not violent: it is soft, a low, continuous, indistinct sound. Is this the murmur of someone talking to themselves, rehearsing the self's experience? It is Creole, perhaps. There are no companions in the windowless room except the taciturn Grace Poole.

Chapter 12: The second reference to Bertha's sounds. Feeling trapped, Jane walks backwards and forwards along the corridor of the third storey, longing for a more expansive life, a strange analogue of the trapped woman guarded by Grace:

> When thus alone, I not unfrequently heard Grace Poole's laugh: the same peal, the same low, slow ha! ha! which, when first heard, had thrilled me: I heard, too, her eccentric murmurs; stranger than her laugh. There were days when she was quite silent; but there were others when I could not account for the sounds she made.

The 'eccentric' murmurs are stranger than the laugh, indicative of mood variation, perhaps, but of the fact that Bertha's language means something to *her*. The 'slow ha! ha!' is surely a sound of recognition, a self-reflexive sound, as before. It is not at all surprising that the laugh occurs when Jane is present because her presence calls it out. There is a strange symbiosis to this episode of sound: both women are, or feel, trapped. Bertha's laugh indicates a recognition of Jane's presence: and surely that double negative – 'not unfrequently', often – denotes a disavowal on Jane's part. She goes to the third floor to hear the laugh and at some level Jane *knows* about Bertha.

Chapter 15 portrays the bed-burning episode: 'Tongues of flame darted round the bed'. Before this Jane hears a 'demoniac laugh – low, suppressed, and deep ... goblin-laughter'. In response to Jane's 'Who is there?': 'Something gurgled and moaned' (p.175). The laughter again – this time demoniac and goblin-like, evil and supernatural – is also 'suppressed', suppressed, presumably, because an early discovery of Rochester's burning bed would save him. None of these sounds is without motivation. The sound is also human. To moan is to utter a long, low sound: a moan expresses physical or mental suffering, even sexual suffering. With the word 'gurgled' Bertha herself seems choked by smoke. Is the candle a warning or a threat? Only Rochester's bed is fired. The students made a metaphorical transfer, as we have seen, from the 'tongues of flame' Bertha created to her own 'tongue' – her own language, which Rochester is incapable of understanding. He is oblivious to these flames though immersed in smoke.

In Chapter 20 Mason is wounded by Bertha. The night silence is 'rent in twain by a savage, a sharp, a shrilly sound' (p. 239). It is not clear *whose* cry this is. But later sounds are less ambiguous. Within Bertha's room a 'snarling, snatching sound, almost like a dog quarrelling' is heard. When Rochester enters the room 'A shout of laughter greeted his entrance; noisy at first, and terminating in Grace Poole's own goblin ha! ha!' (p. 241). We don't know whether this is a shout of fury, defiance, or terror. The attack on Mason, however, is not irrational – it is he, along with Rochester, who is responsible for her fate.

This is the first time that Bertha is explicitly associated with animal life, a dog, 'snatching' – and significantly only when Rochester appears. It is Rochester whose presence endows this woman with bestiality. But at the same time Bertha laughs, asserting her human standing, though Jane chooses the epithet 'goblin', creating an entity at once human and supernatural. It's another symbiosis: Jane the small child in the Red-room sees herself as a ghostly image in the mirror; Jane the young woman when first seen by Rochester is a fairy, a fay figure in the dusk. It is not taxonomically impossible, the novel insists, to put the two categories of human and haunter together, to imagine an entity both sprite and human. In fact, to do so is to *extend* the human imagination. It may be that only human beings can be endowed with a spectrum of elements, sub-human, human, supernatural – from animal to spirit.

Left alone with Mason as Rochester fetches the surgeon, Jane listens for noise from the 'fiend' and hears three sounds at three long intervals – 'a step creak, a momentary renewal of the snarling canine noise, and a deep human groan.' The groan and the moan of Chapter 15 have an affinity. This time we can be sure none of these sounds belong to Mason, since he is in the outer room with Jane (p. 243). Canine, human, goblin – Bertha occupies all these categories. So does Jane. Arguably both are subjected by, assaulted by, the sub-human categories applied to them. 'Bad animal', says John Reed of Jane.

Chapter 26: Jane sees Bertha for the second time. In Chapter 25 Bertha is deemed a 'vampire' by Jane. Here:

> [A] figure ran backwards and forwards. What it was, whether beast or human being, one could not, at first sight, tell: it grovelled, seemingly, on all fours; it snatched and growled like some strange wild animal: but it was covered with clothing. (p. 338)

This being, 'the clothed hyena', utters sounds: 'A fierce cry ... The maniac bellowed': she is tied to a chair 'amidst the fiercest of yells' (p. 339). There is total taxonomic confusion here, caught in the contradiction – a 'strange wild animal ... covered with clothing'. Beasts snatch and growl and bellow but they do not yell. The term 'maniac' is reserved for human madness. Bertha becomes an animal in Rochester's presence either from terror or fury or both. She will think by this time that Rochester is bigamously married. What will he do with her now? To grovel – though the grovelling is qualified by 'seemingly' – in self-defence or in violent aggression actually seems a 'natural' response to Rochester's provocative entry into her prison. Bertha is on display and at his mercy. We can only attribute absolute bestiality to her if we decide to delete any resemblance to human species being from this account. For Rochester this is easy: he demonises her; she is a 'demon' at one time, at another, a 'fierce ragout' (p. 339), a West Indian culinary dish from the colonised other.

Rochester only once calls her by her name, Bertha, when she is on the rooftop of the burning Thornfield, about to throw herself off. We might compare this with the supernatural call, 'Jane'.

What conclusion can be reached through the students' meticulous reading of Bertha's sounds? It does not, perhaps, preclude racist and colonial readings on Brontë's part. When we remember the overt racism of the juvenilia in Charlotte's Angrian sagas interpretative caution must be inevitable.[10] How we see Bertha partly depends on how we understand Rochester's self-serving story of his West Indian adventure and his search for a Creole heiress. If the weight of narrative sympathy is behind him our view of Bertha correspondingly shifts. But isn't he blind before he is blinded? Isn't the extremity of blinding a kind of narrative justice for his degradation of Bertha as well as the bigamous marriage? I think these questions can be endlessly debated, but what is not in question is the text's constant attempt to calibrate Bertha and Jane; its attempt to compare the two women as species being and its understanding that human subjecthood can be arbitrarily taken away from both women. Bertha is not just a Gothic accident however we think of her.

A word about my use of Agamben here. I did not introduce his study, *Homo Sacer: Sovereign Power and Bare Life* (1995) to the students over the course of the summer, but in retrospect it is a useful framework for thinking about dispossession in the nineteenth century, even though Agamben's interest is in the totalitarian regimes of the twentieth century. Nor is he preoccupied with the racism that must enter into a discussion of Bertha – though at the end of his book he does assert that 'today's democratico-capitalist project of eliminating the poor classes through development not only reproduces within itself the people that is excluded but also transforms the entire population of the Third World into bare life'.[11] The importance of Agamben for *Jane Eyre* is that he is interested in analysing the *structures* that bring about the assumption that classes of persons can be defined as not fully human. For 'bare life' is 'a life that may be killed by anyone – an object of violence that exceeds the sphere both of law and of sacrifice',[12] beyond both divine and human law. It comes about as 'the sovereign decision, which suspends law in the state of exception and thus implicates bare life within it'.[13] Agamben traces the origin of bare life to the history of Roman law, the primal system of the Western polis, where the very constitution of the law creates the possibility of excluding persons from it: by declaring a state of exception, a state of exclusion, sovereign power negates any entitlement to the excluded subject, assumes its dehumanisation and allows any kind of violation, assault or abjection to take place; and these assaults do not fall under the rubric of homicide or sacrifice. Bare life's only characteristic is that it is alive, as an animal is alive. Bare life belongs to a 'biopolitics' because it is a politics of the human body, which is under the continual threat of death. For Agamben, the law is complicated by Western metaphysics which assumes a 'pure Being' isolated from the forms of concrete life,[14] thus creating a biopolitical fracture that creates a lesion between the body and Being.

Agamben's example of the state of exception is the concentration camp, but Blake's Chimney Sweepers, slavery, the press gang, factory labour and prostitution in the nineteenth century, and the zero hours contract and rough sleeping today, are all examples of bare life – the not fully human. My complaint against Agamben is that he so normalises the state of exception as to bring about, unintentionally, the inevitability of it. Nevertheless, his work led me to think of the deficit subject (my term, not his) in the nineteenth century because it gets to the root of the outrage and horror of suffering endured by some subjects then. If we turn to Jane, it is clear that as an orphan she is in a state of exception: the law of the family as practiced by Mrs. Reed abjects her; she can be subject to any form of abuse without recourse to justice. Her abusers don't come under the rubric of moral or legal appeal. In her wanderings as a beggar she becomes a Malthusian 'surplus' or unnecessary subject. Similarly, Bertha is declared not fully human by Rochester's law of the father, which both enables the masculine depredation of her West Indian fortune and the abuse of incarceration without end. Throughout, she protests against the taxonomy of bare life that does not grant her human personhood.

The Students

I asked each of the seventeen students to bring to the final meeting 250 words about the rationale for this kind of work. I will use the students' own words. There were three (perhaps four) kinds argument, often so interfused with one another that I find them difficult to separate. Because the writing was informal and personal it was often exclamatory.

1. The emotional, sensory and affective being of the text is released by this method, 'Responding with art', as one student put it.

2. Intense engagement means repeated close reading and searching attention to the words of the text. You become an 'explorer' not a tourist reading an ad from the past. The text is 'multi-faceted and complex'.
3. The end product, physically available for interpretation, is unique. Unlike writing, the concrete artwork is immediate. But it asks for interpretation by others as a condition of its existence.
4. The importance of working in a group.

Here are some of the students' comments:

Student 1
I don't believe imagination and rational intelligence can be bifurcated … I feel like I finally understand what Gertrude Stein was trying to express when she wrote 'a rose is a rose is a rose'. The amount of times I reread the same passages from *Jane Eyre* or sat and repeated single phrases in my head was innumerable, and as a result, the meaning of Brontë's words became something more. Yes, they became abstract art, photography, and performance, but Brontë's words became charged in a way that felt emotional and personal.

Student 2
I experienced how art can serve as its own critical experience of a text. The creative process demands intense close reading, especially of heavily imagistic passages, and not always those that readily appear thematically significant. Responding with art diverges from standard analytical responses by requiring the reader to experience the text in a sensory, non-judgmental way. Personal response drives the process, leading to a more embodied, visceral engagement with the text. Once the art is created, it can be interpreted in its own right. It can speak back to its parent text, highlighting connections that may otherwise go unnoticed.

Student 3
Every single time I went back to reread a passage for artistic inspiration, or for small details during a project, I found something new … Often times when I set out to think critically about a piece of writing, or try and analyze something, I feel frustrated because I am unable to come up with new and original ideas. With *Jane Eyre* my ideas were original every time because they were expressed through artwork that has never been created before.

Student 4
Three headings:

- The text lodges in your brain ... imprinted on my brain because I have looked at it over and over.
- The text brings you together ... the focus moves away from correctness to originality and boldness.
- The text moves you. You are moved by other people's bodies ... the text moves from the purely cerebral to the visceral.

Student 5
Pure language can allow a student to avoid the emotions, the actions, the sounds and textures of a scene ... [the art method involves] Interpretative risks ... and multiple right answers.

This course was an experiment. Teaching comes in many forms. I am not suggesting that this is the invariable way to teach, simply that we can learn how a text opens up when students address it in this way. At least as far as *Jane Eyre* is concerned, the wild racism of the juvenilia and the colonial violence of the Angrian sagas that Charlotte and Branwell made together does not transfer unproblematically to the novel.

Notes

1 See www.middlebury.edu/blse.
2 The course description of the six-week course, five hours of classes a week, was as follows: 'After an in-depth reading of Charlotte Brontë's novel, the class will plan collaborative interpretations of the text through the media of visual art (e.g. painting, puppetry), photography and video, movement (e.g. mime, dance) and sound. Everyone will participate in creative work in each medium, working in groups. The end product will be an exhibition/installation of our work. Concurrently with this work groups will look at Brontë's letters and a selection of criticism. Assessment will be through extracts from your class diary and the study of one chapter in light of your interpretative work. Be prepared to work in groups, to spend time working together after class, and to take risks'.
3 Charlotte Brontë, *Jane Eyre* (1847), Stevie Davies, ed. (London: Penguin, 2006). Subsequent page references in the text.
4 Bertha is minimally clad and fed, but seen as a clothed animal at the time Rochester reveals her to Jane and his brother-in-law after the wedding. Sleep deprivation is her norm.
5 Note that these photographs were an aide memoir to the class event: groups of students composed themselves in living groups/tableaux and the rest of

the class crowded round to view each group and discuss what the assembly suggested.
6 Giorgio Agamben, *Homo Sacer: Sovereign Power and Bare Life* (1995), Daniel Heller-Roazen, trans. (Stanford: Stanford University Press, 1998).
7 This discussion repeats some material from my *Novel Politics: Democratic Imaginations in Nineteenth-Century Fiction* (Oxford: Oxford University Press, 2016), pp. 185–8.
8 Agamben, p. 88.
9 The classic and path-breaking reading of this kind was Sandra Gilbert and Susan Gubar, *The Madwoman in the Attic: The Woman Writer and the Nineteenth-Century Literary Imagination* (New Haven and London: Yale University Press, 1979), pp. 336–71.
10 See Isobel Armstrong, *Charlotte Brontë's City of Glass*, The Hilda Hulme Lecture, 1992 (London: University of London, 1993), note 33, p. 33. I quote a passage from the Angrian sagas where Zamorna vows the 'extirpation' of 'Quashia and his Africans'.
11 Agamben, p. 180.
12 Ibid., at p. 86.
13 Ibid., at p. 83.
14 Ibid., at p. 182.

CHAPTER 12

Fiction as Critique
Postscripts to Jane Eyre *and* Villette

Barbara Hardy

I have never cared for the sequels to nineteenth-century novels except those William Makepeace Thackeray and Anthony Trollope wrote themselves, so it was a surprise to me when I suddenly thought of writing a postscript to George Eliot's *The Mill on the Floss*. I was teaching a seminar on nineteenth-century poetry and fiction, in a course of literary criticism for the Creative Writing MA in Sussex University, and we were discussing the function and success of the character Lucy Deane, Maggie Tulliver's cousin. One student, reluctantly conceding that she had simplified or mis-read this character, concluded, 'But you wouldn't want to write a novel about her, would you?' This remark made me stop to wonder: no, I didn't want to write a novel about Lucy, but how about a short story, concentrating on this relatively minor, but individualised, developing, and understated character? I had once said that George Eliot never creates minor characters, only characters subordinated for her novel but implicitly containing what she called the 'centre of self'. '[B]ut why always Dorothea?' she asks in Chapter XXIX of *Middlemarch*, and I began to ask, 'Why only Maggie? Why not also Lucy Deane?'

I began my book *Dorothea's Daughter and Other Nineteenth-Century Postscripts* because I wanted to revive Lucy Deane, and I went on to write postscripts to *Middlemarch*, and novels by Jane Austen, Charlotte Brontë, Charles Dickens, and Thomas Hardy.[1] I called the stories postscripts rather than sequels because they are highly selective and respect the conclusions – those deaths, marriages and births which form the finales to nineteenth-century novels – only detaining one or two characters for a little while, after the end, to raise questions and dwell on suggestions not fully answered or developed in the novels. I was not making large additions or extensions to the novelists' work, but pulling out loose threads in the original fabric to weave a little new material. Sometimes, as Isobel Armstrong has suggested, finding 'hidden narrative secrets' already embedded in the novel.[2]

The characters I developed, and the novelists I revised, were not all women, but in the course of writing the book a unifying feminist theme emerged, although at no point did I consciously elaborate or sharpen this subject. There also emerged a purpose of making literary judgment and analysis through creative rather than critical discourse, but this became a more conscious process as I wrote, revised and assembled my stories.

'Adèle Varens' was originally intended to dramatise and give a voice to the character of my eponymous heroine, a character I had long thought unimaginatively and disturbingly dismissed from *Jane Eyre*'s happy ending and moral conclusion. In the course of writing the first section of my story, in which Edward and Jane Rochester discuss a visit from Adèle after the birth of their son, I decided to skip the visit and include Adèle by indirection, putting my case against her exclusion from the Rochester family and the novel's Providential ending in the conversation of the happily married couple, that congenial and continuous talk which Jane the narrator emphasises in her last remarks to the Reader. I used Jane to represent the case for Adèle, and against Rochester, with a particular emphasis because the situations of the two women characters, as children, were strikingly similar, though this parallelism is not made explicit in *Jane Eyre*.

I should also say that I had not anticipated the narrative mode of my stories, which turned out to be largely dialogic: I did not use the first person of *Jane Eyre* and *Villette*, or the convention favoured by most nineteenth-century novelists, the authorial or pseudo-authorial narrative. In addition, I did not try to emulate or imitate the style of my subject authors, except in the odd phrase or word, though I re-read carefully, bore language in mind, and tried to avoid modern constructions and lexis.

Charlotte Brontë takes care to include Adèle in Jane's conclusion:

> You have not quite forgotten little Adèle, have you, reader? I had not; I soon asked and obtained leave of Mr Rochester, to go and see her at the school where he had placed her. Her frantic joy at beholding me again moved me much. She looked pale and thin: she said she was not happy. I found the rules of the establishment were too strict, its course of study too severe, for a child of her age: I took her home with me. I meant to become her governess once more; but I soon found this impracticable: my time and cares were now required by another – my husband needed them all. So I sought out a school conducted on a more indulgent system; and near enough to permit of my visiting her often, and bringing her home sometimes. (p. 450)

This is to exclude rather than include, with a revealing tension between the familial 'home' and the last grudging 'sometimes'. My design was to confront that tension, and, as part of a larger intent, to correct another uncertainty or problem in the novel, Rochester's contradictory account of his relation to Adèle when he tells Jane his story – or, as we come to know, a part of his story:

> 'But unluckily the Varens ... had given me this fillette Adèle; who she affirmed was my daughter: and perhaps she may be; though I see no proofs of such grim paternity written in her countenance ... I acknowledged no natural claim on Adèle's part to be supported by me, nor do I now acknowledge any, for I am not her father; but hearing she was quite destitute, I e'en took the poor thing out of the slime and mud of Paris, and transplanted it here, to grow up clean in the wholesome soil of an English country garden.' (p. 144)

These irreconcilable comments, 'perhaps she may be' and 'I am not her father', may come from an intention to show Rochester's feeling as contemptuous, confused, vacillating, and ashamed, as he indulges in his edited confessional: Brontë does not indicate this and never settles the issue. Adèle may or may not be Rochester's illegitimate child, and, of course, he cannot know the truth: the Rochesters' feelings towards her would almost certainly be affected by their marriage and the birth of their child in ways the novelist chooses not to envisage in her conclusion. So, I felt free – indeed, impelled – to imagine that Edward's feeling for his young son James strengthens his rejection of Adèle, while Jane's experience as a mother strengthens her affection and reinforces her sympathy. Her own home-coming, towards which the original novel slowly but surely leads her and us – and which I think is a major theme in *Jane Eyre* – makes her aware of something she doesn't mention, though the sensitive reader will see it – what she and the girl have in common, parentless dependency, the experience of destitution, and a hard school. This was a parallel I wanted to emphasise, making Jane remember her own life-story as we often do when telling another's:

> 'I do love her. I am not jealous of her mother. I can never resent her. Remember that she is like me, a motherless child, and a child who never knew her father, but for some of her childhood, and unlike me in my childhood, she has not been not quite unloved. However, unlike me, she will never know her true parentage. Long ago, when you talked to me about her history and your life with her mother I took her on my knee and caressed her, most fondly, as I had never been in the habit of doing.'[3]

Charlotte Brontë imagined Jane imagining Adèle's physical and emotional response to loneliness, but I wanted to push her still further:

> 'Jane, you have planted the summer flowers in my garden and in my heart, but I do solitude *à deux* by always bringing that brat here.'

'Edward, it is no longer a solitude *à deux*. We are a family now. Adèle is not a brat. I know very well that when she was eight or nine, when I first came to Thornfield, she was vain and spoilt and frivolous, but you know that there was very good reason for it. And she is much changed. Please to give me credit for some improving and steadying influence. Oh Edward – you are wearing your fierce look. She will only be here for a week or two. I have no desire to interrupt her studies but Miss Dalrymple writes that she is doing well, and will lose nothing by a brief holiday. And how can you say I bring her always? I have always visited her more often than she comes here, but as you know very well, I have been unable to do that for the last months. Now I want Adèle to share a little of our new happiness, and meet her brother.'[4]

In the case of Brontë, I was consciously returning to an earlier critical discussion, in this case in *The Appropriate Form,* where I questioned the novel's constraining imposition of a Providential pattern, inherited by Charlotte Brontë from an English tradition in novel-writing beginning with John Bunyan and Daniel Defoe.[5] In my postscript I am keeping the pattern, but questioning its simplifications by having Jane and Rochester argue about his complacent acceptance and interpretation of a kind and easy Providence and uncomplicated reward. This is how I make Jane scrutinise the matter:

'We have come to think that our freedom to marry was a Providential gift. But perhaps my idea of Providence is not yours. My Providence is merciful but not gentle. I believe we are not allowed to escape entirely the shadow of our past life and actions. That would make Providence too soft in its forgiving. Providence does not bestow a happy ever-after, like an author of novels. It imposes hard discipline. You have ease and happiness, our marriage, our son, your partly restored sight, but Adèle presents you with a pain, a problem, a discord, a demand, not a healing, not a solution, not a harmony, not a gift. It is different for me; it is easier for me.'[6]

I also remembered a smaller problem – apparently an oversight – in Jane's home-coming: Rochester settles in Ferndean after Thornfield is burnt down, and after the marriage he and Jane and their child live in the house he had scrupulously thought too unhealthy for his mad wife Bertha, preferring to install her in the airier attic in Thornfield. In my postscript I took the liberty of making some sanitary improvements to the damp house and grounds of Ferndean.

The characters and destinies of Lucy Snowe and Paulina Bretton, unlike those of Jane Eyre and Adèle Varens, are fixed by Charlotte Brontë in a highly conspicuous structural relation of comparison and contrast, but in my story 'Lucy and Paulina: the Conversation of Women', I have tried to complicate or blur the polarity, to glimpse Lucy's life without Paul Emmanuel, and to imagine something of Lucy's life as teacher and spinster and Paulina's life as wife and mother.

Once again, I did not start with a critical judgment and commentary: I wanted to bring the two complex young women in *Villette* together again, but I developed a critique of the dogmatic antithesis which also involved – I can't honestly say whether this was a cause or an effect – seeing Paulina as more than a conventionally contented wife and mother, but open to self-reflection and social analysis. I have imagined some development for her: she marries when she is still very young and she always shows signs of considerable intelligence and imagination, in her talk and in Lucy's judgment.

Villette has one of the most startling and original conclusions in Victorian fiction. I have always been one of those readers who see that ending as less open than it may appear, for several reasons: the conclusion completes the sea, storm and weather imagery that runs through the novel, disturbingly and subtly merging metaphor and actuality – it is a rhetorical rather than a literal merging, since, of course, the level of 'actuality' is fictional; the narrator is called Lucy Snowe, is still Lucy Snowe; the narrative, tone, and language at the end are faithful to the affective form of serious teasing present at the beginning of the book, in the references to Lucy's early life and its losses, references which are reticences. So, I believe the novelist meant what she made the narrator say, that readers who dislike tragic endings won't have an unhappy ending – Paul Emmanuel's drowning and Lucy's heart-breaking bereavement – forced directly on them. I have left the superficial ambiguity untouched, seeing the so-called open ending as a dark play, an ironic veiled closure. I had not wanted to undermine the providential pattern and conclusion of *Jane Eyre* and I did not want to undermine the brilliantly subtle indirection and narrative wit of *Villette's* ending.

Lucy Snowe has come into her professional inheritance, living in the house her lover and teacher took and furnished for her, and prospering as he had hoped and predicted: it is important that the house where she lives alone is an imaginative donation, a betrothal gift from a man who loved her and whom she loved, whose habitual gifts – flowers, sweet titbits, books – have long graced their friendship. The fact that he did not buy her the house, which she has to rent, severs the link with gross transaction, is choice not purchase, liberation not condescension, emphasising the emotional significance and making the gift purely sacramental. The description of the house Paul chose, and his welcome to Lucy, imagined out of Brontë's loss, love, and wish-fulfilment, is not floridly romantic: its poetry is that of the world at its everyday best, the poetry of work, moderate pleasure, relaxation, air and space. Its objects are functional and ornamental, lightly brushed by the pathetic fallacy. It makes the most beautiful and touching image in *Villette*, and one of the most moving in fiction, a domestic and hospitable love-scene preparing for solitude, but a solitude which is to be occupied and healthy:

> The vestibule was small, like the house ... a French window with vines trained above the panes, tendrils and green leaves kissing the glass... M. Paul disclosed a parlour, or salon ... Its delicate walls were tinged like a blush; its floor was waxed; a square of brilliant carpet ... a little couch ... a lamp ... the recess of the single ample window was filled with a green stand, bearing three green flower-pots, each filled with a fine plant glowing in bloom ... violets in water ... The lattice of this room was open; the outer air breathing through. (p. 485)

The emphasis is placed on the parlour or salon – Lucy's room – but the lovers go through the kitchen, sleeping-rooms and class-room, then eat on the balcony, in rites of home-coming and valediction. (Perhaps the novelist's intuition kept them from eating in the parlour, so that it could not be haunted.)

Starting from the objects, colour, light, air and space Paul and the novelist provide, I needed few additions – a Dickens novel, Schiller and Mozart – to animate and particularise Lucy's school, and the room of her own, as a sacred space enriched by art and nature, not a place of barren or narrow spinsterhood. If my postscript exaggerates the solaces of her survival and solitude, I hope it is a legitimate, if not the only, reading; returning to a scene is in danger of over-interpretation, but I am developing an idea rather than complaining that it is undeveloped.

Like Brontë, I have invoked benign weather for the occasion, and I have invented Paulina's response to Lucy's home – but what else could it be for a woman of her history, intelligence, and sensibility but appreciative and enlightened? I imagined that she expected to feel pity for her solitary and bereaved friend: readers will remember that when she first heard from truth-telling Lucy that she works as a teacher to earn her living, she pities her. In my scene and dialogue, she moves from pity to admiration. Experienced in love, marriage, maternity, and bereavement, she has grown and is growing, and understands something of Lucy's lonelier, but more creative growth. Her feeling for her friend goes back to the night Lucy took the little girl into her own bed to warm and comfort her.

Lucy earns her home-coming; in the novel she ends her narrative sadly, with a bitter irony, but I think may be imagined living on energetically and contentedly – like Jane in her village school – of course lamenting the drowned Paul, the loss of love, the absence of elective affinity, but at work in her own space and home. To put a feminist twist on it, she lives without a husband and a family, in what I call, without anachronism, a room of her own – perhaps the first of its kind in literature.

Once more, I rely on dialogue, on a narrative and speculative conversation which converts, exchanges and changes by imaginative and sympathetic telling and listening. (My interest in forms of 'natural' narrative sharpened my making of this story.[7]) This is Paulina confiding to Lucy, spontaneously, in response to her old friend and new impressions. I have given her speech the sea-imagery prevalent in the first-person narrative of *Villette*:

> 'But now Lucy, I am pleased that yours is not a sad solitude, is it? I asked you if you minded solitude, but I can see for myself. Here we are – I am visiting you on a spring morning and see you have your work and other pleasures, soft breezes, sunlight in a green garden, music and books to lighten the heart. It has not been all tempest and shipwreck for you. Do not look so surprised that I recall your own images. They were strong and vivid.'[8]

> '[T]hese are things I have not thought out in advance – or have only half-thought. I am saying some of these things for the first time. My life – my life with Graham and the children – is the life of a loving, a companionable, and a valuable wife and mother, and in it I feel fulfilled. Not only by its joys and rewards but by its pains, its shared pains, its anxieties, its endurance. But I see also that for many women, with less fortunate marriages, less wealth, less of a true and equal partnership, there may be frustration and loss, faculties rusted, dreams not realised. Even for me, from my safe harbour, there is the sense – and Lucy, this is knowledge, not fancy – that putting out to sea alone, without another firm hand at the helm, can bring its boons and benisons.'[9]

In the novel, Lucy articulates the novel's dominant structure, what in *The Appropriate Form* I called 'dogmatic form', a version of the Providential pattern of *Jane Eyre*. This I have also attempted to modify, through Paulina's developed consciousness. In using Paulina, described by Lucy in Chapter 32 of *Villette* as 'delicate, intelligent, and sincere', to offer her own articulation and modification, I have tried to do so dramatically, through a character I thought capable of more growth and complexity than the constraining antithetical and simple plot-pattern of the original novel allowed or showed:

> 'I pitied you for having to work as a teacher ... to earn your bread. How young and ignorant I was then! I do not envy you your life, Lucy, but I honour it, because I can imagine its powers and its rewards. When you

said just now that you were a working woman, you looked confident, proud, even exalted. You earn your living, every hour of your day and every day of your life, and with all your resources, your busy hands and your mind also. I see that a life like yours, not shaped in the ordinary mould for women of our time, not crowned with the blessing – as it is generally thought a crown and a blessing – of family, a good husband and healthy offspring – such a life can be creative in the way a man's life can be rich and creative. It can even be enviable, in a way, by women like me, who take the conventional path, whose lives are sheltered and cast in the usual mould. But not sheltered from pain and death, Lucy.'[10]

'I give everything I can to my husband and my children, but at times – for example, today, as I think about you and your work and your precious solitude – I know I am not using all my powers. And when you speak of a room of your own I suddenly realise that I am never – ever – alone. It is not that I am discontented or pine for solitude but now and then I seem to glimpse the possibilities of another kind of life. I do not know why I am saying all this. I seem to see my life through your eyes, as they might see mine, or perhaps because I can see – or imagine – your life.'[11]

In these two speeches I have attempted what I hope is an appropriately muted feminism, a recognition and celebration of Brontë's feminism.

I want to mention one aspect of the story which it shares with my postscripts to Austen's *Mansfield Park*, Dickens's *Little Dorrit*, and Eliot's *Middlemarch*: the presence and pressure of memory. In returning to – and in a sense revising – psychological novels with a happy ending, one confronts the characters' memory of past feelings and events that preceded that ending. We do not live only in the present. When talking affectionately but reticently to Paulina, Lucy would inevitably remember her own past love for Graham Bretton, unknown to his wife, and their brief but vital friendship memorialised in his letters, secrets sealed and buried deep in the garden of Madame Beck's school, so I have let this memory rise 'naturally' in Lucy's thought as the two women look out of the window and talk, a sub-text that cannot spoken or shared.

Like George Eliot, Charlotte Brontë was too imaginative a novelist to create minor characters: never trust the artist, trust the tale – but read between the lines. I have written these tales as postscripts, as critique in the form of fiction, and as *hommage*.

Notes

1 Barbara Hardy, *Dorothea's Daughter and Other Nineteenth-Century Postscripts* (Brighton: Victorian Secrets Ltd., 2011).
2 Isobel Armstrong, 'Introduction', in *Form and Feeling in Modern Literature: Essays in Honour of Barbara Hardy*, William Baker and Isobel Armstrong, eds. (London: Legenda, 2013), p. 2.

3 Hardy, *Dorothea's Daughter*, p. 54.
4 Ibid., at p. 55.
5 Barbara Hardy, *The Appropriate Form: An Essay on the Novel* (London: Athlone Press, 1964).
6 Hardy, *Dorothea's Daughter*, p. 55.
7 Barbara Hardy, *Tellers and Listeners: The Narrative Imagination* (London: Athlone Press, 1974).
8 Hardy, *Dorothea's Daughter*, p. 66.
9 Ibid., at p. 67.
10 Ibid., at p. 68.
11 Ibid., at p. 70.

CHAPTER 13

We Are Three Sisters
The Lives of the Brontës as a Chekhovian Play
Blake Morrison

In 1834 Charlotte Brontë wrote as follows to Ellen Nussey, after Ellen had visited London:

> I was greatly amused at the tone of nonchalance which you assumed while treating of London, and its wonders, which seem to have excited anything rather than surprise in your mind: did you not feel awed while gazing at St Paul's and Westminster Abbey? had you no feeling of intense, and ardent interest, when in St James' you saw the Palace, where so many of England's Kings, had held their courts? ... You should not be too much afraid of appearing country-bred, the magnificence of London has drawn exclamations of astonishment from travelled men, experienced in the World, its wonders and its beauties.[1]

We know Charlotte felt more ambivalence towards London than this suggests. Nevertheless, it's hard not to be reminded of the way that Olga, Masha and Irina – Anton Chekhov's three sisters, stuck in the provinces, far from the capital – romanticise and yearn for faraway Moscow. And it's one of many conjunctions, parallels or suggestive overlaps which made me persevere with what might otherwise seem a faintly ludicrous or bonkers enterprise: to rework Chekhov's play *Three Sisters* (1900) so that the three sisters become the Brontë sisters and their grand house in the Russian countryside the parsonage at Haworth.

The idea was first suggested to me by the theatre critic Susannah Clapp, who'd seen a couple of other adaptations I'd done for the theatre company Northern Broadsides and thought this might make another, not least since the company is based in Halifax, quite close to Haworth. Susannah was looking at it from the point of view of a Chekhovian, not a Brontëite; she was thinking simply of the fact that Chekhov's play has three sisters and a wayward brother and wondered whether something could be made of that. But when I began to research the possibilities ten years ago, with Chekhov on one side of my desk and the Brontës' poems and novels along with Juliet Barker's biography on the other, I was encouraged to see how many of the ideas or themes that preoccupied the Brontës also feature in Chekhov's play: work, love, education, marriage, the role of women, the dangers of addiction to drugs or alcohol, the rival claims of country and city, a background of social change and political unrest. And there is evidence that the parallels are no mere coincidence. According to his biographer, Donald Rayfield, one of the books Chekhov ordered for the library of his home town, Taganrog, and which he kept for nearly a month before sending it on, was an account of the Brontës by Olga Peterson (a Russian married to an Englishman, he infers). The fact that Chekhov's dancing teacher at school was a Greek called Vrondi, and in demotic Greek (which Chekhov knew a little) Brontë and Vrondi are virtual homonyms, may have tickled Chekhov's fancy still further. I know nothing of the Peterson biography; perhaps it's just a translation or rip-off of Elizabeth Gaskell's *The Life of Charlotte Brontë* (1857). But in any case, Chekhov wouldn't have had to read a biography to be aware of the Brontës; by the end of the nineteenth century they were internationally famous. It's enough to say that a rough outline of their story might have lodged at the back of his mind.

Encouraged, I began to fill a notebook with ideas for an adaptation and noted further parallels between Chekhov's text and the lives of the Brontës: the presence of a long-suffering and faithful old servant, for instance (Anfisa in Chekhov, Tabby Aykroyd in Haworth) and the damaging effect on the only son or brother in the family of a woman whom the sisters deeply resent (Natasha in Chekhov, Lydia Robinson for the Brontës). But there were also differences and disjunctions so overwhelming that I began to despair. Masha – the sister most obviously resembling Emily – is married; Emily wasn't. Andrei the brother is married and has children; Branwell wasn't and hadn't, or none that we know of for sure. The father in the Chekhov is dead; Patrick survived to a good age, outliving all his children. The Prozorovs' house is vast and full of soldiers; the parsonage was cramped, the Brontës poor and no soldier ever crossed the threshold. There is a duel at the end of Chekhov; there wouldn't have been a duel in Haworth (Patrick, we know, felt strongly about duelling and wrote letters condemning the practice). Most important of all: the three sisters aren't writers; the Brontës were.

Unsurprisingly, perhaps, I set the project aside. But when I came back to it, a couple of years ago, I began to see a way in which the adaptation might work – with the disjunctions embraced and made part of the text rather than denied. So, while the Chekhov play begins with the three sisters thinking back to the death of their father; in my version

Charlotte – the only one old enough to remember – recalls her mother's death and there's reference to the deaths of Elizabeth and Maria. In the Chekhov play, it's the Saint's Day of the youngest sister, Irena; so, in our version it's Anne's birthday and she is looking back at her diary paper from four years previously and writing a new one. The girls discuss work: Irena longs to do something useful, or rather Anne does, and talks about that idea they had for starting a school – could they not revive it? Enter, after much anticipation, Vershinin, the charismatic and flirtatious lovesick major. There was no lovesick major in the lives of the Brontës, but there was for a short time a lovesick curate, the flirtatious William Weightman, whom they found ridiculous yet also warmed to. He tells the three sisters how marvellous he thinks London is but how attractive the local Yorkshire landscapes are too. He's just puzzled by the remoteness of the nearest station – as those coming to Haworth were, since the line stopped at Keighley. In Chekhov there are three or more other soldiers onstage in the first act. Here they become a doctor, clearly well-known to the Brontës, and Patrick Brontë, the girls' father, who banter away in the background about guns – Patrick, we know, used to keep a pistol by his bed, and discharge it every morning, so the image of a shot ringing out – crucial to the denouement of the Chekhov – is worked in right at the start. As to soldiers, well Anne tells Weightman, 'This house was once full of soldiers, you know', explaining that when they were children they used to play with toy ones and make up stories about them.

At this point the action has barely begun, but I hope it's clear already how the stuff of the Brontës' lives make sense in the context of Chekhov's play. Of course, Charlotte liked to pretend that no action ever really took place in their lives, speaking of 'the torpid retirement where we lie like dormice'.[2] But it seems to me that the years 1846–1848 were active – and dramatic – in the extreme, with the publication of Charlotte, Emily and Anne's first novels, the fallout from Branwell's affair with Lydia Robinson wrecking his life and Patrick's cataract operation for his eyes. Of course William Weightman was long dead by then; I've allowed myself some licence with chronology. Chekhov's play stretches over four years; in the adaptation, the action takes place over a matter of a few weeks in early 1848, with Chartist riots brewing in the background, and the deaths of Branwell, Emily and Anne looming ahead.

Mrs. Robinson – as even the name suggests – was too much of a gift to leave out and in the play she assumes the role of Natasha, Andrei's wife, whose vulgar bright-green dress so offends the sisters when she first appears and who, in effect, takes over the house, literally dislodging them from their rooms. Mrs. Robinson would never have visited Haworth, of course, and that's the cheekiest of all the liberties I've taken in the play. A film version wouldn't have required this, but onstage, with the action all confined to one set, the Parsonage, there's no alternative. So the unlikelihood of it becomes a sort of joke. 'We never thought to see you here,' Anne says, when Mrs. R arrives and when Mrs. R explains she just called in on her way to Scarborough, Charlotte responds, 'You'll not get there from here'. Rebuffed, Mrs. R complains what an ugly little place Haworth is when she'd been expecting something more like Harrogate. The way she outstays her welcome may be comic, but the effect on Branwell of her departure (and of his dismissal from the post of tutor to her family) makes for tragedy.

Chekhov's play is a tragic-comedy. That's the tone I've tried to strike too. I was keen to disperse the gloom surrounding the Brontës, to challenge the stereotypes of them as repressed, miserable and unworldly, and to show Haworth not as a desolate, dead-end spot cut off from the world, but as a town or village to which educated people came and where enlightened ideas penetrated. So, there's a reference from Anne to hearing Paganini play in Halifax and to the well-stocked shelves of Keighley library. And we see Patrick campaigning for better sanitation, education and living conditions for the poor. And there's humour and light as well death and darkness.

The ending was a problem. In the Chekhov, Toozenbach, the Baron, whom the youngest sister Irena has agreed to marry even though she doesn't love him, dies in a duel. Even with the equivalent of the Baron introduced into the play – a complete invention this – as a doctor who loves Anne but is rejected by her, and even with Patrick's pistol present as a prop, the remoteness of that death by duelling from the reality of the Brontës' lives seemed too much to accommodate. I toyed with an open ending: a shot rings out and there are three possibilities for who is dead – Branwell, the doctor or the lovesick curate. This open ending was still in the text at a read-through at the Baptist chapel in Haworth in July, just a few weeks before rehearsals began. The audience – composed in part of Brontë enthusiasts and professionals, including Juliet Barker – all inferred it was Branwell who had died. But I didn't feel I could allow that ending to stand. So the text was changed so that the gunshot becomes a red herring and the actual circumstances leading up to Branwell's death – his collapse in the street – are now acknowledged.

The way I changed the ending is indicative of the process generally as I worked through different drafts, seven in all. The more I went on, the more it mattered to be true to the Brontës' lives and to alter the text accordingly – with the Chekhov used as a template, or launch pad, rather than slavishly adhered to. So, for example, in Act 2 of his play, some musicians arrive, but are turned away by Natasha because she thinks they'll wake the children. The Brontës would surely never have invited musicians to turn up at the Parsonage, least of all at night. But Mrs. Robinson might, to provide some entertainment and stop their annoying habit of spending all their time reading and writing. So I've reversed the premise here: she encourages them to come and it's the three sisters who send them packing. Likewise in Act 3 of Chekhov there's a fire in town, with the local population much affected. There's no record of a fire in Haworth. For a time I toyed with having a gas explosion – there was one in 1857, twelve years after gas lighting had been introduced in the town. But in the end I went for the bog burst or mud-slide instead – an event that occurred much earlier than 1848, when the sisters were children, but which did wreak havoc.

Throughout the play I've accommodated lines both from the novels and the letters. 'Do you think I'm an automaton, that I have no feelings' comes from *Jane Eyre*, but here Charlotte uses it in an exchange with Anne. The letters are perhaps the resource I've exploited more often, because in them it's the sisters speaking directly, rather than one of their characters speaking – artistic freedom's well and good, but I think it's unsafe to equate what fictional characters say with what their creators thought and believed. The Chekhov play ends with the three sisters stoically trying to make sense of life and go on with it in the shadow of death – an opportunity, I thought, to include Anne's moving words in a letter, shortly before her death, 'If I knew I was dying, I think I could resign myself to it. Only ... I long to do some good in the world before I leave it. I've all these schemes in my head and I'd hate them to come to nothing'.[3] Most of the letters are Charlotte's, of course, so she is the sister who gets most of her own lines, as it were. Famously, she described the visit to London with Anne, to meet their publishers and reveal who they really were – three women called Brontë, not three men called Bell. That episode not only gives the play its title, but makes such a powerful scene that I added a new, short extra act in which Charlotte and Anne return from London and tell Emily of all that took place (the lines shared out between them) as they would have done – with Emily's fury at being betrayed being an integral part.

CHARLOTTE: Smith & Elder's is also a bookshop. So we went to the counter and asked for Mr. Smith. The lad seemed surprised – two frumpy women with northern accents wanting his boss – but he went to fetch him. Then he came back on his own saying Mr. Smith wanted to know our names.
ANNE: We could have said Bell but all we said was it was private. So the lad went off again. Then this tall youngish gentleman appeared from the back of the shop. 'Did you wish to see me?' he said. 'Is it Mr. Smith?' Charlotte said. 'It is indeed,' he said. Then Charlotte handed him his letter.
CHARLOTTE: The one he'd just sent me, addressed to Currer Bell.
ANNE: 'Where'd you get this?' he said, and stared at us, baffled, and looked at the letter, and looked at us, and then it dawned on him and we all smiled.
EMILY: *What* dawned? What did you tell him?
CHARLOTTE: We didn't say anything … not then.
ANNE: But we knew he knew. He took us to a back room, and introduced us to Mr. Williams, who's older, about fifty, and we chatted, or *they* chatted, and Mr. Smith said we must stay a few days and meet his two sisters, and if Mr. Thackeray was in town maybe we'd like to meet him too, because he'd read our books and admired us and…
EMILY: But what did you say about our names?
ANNE: Well, Charlotte stopped him then, you see, and went all serious, and told him straight.
CHARLOTTE: 'We are three sisters,' I said.
EMILY: Three! So he knows there's no Ellis. How could you?
CHARLOTTE: I'm sorry. It just slipped out.

The other scene which I felt could be imported almost word for word was that in which Charlotte tells her father – in the wake of *Jane Eyre*'s astonishing and almost immediate success – that she has written and published a book:

CHARLOTTE: Books. We couldn't live without books.
PATRICK: True enough. We're all readers in this house.
CHARLOTTE: Not just readers. Every night when you're in bed we write.
PATRICK: Oh I know you're always writing letters…
CHARLOTTE: Not just letters. Stories. Poems…
PATRICK: I was a bit of a poet myself, in my youth.
CHARLOTTE: What I'm trying to say is that I've written a book.
PATRICK: Have you, my dear?
CHARLOTTE: Yes, and I'd like you to read it. [*hands him book*]
PATRICK: I can't – it'll be wasted on me.
CHARLOTTE: But your eyesight's good now. You said it yourself.
PATRICK: Even so, I can never make out your writing.
CHARLOTTE: It's not a manuscript. It's printed.

PATRICK: But the cost! How could you afford it?
CHARLOTTE: I didn't pay for it to be printed, a publisher did.
PATRICK: They must be daft. Nobody's heard of you.
CHARLOTTE: I've chosen a different name, Currer Bell, see.
PATRICK: What good will that do? No one's heard of him, either.
CHARLOTTE: No. But the book seems to be selling. I've brought you some reviews.
PATRICK: All right, I'll try a page or two. But you mustn't forget your duties. Books can't be the business of a woman's life.

The last line comes from Robert Southey in his response to the poems Charlotte had sent him. But most of the rest is straight from Gaskell. And it was Gaskell who provided another haunting image which it was impossible not to want to use in the production – the image of the three sisters circling a table at night after their father has gone to bed, reading from and discussing their work.

In this chapter I've concentrated on my own role in preparing for the production. But ultimately the role of the writer or adaptor is a limited one: it's the director, actors, lighting technicians and set designers who bring the play to life. In fact, I was extremely lucky in having a designer, Jessica Worrall, who has been a Brontë fan since childhood and didn't just come up with a brilliant set – into which the graves of Haworth churchyard, and the pillar portrait of the sisters (with Branwell airbrushed out) find their way – but who also had many useful suggestions to make about the text. As to the actors, well, they all play their part, but it's when the three sisters are alone together onstage that the real chemistry takes place. Which is as it should be.

For me writing the play was a kind of homecoming as I grew up within striking distance of Haworth in an old rectory at the top of a village with a view out onto moors – and with an Irish parent. And even before this play, I'd also had some history with the Brontës, the composer Howard Goodall having approached me with the idea of collaborating on a musical of *Wuthering Heights* back in the 1980s. (Leicester Haymarket Theatre seemed interested in staging it. But there were five musical versions of *Wuthering Heights* doing the rounds at the time and, in the end, it was Tim Rice's *Heathcliff*, starring Cliff Richard, that prevailed.) I'm conscious of how many theatric, filmic, operatic, choreographic and televisual adaptations there have been, not just of the Brontës' novels, but of their lives. But I hope that by finding this unusual route into the Brontës, via a play written by a Russian in 1904, audiences might be moved to learn things about them they didn't know, or if not learn, then at least be entertained.

Notes

1 Charlotte Brontë to Ellen Nussey, Haworth, 20 February 1834, in Juliet Barker, *The Brontës: A Life in Letters* (London: Penguin, 1997), p. 28.
2 Charlotte Brontë to William Smith Williams, Haworth, 21 December 1847, in Barker, p. 174.
3 This paraphrases Anne's words to Ellen Nussey, 5 April 1849, in Barker, p. 228.

Bibliography

Agamben, Giorgio, *Homo Sacer: Sovereign Power and Bare Life* (1995), Daniel Heller-Roazen, trans. (Stanford: Stanford University Press, 1998).
Alexander, Christine and Jane Sellars, *The Art of the Brontës* (Cambridge: Cambridge University Press, 1995).
Allott, Miriam, ed., *The Brontes: The Critical Heritage* (London: Routledge, 1974).
Altholz, Josef, 'The Warfare of Conscience with Theology', in *The Mind and Art of Victorian England,* Josef Altholz, ed. (Minneapolis: University of Minnesota Press, 1976), pp. 58–77.
Anon., 'On Hypochondriasis', *Journal of Psychological Medicine and Mental Pathology*, 3 (1850), 1–14.
  'The Passions', review of scientific works, *Journal of Psychological Medicine and Mental Pathology*, 3 (1850), 141–64.
  'Vivisection', *Times*, August 19, 1843, 7.
Arata, Stephen, *Fictions of Loss in the Victorian Fin de Siècle* (Cambridge: Cambridge University Press, 1996).
Arendt, Hannah, *Men in Dark Times* (London: Jonathan Cape, 1970).
Armstrong, Isobel, *Charlotte Brontë's City of Glass*, The Hilda Hulme Lecture, 1992 (London: University of London, 1993).
  'Introduction', in *Form and Feeling in Modern Literature: Essays in Honour of Barbara Hardy*, William Baker and Isobel Armstrong, eds. (London: Legenda, 2013).
  Novel Politics: Democratic Imaginations in Nineteenth-Century Fiction (Oxford: Oxford University Press, 2016).
Austen, Jane, *Emma*, Richard Cronin and Dorothy McMillan, eds. (Cambridge: Cambridge University Press, 2005).
Bain, Alexander, *The Senses and the Intellect* (London: John W. Parker and Son, 1855).
Bakhtin, M.M., *The Dialogic Imagination: Four Essays*, Carl Emerson and Michael Holquist, trans. (Austin: University of Texas Press, 1981).
Baldick, Chris, ed., *The Oxford Book of Gothic Tales* (Oxford: Oxford University Press, 1992).
Barker, Juliet, *The Brontës* (London: Weidenfeld and Nicolson; New York: St. Martin's Press, 1994).

The Brontës: A Life in Letters (London: Penguin, 1997).
The Brontës: Wild Genius on the Moors (2010; New York & London: Pegasus Books, 2013).
Barlow, John, *On Man's Power Over Himself to Prevent or Control Insanity* (1843), 2nd edn. (London: William Pickering, 1849).
Barthes, Roland, 'The Death of the Author,' in *Image Music Text*, Stephen Heath, ed. (London: Fontana Press, 1977), pp. 142–8.
Bate, Jonathan, *Romantic Ecology: Wordsworth and the Environmental Tradition* (London: Routledge, 1991).
Bayley, John, 'Kitchen Devil' [review of Frank, *Chainless Soul*], *London Review of Books*, 12:24 (20 December 1990), p. 16.
Beaumont, Matthew, 'Heathcliff's Great Hunger: The Cannibal Other in *Wuthering Heights*', *Journal of Victorian Culture*, 9:2 (2004), 137–163.
Beer, Gillian, *Open Fields: Science in Cultural Encounter* (Oxford: Clarendon, 1996).
Beer, John, *Providence and Love: Studies in Wordsworth, Channing, Myers, George Eliot, and Ruskin* (Oxford: Clarendon Press, 1998).
Beirne, Piers, *Hogarth's Art of Animal Cruelty: Satire, Suffering and Pictorial Propaganda* (Basingstoke: Palgrave Macmillan, 2015).
Benckhuysen, Amanda, 'The Prophetic Voice of Christina Rossetti', in *Recovering Nineteenth-Century Women Interpreters of the Bible*, Christiana de Groot and Marian Ann Taylor, eds. (Atlanta: Society of Biblical Literature, 2007), pp. 165–80.
Benjamin, Walter, 'The Storyteller: Reflections on the Work of Nikolai Leskov', in *Illuminations*, Hannah Arendt, ed., Harry Zohn, trans. (New York: Schocken, 1968), pp. 83–109.
Berg, Maggie, 'Hapless Dependents: Women and Animals in Anne Brontë's *Agnes Grey*', *Studies in the Novel*, 34:2 (2002), 177–97.
'"Let me have its bowels then": Silence, Sacrificial Structure, and Anne Brontë's *The Tenant of Wildfell Hall*', *LIT: Literature Interpretation Theory*, 21:1 (2010), 20–40.
Berman, Carolyn Vallenga, *Creole Crossings: Domestic Fiction and the Reform of Colonial Slavery* (Ithaca, NY: Cornell University Press, 2006).
Berry, Elizabeth Hollis, *Anne Brontë's Radical Vision: Structures of Consciousness* (Victoria: University of Victoria, English Literary Studies, 1994).
Bersani, Leo, 'Desire and Metamorphosis', in *A Future for Astyanax: Character and Desire in Literature* (London: Marion Boyars, 1978), pp. 189–219.
Blasing, Mutlu Konuk, *Lyric Poetry: The Pain and the Pleasure of Words* (Princeton: Princeton University Press, 2007).
Blouet, Olwyn, 'Slavery and Freedom in the British West Indies, 1823–1833: The Role of Education', *History of Education Quarterly*, 30:4 (1990), 625–43.
Blumberg, Ilana, *Victorian Sacrifice: Ethics and Economics in Mid-Century Novels* (Columbus: Ohio State University Press, 2013).
Bourke, Joanna, *What it Means to be Human: Reflections from 1791 to the Present* (London: Virago, 2011).

Bowen, John, 'The Brontës and the Transformations of Romanticism', in *The Nineteenth-Century Novel, 1820–1880, Volume 3 of The Oxford History of the Novel in English*, John Kucich and Jenny Bourne Taylor, eds. (Oxford: Oxford University Press, 2011), pp. 203–19.
Boxall, Peter, 'The Limits of the Human', in *Twenty-First-Century Fiction: A Critical Introduction* (Cambridge: Cambridge University Press, 2013), pp. 84–122.
Brantlinger, Patrick, *Taming Cannibals: Race and the Victorians* (Ithaca, NY: Cornell University Press, 2011).
Bristow, Joseph, 'Reforming Victorian Poetry: Poetics after 1832', in *The Cambridge Companion to Victorian Poetry*, Joseph Bristow, ed. (Cambridge University Press, 2000), pp. 1–24.
Brontë, Charlotte, *An Edition of the Early Writings of Charlotte Brontë*, Christine Alexander, ed., 2 vols. (Oxford: Basil Blackwell, 1987–91).
  The Letters of Charlotte Brontë with a selection of letters by her family and friends, Margaret Smith, ed., 3 vols. (Oxford: Oxford University Press, 1995–2004).
Brontë, Charlotte and Emily Brontë, *The Belgian Essays*, Sue Lonoff, ed. and trans. (New Haven, CT: Yale University Press, 1997).
Brontë, Charlotte, Emily Brontë and Anne Brontë, *Poems by Currer, Ellis, and Acton Bell* (London: Aylott and Jones, 1846).
Brontë, Emily, *Emily Brontë: The Complete Poems*, Janet Gezari, ed. (London: Penguin, 1992).
  Five Essays Written in French, trans. Lorine White Nagel, Fannie E. Ratchford, ed. (Austin: University of Texas Press, 1948).
  The Poems of Emily Brontë, Derek Roper and Edward Chitham, eds. (Oxford: Clarendon Press, 1995).
Brooke, John Hedley, *Science and Religion: Some Historical Perspectives* (Cambridge: Cambridge University Press, 1991).
Brown, Laura, *Homeless Dogs and Melancholy Apes: Humans and Other Animals in the Modern Literary Imagination* (Ithaca, NY: Cornell University Press, 2010).
Brown, Thomas, *Lectures on the Philosophy of the Human Mind*, 4 vols. (Edinburgh: W. and C. Tait; London: Longman, Hurst, Rees, Orme, and Brown, 1820).
Burrows, George Man, *Commentaries on the Causes, Forms, Symptoms, and Treatment, Moral and Medical, of Insanity* (London: Thomas and George Underwood, 1828; repr. New York: Arno Press, 1976).
Byron, George Gordon, Lord, *The Complete Poetical Works of Lord Byron*, Paul Elmer More, ed. (Boston, MA: Houghton Mifflin, 1905).
Byron, Glennis, *Dramatic Monologue* (London: Routledge, 2003).
Carlisle, Janice, 'The Face in the Mirror: *Villette* and the Conventions of Autobiography', *ELH*, 46 (1979), 262–89.
Carlyle, Thomas, *On Heroes, Hero-Worship and the Heroic in History* (London: Chapel & Hall Ltd, 1897).

Carnell, Rachel, 'Feminism and the Public Sphere in Anne Brontë's *The Tenant of Wildfell Hall*', *Nineteenth-Century Literature*, 53:1 (1998), 1–24.
Carpenter, William Benjamin, *Principles of Mental Physiology* (1874), 4th edn. (London: Henry S. King & Co., 1876; repr. 1877).
Carroll, Joseph, 'The Cuckoo's History: Human Nature in *Wuthering Heights*', *Philosophy and Literature*, 32:2 (2008), 241–57.
Chadwick, Ellis H., *In the Footsteps of the Brontës* (London: Isaac Pitman, 1914).
Channing, William Ellery, *Self-Culture* (London: 1838).
  Slavery (London: Hunter, 1836).
Chez, Keridiana, *Victorian Dogs, Victorian Men: Affect and Animals in Nineteenth-Century Literature and Culture* (Columbus: Ohio State University Press, 2017).
Chitham, Edward, *The Birth of Wuthering Heights: Emily Brontë at Work* (Basingstoke: Palgrave, 1998).
Clarke, Micael M., 'Emily Brontë's "No Coward Soul" and the Need for a Religious Literary Criticism', *Victorians Institute Journal*, 37 (2009), 195–223.
Cobbe, Frances Power, *Darwinism in Morals and Other Essays* (London: Williams and Norgate, 1872).
Cohen, William A., *Embodied: Victorian Literature and the Senses* (Minneapolis: University of Minnesota Press, 2008).
Coleridge, Samuel Taylor, *Biographia Literaria*, 2 vols. (New York: Kirk and Mercein, 1817).
  The Poetical Works of Samuel Taylor Coleridge, James Dykes Campbell, ed. (London: Macmillan, 1905).
Collini, Stefan, *Public Moralists: Political Thought and Intellectual Life in Britain, 1850–1930* (Oxford: Clarendon Press, 1991).
Conolly, John, *An Inquiry Concerning the Indications of Insanity* (London: John Taylor, 1830; repr. London: Dawsons, 1964).
Cooper, Isabella, 'The Sinister Menagerie: Animality and Antipathy in *Wuthering Heights*', *Bronte Studies*, 40:1 (2015), 252–62.
Cormie, Lee, 'The Hermeneutical Privilege of the Oppressed: Liberation Theologies, Biblical Faith, and Marxist Sociology of Knowledge', *Proceedings of the Annual Convention of the Catholic Theological Society of America*, 32 (1978), 155–81.
Cosslett, Tess, *Talking Animals in British Children's Fiction* (Aldershot: Ashgate, 2006).
Cowper, William, *Life and Works of William Cowper*, Revd. T. S. Grimshawe, ed. (London: Saunders and Otley, 1836).
  The Poems of William Cowper, 3 vols., John D. Baird and Charles Ryskamp, eds. (Oxford: Clarendon Press, 1980–95).
Cunningham, Valentine, *In the Reading Gaol: Postmodernity, Texts, and History* (Oxford: Blackwell, 1994).
Curran, Stuart, 'Romantic Poetry: The I Altered', in *Romanticism and Feminism*, Anne K. Mellor, ed. (Bloomington: Indiana University Press, 1988), pp. 185–203.

Dallas, E. S., *The Gay Science*, 2 vols. (London: Chapman and Hall, 1866; New York: Johnson Reprint Corporation, 1969).
Dart, Gregory, *Rousseau, Robespierre, and English Romanticism* (Cambridge: Cambridge University Press, 1999).
Davies, Stevie, *Emily Brontë: Heretic* (London: Women's Press, 1994).
Davion, Victoria, 'Is Ecofeminism Feminist?', in *Ecological Feminism*, Karen J. Warren, ed. (London: Routledge, 1994), pp. 8–29.
De Bolla, Peter, *The Architecture of Concepts: The Historical Formation of Human Rights* (New York: Fordham University Press, 2013).
De Quincey, Thomas, 'Suspiria de Profundis: Being a Sequel to the Confessions of an English Opium-Eater', *Blackwood's Magazine*, 57 (June 1845), 739–51.
Derrida, Jacques, 'The Animal That Therefore I Am (More to Follow)', David Wills, trans. *Critical Inquiry*, 28:2 (2002), 369–418.
Dickens, Charles, *Our Mutual Friend* (1864–1865) (Harmondsworth: Penguin, 1985).
Dolan, Elizabeth A., *Seeing Suffering in Women's Literature of the Romantic Period* (Aldershot: Ashgate, 2008).
Douglas, James, *On the Philosophy of the Mind* (Edinburgh: Adam and Charles Black; London: Longman, Orme, Brown, Green, & Longmans, 1839).
Drummond, William Hamilton, *The Rights of Animals, and Man's Obligation to Treat them with Humanity* (London: John Mardon, 1838).
Dumas, Paula E., *Proslavery Britain* (London: Palgrave Macmillan, 2016).
Eisenman, Stephen F., *The Cry of Nature: Art and the Making of Animal Rights* (London: Reaktion Books, 2013).
Eliot, George, *Daniel Deronda* (1876) (Harmondsworth: Penguin, 1986).
The Mill on the Floss (1860) (Harmondsworth: Penguin, 1985).
Elliotson, John, *Human Physiology* (London: Longman, Rees, Orme, Brown, Green and Longman, 1835).
Elliott, Mary, *The History of a Goldfinch* (London: W. and T. Darton, 1807).
The History of a Goldfinch: Addressed to those Children who are Dutiful to their Parents, and Humane to their Fellow Creatures (Philadelphia: B. and T. Kite, 1807).
Ellis, Sarah Stickney, *Mothers of England, their Influence and Responsibility* (London: Fisher, Son and Co., 1843).
Evans, G. R., *Augustine on Evil* (Cambridge: Cambridge University Press, 1982).
Ferguson, Christine, *Language, Science and Popular Fiction in the Victorian Fin-de-Siècle: The Brutal Tongue* (Aldershot: Ashgate, 2006).
Ferguson, Moira, *Animal Advocacy and Englishwomen, 1780–1900: Patriots, Nation and Empire* (Ann Arbor: University of Michigan Press, 1998).
Fermi, Sarah, 'A Question of Colour', *Brontë Studies*, 40:4 (2015), 334–42.
Fiddes, Paul S., *The Promised End: Eschatology in Theology and Literature* (Oxford: Blackwell, 2000).
Fike, Francis, 'Bitter Herbs and Wholesome Medicines: Love as Theological Affirmation in *Wuthering Heights*', *Nineteenth-Century Fiction*, 23 (1968), 127–49.

Frank, Katherine, *Emily Brontë: A Chainless Soul* (London: Hamish Hamilton, 1990).
French, Richard, *Antivivisection and Medical Science in Victorian Society* (Princeton: Princeton University Press, 1975).
Gallant, Christine, 'The Archetypal Feminine in Emily Brontë's Poetry', in *Victorian Women Poets,* Tess Cosslett, ed. (London: Longman, 1996), pp. 44–54.
Gaskell, Elizabeth, *The Life of Charlotte Brontë (1857)*, Angus Easson, ed. (Oxford: Oxford University Press, 1996).
Gazzaniga, Michael S., *The Ethical Brain* (New York: Dana Press, 2005).
Gezari, Janet, *Last Things: Emily Brontë's Poems* (Oxford: Oxford University Press, 2007).
Gigerenzer, Gerd and Daniel G. Goldstein, 'Reasoning the fast and frugal way: Models of bounded rationality', *Psychological Review,* 103 (1996), 650–69.
Gilbert, Sandra and Susan Gubar, *The Madwoman in the Attic: The Woman Writer and the Nineteenth-Century Literary Imagination* (New Haven, CT: Yale University Press, 1979).
Gladwell, Malcolm, *Blink: The Power of Thinking Without Thinking* (New York: Little, Brown and Company, 2005).
Glen, Heather, *Charlotte Brontë: The Imagination in History* (Oxford: Oxford University Press, 2002).
Goff, Barbara Munson, 'Between Natural Theology and Natural Selection: Breeding the Human Animal in *Wuthering Heights*', *Victorian Studies,* 27:4 (1984), 477–508.
Graham, Thomas John, *Modern Domestic Medicine* (London: Simpkin and Marshall et al., 1826).
Greg, William Rathbone, 'Why are Women Redundant?', *National Review,* 14 (April 1862), 434–60.
Hagan, Sandra and Juliette Wells, eds., *The Brontës in the World of the Arts* (Aldershot: Ashgate, 2008).
Hague, Angela, 'Charlotte Brontë and Intuitive Consciousness', *Texas Studies in Language and Literature,* 32:4 (1990), 584–601.
Hague, William, *William Wilberforce: The Life of the Great Anti-Slave Trade Campaigner* (New York: Harcourt, 2008).
Hamilton, William, *Lectures on Metaphysics and Logic*, Henry L. Mansel and John Veitch, eds., 2 vols. (Boston, MA: Gould and Lincoln, 1871).
Harari, Yuval Noah, *Sapiens: A Brief History of Humankind* (2011; London: Vintage, 2014).
Hardy, Barbara, *The Appropriate Form: An Essay on the Novel* (London: Athlone Press, 1964).
  Dorothea's Daughter and Other Nineteenth-Century Postscripts (Brighton: Victorian Secrets Ltd., 2011).
  Tellers and Listeners: The Narrative Imagination (London: Athlone Press, 1974).
Harman, Claire, *Charlotte Brontë* (London: Viking, 2015; New York: Alfred A. Knopf, 2016).

Hayles, Katherine N., *How We Became Posthuman: Virtual Bodies in Cybernetics, Literature, and Informatics* (Chicago: University of Chicago Press, 1999).
H. D., *Pilate's Wife*, Joan A. Burke, ed. (New York: New Directions, 2000).
Heilman, Robert, 'Charlotte Brontë's New Gothic', in *Jane Austen to Joseph Conrad: Essays Collected in Memory of James T. Hillhouse*, Robert C. Rathburn and Martin Steinmann, Jr., eds. (Minneapolis: University of Minnesota Press, 1958), pp. 118–32.
Henck, K. Cubie, '"That Peculiar Voice": Jane Eyre and Mary Bosanquet Fletcher, an Early Wesleyan Female Preacher', *Brontë Studies*, 35:1 (2010), 7–22.
Hilton, Boyd, *The Age of Atonement: The Influence of Evangelicalism on Social and Economic Thought 1785–1865* (Oxford: Clarendon Press, 1988).
Holland, Henry, *Chapters on Mental Physiology*, 2nd edn., revised and enlarged (London: Longman, Brown, Green, Longmans, & Roberts, 1858).
Medical Notes and Reflections (1839), 2nd edn., rev. (London: Longman, Orme, Brown, Green, and Longmans, 1840).
Medical Notes and Reflections, 3rd edn. (London: Spottiswoode, 1855).
Homans, Margaret, *Women Writers and Poetic Identity: Dorothy Wordsworth, Emily Brontë, and Emily Dickinson* (Princeton: Princeton University Press, 1980).
Hornosty, Janine, 'Let's Not Have Its Bowels So Quickly Then: A Response to Maggie Berg', *Brontë Studies*, 39:2 (2014), 130–40.
Houston, Gail Turley, *Victorian Women Writers, Radical Grandmothers and the Gendering of God* (Columbus: Ohio State University, 2013).
Hunt, Lynn, *Inventing Human Rights: A History* (New York: Norton, 2007).
Jenkins, Ruth Y., *Reclaiming Myths of Power: Women Writers and the Victorian Spiritual Crisis* (Lewisburg: Bucknell University Press, 1995).
Kay, James Phillips, *The Moral and Physical Condition of the Working Classes Employed in the Cotton Manufacture in Manchester* (London: Ridgway, 2nd edn., 1832).
Kosslyn, Stephen M., *Image and Brain: The Resolution of the Imagery Debate* (Cambridge, Mass.: MIT Press, 1994).
Kreilkamp, Ivan, 'Petted Things: *Wuthering Heights* and the Animal', *Yale Journal of Criticism*, 18:1 (2005), 87–110.
Voice and the Victorian Storyteller (Cambridge: Cambridge University Press, 2005).
Krueger, Christine L., *The Reader's Repentance: Women Preachers, Women Writers and Nineteenth-Century Social Discourse* (London: University of Chicago Press, 1992).
Reading for the Law: British Literary History and Gender Advocacy (Charlottesville: University of Virginia Press, 2010).
Lamonica, Drew, *We Are Three Sisters: Self and Family in the Writing of the Brontës* (Columbia: University of Missouri Press, 2003).
Lamonica, Maria, 'Jane's Crown of Thorns: Feminism and Christianity in *Jane Eyre*', *Studies in the Novel*, 34:3 (2002), 245–63.
Langland, Elizabeth, 'The Voicing of Feminine Desire in Anne Brontë's *The Tenant of Wildfell Hall*', in *Gender and Discourse in Victorian Literature and*

Art, Anthony H. Harrison and Beverly Taylor, eds. (DeKalb, IL: Northern Illinois University Press, 1992), pp. 111–23.

LaPorte, Charles, 'George Eliot, The Poetess as Prophet', *Victorian Literature and Culture*, 31:1 (2003), 159–79.

Larson, Janet L., '"Who Is Speaking?" Charlotte Brontë's Voices of Prophecy', in *Victorian Sages and Cultural Discourse: Renegotiating Gender and Power*, Thais E. Morgan, ed. (New Brunswick, NJ: Rutgers University Press, 1990), pp. 66–86.

Laycock, Thomas, *A Treatise on the Nervous Diseases of Women; Comprising an Inquiry into the Nature, Causes, and Treatment of Spinal and Hysterical Disorders* (London: Longman, Orme, Brown, Green, and Longmans, 1840).

Lecaros, Cecilia Wadsö, *The Governess Novel* (Lund: Lund University Press, 2001).

Lee, Julia Sun-Joo, *The American Slave Narrative and the Victorian Novel* (Oxford: Oxford University Press, 2010).

Lee, Monika Hope, '"A Mother Outlaw Vindicated": Social Critique in *The Tenant of Wildfell Hall*', *Nineteenth-Century Gender Studies*, 4:3 (2008).

Levine, George, *Realism, Ethics and Secularism: Essays on Victorian Literature and Science* (Cambridge: Cambridge University Press, 2008).

Lewes, George Henry, *The Physiology of Common Life*, 2 vols. (Edinburgh: Blackwood and Sons, 1859–1860).

Problems of Life and Mind, 2 vols. (London: Trubner & Co., 1874–1879).

'Recent Novels: French and English', *Fraser's Magazine*, 36 (December 1847), 686–95.

Lewis, Alexandra, 'The Ethics of Appropriation; Or, the "mere spectre" of Jane Eyre: Emma Tennant's *Thornfield Hall*, Jasper Fforde's *The Eyre Affair* and Gail Jones's *Sixty Lights*', in *Charlotte Brontë: Legacies and Afterlives*, Amber Regis and Deborah Wynne, eds. (Manchester: Manchester University Press, 2017), pp. 197–220.

'Memory Possessed: Trauma and Pathologies of Remembrance in Emily Brontë's *Wuthering Heights*', in *Acts of Memory: The Victorians and Beyond*, Ryan Barnett and Serena Trowbridge, eds. (Newcastle upon Tyne: Cambridge Scholars Publishing, 2010), pp. 35–53.

'Psychology and Psychiatry', in *The Blackwell Encyclopedia of Victorian Literature*, 4 vols., Dino F. Felluga, Pamela K. Gilbert and Linda K. Hughes, eds. (Hoboken, NJ: Wiley-Blackwell, 2015).

'Stagnation of Air and Mind: Picturing Trauma and Miasma in Charlotte Brontë's *Villette*', in *Picturing Women's Health*, Francesca Scott, Kate Scarth and Ji Won Chung, eds. (London: Pickering and Chatto, 2014), pp. 59–76.

'"Supposed to be very calm generally": Anger, Narrative, and Unaccountable Sounds in Charlotte Brontë's *Jane Eyre*', in *Feminist Moments: Reading Feminist Texts*, Susan Bruce and Kathy Smits, eds. (London: Bloomsbury, 2016), pp. 67–74.

Locke, John, *Some Thoughts Concerning Education*, J. W. Yolton and J. S. Yolton, eds. (Oxford: Clarendon Press, 1989).

Lonoff, Sue, 'Brontë Scholarship: Retrieval, Criticism, Pedagogy', *Victorian Studies*, 39:1 (1995), 55–63.
  'The Education of Charlotte Brontë: A Pedagogical Case Study', *Pedagogy*, 1:3 (2001), 457–77.
Lutz, Deborah, *The Brontë Cabinet: Three Lives in Nine Objects* (New York: Alfred A. Knopf, 2015).
Madden, William A., '*Wuthering Heights*: The Binding of Passion', *Nineteenth-Century Fiction*, 27:2 (September 1972), 127–54.
Marsden, Simon, *Emily Brontë and the Religious Imagination* (London: Bloomsbury, 2014).
Marsh, Joss, *Word Crimes: Blasphemy, Culture, and Literature in Nineteenth-Century England* (Chicago: University of Chicago Press, 1998).
Martineau, Harriet, *Autobiography*, 2 vols. (London: Virago, 1983).
  Household Education (London: Edward Moxon, 1849).
Mason, Emma, 'The Key to the Brontës? Methodism and *Wuthering Heights*', in *Biblical Religion and the Novel, 1700–2000*, Mark Knight and Thomas Woodman, eds. (Aldershot: Ashgate, 2006), pp. 69–77.
  '"Some god of wild enthusiast's dreams": Emily Brontë's Religious Enthusiasm', *Victorian Literature and Culture*, 31:1 (2003), 263–77.
Mathewes, Charles, *Evil and the Augustinian Tradition* (Port Chester, NY: Cambridge University Press, 2001).
Mazzeno, Laurence W. and Ronald D. Morrison, eds., *Animals in Victorian Literature and Culture* (London: Palgrave Macmillan, 2017).
McMaster, Juliet, 'Imbecile laughter and desperate earnest in *The Tenant of Wildfell Hall*', *Modern Language Quarterly*, 43:4 (1982), 352–68.
Mellor, Anne K., *Mothers of the Nation: Women's Political Writing in England 1780–1830* (Bloomington: Indiana University Press, 2002).
  Romanticism and Gender (London: Routledge, 1993).
Meyer, Susan L., *Imperialism at Home: Race and Victorian Women's Fiction* (Ithaca, NY: Cornell University Press, 1996).
Milbank, Alison, *Daughters of the House: Modes of the Gothic in Victorian Fiction* (London: Macmillan, 1992).
Milbank, John, *Being Reconciled: Ontology and Pardon* (London: Routledge, 2001).
Mill, John Stuart., *Autobiography* (1873; New York: Liberal Arts Press, 1957).
  'On the Subjection of Women' (1869), in *On Liberty and Other Writings*, Stefan Collini, ed. (Cambridge: Cambridge University Press, 1989), pp. 117–218.
  A System of Logic: Ratiocinative and Inductive (Cambridge: Cambridge University Press, 2011).
  'Utilitarianism' (1861), in *Essays on Ethics, Religion, and Society*, J. M. Robson, ed. (Toronto: University of Toronto Press, 1969).
Miller, Andrew H., *The Burdens of Perfection: On Ethics and Reading in Nineteenth-Century British Literature* (Ithaca, NY: Cornell University Press, 2008).
Miller, J. Hillis, *The Disappearance of God: Five Nineteenth-Century Writers*, 3rd edn. (Urbana: University of Illinois Press, 2000).
  Fiction and Repetition: Seven English Novels (Oxford: Basil Blackwell, 1982).

Miller, Lucasta, *The Brontë Myth* (London: Vintage, 2002).
Millingen, J. G., *The Passions; or, Mind and Matter. Illustrated by Considerations on Hereditary Insanity* (London: John and Daniel A. Darling, 1848).
Mitchell, W. J. T., 'Visible Language: Blake's Wond'rous Art of Writing,' in *Romanticism and Contemporary Criticism*, Morris Eaves and Michael Fischer, eds. (Ithaca, NY: Cornell University Press, 1986), pp. 46–95.
More, Hannah, *Moral Sketches of Prevailing Opinions and Manners, Foreign and Domestic: With Reflections on Prayer* (London: Cadell & Davies, 1819).
  Poems (1816), Introd. Caroline Franklin (London: Routledge/Thoemmes Press, 1996).
  Strictures on the Modern System of Female Education, with a View of the Principles and Conduct Prevalent Among Women of Ranks and Fortune, 2 vols. (London, 5th edn., 1799).
Morrison, Blake, *We Are Three Sisters* (Nick Hern Books, 2011).
Morse, Deborah Denenholz, 'Charlotte Brontë' and 'Emily Brontë', in *The Blackwell Encyclopedia of Victorian Literature*, 4 vols., Dino F. Felluga, Pamela K. Gilbert and Linda K. Hughes, eds. (Hoboken, NJ: Wiley-Blackwell, 2015).
  'The Forest Dell, the Attic, and the Moorland: Animal Places in *Jane Eyre*,' in *Time, Space, and Place in Charlotte Brontë*, Diane Long Hoeveler and Deborah Denenholz Morse, eds. (New York: Routledge, 2016).
  '"I Speak of Those I Do Know": Witnessing as Radical Gesture in Anne Brontë's *The Tenant of Wildfell Hall*', in *New Approaches to the Literary Art of Anne Bronte*, Julie Nash and Barbara A. Suess, eds. (Aldershot: Ashgate, 2001), pp. 103–26.
  'The Mark of the Beast: Animals as Sites of Imperial Encounter', in *Victorian Animal Dreams*, Deborah Denenholz Morse and Martin Danahay, eds. (Aldershot: Ashgate Press 2007).
Morse, Deborah Denenholz and Martin A. Dananhay, eds., *Victorian Animal Dreams: Representations of Animals in Victorian Culture and Literature* (Aldershot: Ashgate, 2007).
Mousley, Andy, *Literature and the Human: Criticism, Theory, Practice* (London: Routledge, 2013).
Moyn, Samuel, *Human Rights and the Uses of History* (London: Verso, 2014).
Mushet, David, *The Wrongs of the Animal World. To Which is Subjoined the Speech of Lord Erskine on the Same Subject* (London: Hatchard and Son, 1839).
Neill, Anna, *Primitive Minds: Evolution and Spiritual Experience in the Victorian Novel* (Columbus: Ohio State University Press, 2013).
Oates, Joyce Carol, 'The Magnanimity of *Wuthering Heights*', *Critical Inquiry* 9:2 (1982), 435–49.
Ong, Walter J., *Orality and Literacy* (London: Routledge, 1982).
Otis, Laura, *Rethinking Thought: Inside the Minds of Creative Scientists and Artists* (Oxford: Oxford University Press, 2015).
Parker, Theodore, 'The Public Function of Women', in *Collected Works*, 12 vols., Frances Power Cobbe, ed. (London: Trubner and Co., 1863), VIII, p. 103.

Peacock, Thomas Love, *The Works of Thomas Love Peacock*, H. F. B. Brett-Smith and Clifford Ernest Jones, eds., 10 vols. (London: Constable & Co. Ltd, 1924–34).
Pearce, Lynne, *Romance Writing* (Cambridge: Polity, 2007).
Pemberton, Neil and Michael Worboys, *Rabies in Britain: Dogs, Disease and Culture, 1830–2000* (Basingstoke: Palgrave Macmillan, 2007).
Penner, Louise, '"Not yet settled": Charlotte Brontë's Anti-Materialism', *Nineteenth-Century Gender Studies*, 4:1 (2008), 1–27.
Perkins, David, *Romanticism and Animal Rights* (Cambridge: Cambridge University Press, 2003).
Pickstock, Catherine, *After Writing: On the Liturgical Consummation of Philosophy* (Oxford: Blackwell, 1998).
Repetition and Identity (Oxford: Oxford University Press, 2013).
Pike, Judith, 'Breeching Boys: Milksops, Men's Clubs, and the Modelling of Masculinity in Anne Brontë's *Agnes Grey* and *The Tenant of Wildfell Hall*', *Brontë Studies*, 37:2 (2012), 112–24.
Plasa, Carl, *Charlotte Brontë (Critical Issues)* (Basingstoke: Palgrave Macmillan, 2004).
Textual Politics from Slavery to Postcolonialism: Race and Identity (London: Palgrave Macmillan, 2000).
Poovey, Mary, *Uneven Developments: The Ideological Work of Gender in Mid-Victorian England* (Chicago: Chicago University Press, 1988).
Pratt, Samuel J., *Pity's Gift: A Collection of Interesting Tales to Excite the Compassion of Youth for the Animal Creation*. Selections from Pratt by a Lady (London: printed for T. N. Longman and E. Newbery, 1798).
Prichard, James Cowles, *A Treatise on Insanity and Other Disorders Affecting the Mind* (London: Sherwood, Gilbert, and Piper, 1835).
Rawls, John, *A Theory of Justice* (Cambridge, MA: Harvard University Press, 1971, rev. edn. 1999).
Reardon, Bernard M. G., *Religion in the Age of Romanticism* (Cambridge: Cambridge University Press, 1985).
Reid, John, *Essays on Hypochondriasis, and Other Nervous Affections* (1821), 3rd edn., rev. (London: Longman, Hurst, Rees, Orme, and Brown, 1823).
Rhys, Jean, *Wide Sargasso Sea* (1966) (Harmondsworth: Penguin, 2011).
Richardson, Alan, *The Neural Sublime: Cognitive Theories and Romantic Texts* (Baltimore, MD: Johns Hopkins University Press, 2010).
Ritvo, Harriet, *The Animal Estate: The English and Other Creatures in the Victorian Age* (Cambridge, MA: Harvard University Press, 1987).
Roper, Derek, 'The Revision of Emily Brontë's Poems of 1846', *The Library*, 6th series 6 (1984), 153–67.
Rorty, Amélie Oksenberg, 'A Literary Postscript: Characters, Persons, Selves, Individuals', in *The Identities of Persons*, Rorty, ed. (Berkeley, CA: University of California Press, 1969), pp. 301–23.
Roscoe, William Caldwell, 'Art. VI – Miss Brontë', *The National Review* (July 1857), 127–64.

Ruether, Rosemary Radford, 'Feminist Interpretation: A Method of Correlation', in *Feminist Interpretation of the Bible,* Letty M. Russell, ed. (Oxford: Blackwell, 1985), pp. 111–34.
Ryan, Robert, *The Romantic Reformation: Religious Politics in English Literature 1789–1824* (Cambridge: Cambridge University Press, 1997).
Rylance, Rick, *Victorian Psychology and British Culture, 1850–1880* (Oxford: Oxford University Press, 2000).
Scarry, Elaine, 'Donne: "But yet the body is his booke"', in *Literature and the Body: Essays on Populations and Persons,* Elaine Scarry, ed. (Baltimore, MD: Johns Hopkins University Press), pp. 70–105.
Schaffer, Talia, *Romance's Rival: Familiar Marriage in Victorian Fiction* (New York: Oxford University Press, 2016).
Scheinberg, Cynthia, '"Measure to yourself a prophet's place": Biblical Heroines, Jewish Difference and Women's Poetry', in *Women's Poetry, Late Romantic to Late Victorian: Gender and Genre 1830–1900,* Isobel Armstrong and Virginia Blain, eds. (Basingstoke: Macmillan, 1999), pp. 236–91.
Schmitt, Cannon, *Darwin and the Memory of the Human: Evolution, Savages, and South America* (Cambridge: Cambridge University Press, 2009).
Schramm, Jan-Melissa, *Atonement and Self-Sacrifice in Nineteenth-Century Narrative* (Cambridge: Cambridge University Press, 2012).
  'Towards a Poetics of (Wrongful) Accusation: Innocence and Working-Class Voice in Mid-Victorian Fiction', in *Fictions of Knowledge: Fact, Evidence, Doubt,* Yota Batsaki, Subha Mukherji, and Jan-Melissa Schramm, eds. (Basingstoke: Macmillan, 2011), pp. 193–212.
Sellars, Jane, 'Art and the Artist as Heroine in the Novels of Charlotte, Emily, and Anne Brontë', *Brontë Studies*, 20:2 (1990), 57–76.
Senf, Carol, '*The Tenant of Wildfell Hall*: Narrative Silences and Questions of Gender', *College English*, 52:4 (April 1990), 446–456.
Shelley, Percy Bysshe, *The Complete Poetical Works of Percy Bysshe Shelley,* Thomas Hutchinson, ed. (Oxford: Oxford University Press, 1905).
Sherwood, Yvonne, *Biblical Blaspheming: The Trials of the Sacred in a Secular Age* (Cambridge: Cambridge University Press, 2012).
Shires, Linda, 'Victorian Women's Poetry', *Victorian Literature and Culture*, 27:2 (1999), 601–9.
Shuttleworth, Sally, *Charlotte Brontë and Victorian Psychology* (Cambridge: Cambridge University Press, 1996).
Simon, Bart, 'Introduction: Toward a Critique of Posthuman Futures', *Cultural Critique*, 53 (Winter 2003), 1–9.
Sinclair, May, *The Three Brontës* (London: Hutchinson, 1912).
Slaughter, Joseph R., *Human Rights, Inc: The World Novel, Narrative Form, and International Law* (New York: Fordham University Press, 2007).
Smith, Adam, *The Theory of Moral Sentiments* (1759, rev. 1790), Knud Haakonssen, ed. (Cambridge: Cambridge University Press, 2002).
Snodgrass, Mary Ellen, ed., *Encyclopedia of Feminist Literature* (New York: Infobase Publishing, 2006).

Soper, Kate, *What is Nature? Culture, Politics and the Non-Human* (Oxford: Blackwell, 2004).
Spivak, Gayatri, '*Jane Eyre* and Three Texts of Imperialism', *Critical Inquiry*, 12:1 (1985), 243–61.
Spolsky, Ellen, *Gaps in Nature: Literary Interpretation and the Modular Mind* (Albany: State University of New York Press, 1993).
Spurgeon, Caroline F. E., *Mysticism in English Literature* (Cambridge: Cambridge University Press, 1913).
Sterne, Jonathan, *The Audible Past: Cultural Origins of Sound Reproduction* (Durham: Duke University Press, 2003).
Stevens, Jane, ed., *Mary Taylor: Friend of Charlotte Brontë: Letters from New Zealand & Elsewhere* (Oxford: Oxford University Press, 1972).
Stewart, Garrett, *Reading Voices: Literature and the Phonotext* (Berkeley: University of California Press, 1990).
Stoneman, Patsy, *Jane Eyre on Stage, 1848–1898: An Illustrated Edition of Eight Plays with Contextual Notes* (Aldershot: Ashgate, 2007).
Styler, Rebecca, *Literary Theology by Women Writers of the Nineteenth Century* (Farnham: Ashgate, 2010).
Styles, Rev. John, *The Animal Creation: Its Claims on our Humanity, Stated and Enforced* (London: Thomas Ward and Co, 1839).
Surridge, Lisa, 'Animals and Violence in *Wuthering Heights*', *Brontë Society Transactions*, 24:2 (1999), 161–73.
   Bleak Houses: Marital Violence in Victorian Fiction (Athens, OH: Ohio University Press, 2005).
Sutherland, John, *Lives of the Novelists: A History of Fiction in 294 Lives* (New Haven, CT: Yale University Press, 2012).
Swaminathan, Srividhya, *Debating the Slave Trade: Rhetoric of British National Identity, 1759–1815* (Aldershot: Ashgate, 2009).
Talley, Lee A., 'Anne Bronte's Method of Social Protest in *The Tenant of Wildfell Hall*', in *New Approaches to the Literary Art of Anne Bronte,* Julie Nash and Barbara A. Suess, eds. (Aldershot: Ashgate, 2001), pp. 127–51.
Tayler, Irene, *Holy Ghosts: The Male Muses of Emily and Charlotte Brontë* (New York: Columbia University Press, 1990).
Taylor, Beverly, 'Race, Slavery, and the Slave Trade', in *A Companion to the Brontës*, Diane Long Hoeveler and Deborah Denenholz Morse, eds. (London: Blackwell, 2016), pp. 339–53.
Taylor, Charles, *Sources of the Self: The Making of the Modern Identity* (Cambridge, MA: Harvard University Press, 1989).
Taylor, Jenny Bourne, 'Forms and Fallacies of Memory in 19th-Century Psychology: Henry Holland, William Carpenter and Frances Power Cobbe', *Endeavour*, 23:2 (1999), 60–4.
Taylor, Jenny Bourne and Sally Shuttleworth, eds., *Embodied Selves: An Anthology of Psychological Texts, 1830–1890* (Oxford: Clarendon Press, 1998).
Thomas, Sue, *Imperialism, Reform, and the Making of Englishness in Jane Eyre* (Basingstoke: Palgrave Macmillan, 2008).

Thormählen, Marianne, *The Brontës and Education* (Cambridge: Cambridge University Press, 2007).
  The Brontës and Religion (Cambridge: Cambridge University Press, 1999).
  ed., *The Brontës in Context* (Cambridge: Cambridge University Press, 2012).
Torgerson, Beth, *Reading the Brontë Body: Disease, Desire and the Constraints of Culture* (New York: Palgrave Macmillan, 2005).
Tuer, Andrew W., *Stories from Old-Fashioned Children's Books* (London: Leadenhall Press, 1899–1900).
Turner, James, *Reckoning with the Beast: Animals, Pain, and Humanity in the Victorian Mind* (Baltimore: Johns Hopkins University Press, 1980).
Veith, Ilza, *Hysteria: The History of a Disease* (Chicago: University of Chicago Press, 1965).
Vincent, David, *Literacy and Popular Culture: England 1750–1914* (Cambridge: Cambridge University Press, 1989).
Von Schneidern, Maja-Lisa, '*Wuthering Heights* and the Liverpool Slave Trade', *ELH*, 62:1 (1995), 171–96.
Vrettos, Athena, *Somatic Fictions: Imagining Illness in Victorian Culture* (Stanford: Stanford University Press, 1995).
Wallace, Diana, 'Uncanny Stories: The Ghost Story as Female Gothic', *Gothic Studies*, 6:1 (2004), 57–68.
Wallace, Jeff, 'Literature and Posthumanism', *Literature Compass*, 7/8 (2010), 692–701.
Ward, Keith, *Christianity: A Short Introduction* (Oxford: Oneworld, 2000).
Westcott, Andrea, '"A Matter of Strong Prejudice": Gilbert Markham's Self-Portrait', in *New Approaches to the Literary Art of Anne Brontë*, Julie Nash and Barbara A. Suess, eds. (Aldershot: Ashgate, 2001), pp. 213–25.
Whone, Clifford, 'Where the Brontës Borrowed Books: The Keighley Mechanics' Institute Library Catalogue of 1841', *Brontë Society Transactions*, 11 (1951), 344–58.
Williams, Rowan, *The Edge of Words: God and the Habits of Language* (London: Bloomsbury, 2014).
Wilson, Timothy D., *Strangers to Ourselves: Discovering the Adaptive Unconscious* (Cambridge, MA: Belknap Press of Harvard University Press, 2002).
Winslow, Forbes, *On Obscure Diseases of the Brain, and Disorders of the Mind* (Philadelphia, PA: Blanchard & Lea, 1860).
Wise, T. J. and J. A. Symington, eds., *The Brontës: Their Lives, Friendships and Correspondence*, 4 vols. (Oxford: Shakespeare Head, 1932).
Wolfe, Cary, *Animal Rites: American Culture, the Discourse of Species, and Posthumanist Theory* (Chicago: University of Chicago Press, 2003).
  'Is Humanism Really Humane?', interview with Natasha Lennard, *The New York Times*, January 9, 2017.
Wollstonecraft, Mary, *Original Stories from Real Life: With Conversations Calculated to Regulate the Affections, and Form the Mind to Truth and Goodness* (1788; new edn., London: J. Johnson, 1796).
  Thoughts on the Education of Daughters: With Reflections on Female Conduct, in the More Important Duties of Life (London: Joseph Johnson, 1787).

A Vindication of the Rights of Woman and *A Vindication of the Rights of Men*, Janet Todd, ed. (Oxford: Oxford University Press, 1993).
Woloch, Alex, *The One vs. the Many: Minor Characters and the Space of the Protagonist in the Novel* (Princeton: Princeton University Press, 2003).
Woolf, Virginia, '*Jane Eyre* and *Wuthering Heights*', in *The Common Reader* (New York: Harcourt and Brace, 1925).
A Room of One's Own (New York: Harcourt Brace Jovanovich, 1991).
Yamaguchi, Midori, *Daughters of the Anglican Clergy: Religion, Gender and Identity in Victorian England* (Basingstoke: Palgrave Macmillan, 2014).

Index

adaptive unconscious, 77–78
Agamben, Giorgio, 21, 155, 233, 237, 246–247
   Homo Sacer: Sovereign Power and Bare Life, 7, 246
The American Declaration of Independence, 172
Anglicanism, 221
animal–human relations, 27, 32, 35, 39, 219,
   see also violence against animals
antivivisection campaigns, 32
Arendt, Hannah
   The Human Condition, 7
Aristotle, 10
Armstrong, Isobel, 2, 20–21, 22, 26*n*32, 155, 222*n*3, 226, 250*n*10, 251, 258*n*2
   Novel Politics, 21
Arnold, Andrea, 22, 129
Athénée Royal School, 59
Austen, Jane, 52, 251
   Mansfield Park, 131–132, 258

Bain, Alexander, 67–68, 114, 122
   The Emotions and the Will, 113
   Mind and Body: The Theories of Their Relation, 67
   The Senses and the Intellect, 113
Barbauld, Anna, 207
Barker, Juliet, 260, 267*n*1
Barlow, John, 93, 103, 105*n*41
Barthes, Roland, 108, 122*n*2
Bayley, John, 157, 165*n*23
Beer, Gillian, 8
Benjamin, Walter, 108, 122*n*2
Berman, Carolyn Vallenga, 129, 143*n*16
Bildungsroman, 173, 181–183
Birch, Dinah, 12, 22, 48
Blackwell, Su, 23, 26*n*37
Blackwood's Edinburgh Magazine, 45*n*8, 55, 68, 70
Blake, Charlie
   Beyond Human: From Animality to Transhumanism (with Molloy, Claire and Shakespeare, Steven), 7

Blouet, Olwyn, 130
Bourke, Joanna
   What it Means to be Human: Reflections from 1791 to the Present, 4, 7
Bowen, John, 6, 25*n*23
Boxall, Peter, 2
Bristow, Joseph, 207, 222*n*1
Brontë, Anne, 1, 23, 51, 61, 64, 207, 262, 265
   abuse of animals, 125–129
   Agnes Grey, 12, 27, 39, 41, 43, 58, 61–62, 127, 168, 172, 179, 181, 184
   Biblical quotation in works, 18, 184
   critique of British mastery over wives and slaves, 130
   development as a writer, 61
   1846 *Poems*, 210, 218, 220, 222
   home education, 54–56
   prophetic sublime of, 220–222
   slavery and abolitionist discourse, 129–135, 141
   The Tenant of Wildfell Hall, see *The Tenant of Wildfell Hall*
Brontë, Branwell, 56–57, 156
   'Branwell's Blackwood's Magazine', 56
Brontë, Charlotte, 48, 51, 64, 192, 251, 265
   anti-colonial poems, 212
   attitude towards learning, 56
   Biblical quotation in works, 18, 184
   construction of masculinity, 212
   devotion to father, 54
   evocation of biblical male prophetic models, 212
   experiences as private governess, 54–55
   experiences at Clergy Daughters' School, 54
   female prophetic voice, use of, 19, 210–216, 222
   female rebellion in, 133
   fictional autobiography, use of, 91
   first-person narration, use of, 113–114, 117, 119, 121, 171
   Gaskell's vivid portrait of, 112

Brontë, Charlotte *(cont.)*
  gendered language of nervous disease, use of, 86
  home education, 54–56
  imaginative experiences, 67–70, 73–74, 78
  influence of materialist thought, 68
  Jane Eyre, see *Jane Eyre*
  phrenology aspects in works, 68, 71, 88, 118
  'Pilate's Wife's Dream', 15, 19, 121–122, 210–213
  presentation of internal conflicts, 89
  The Professor, 5, 11, 41, 43, 59
  as pupil and teacher, 54, 59–60, 63
  on realism, 72–73
  relation between teacher and pupil, depiction of, 61
  Roe Head writings, 68
  Shirley, 5, 27, 59, 110–111, 171, 181
  Villette, see *Villette*
  voices and sounds in novels, use of, 107, 113–122
  workings of mind and memory, 85
Brontë, Emily, 15, 16, 17, 18, 42, 44, 51, 63, 64, 86, 147, 150, 154, 155, 192, 261–262, 264–265
  devotion to animals, 152
  female earth spirit, use of, 219
  ghostly nature-spirit, use of, 219
  home education, 54–56
  naturalism in works, 154
  prophetic visionary, 216–220
  repetitions, use of, 197–204
  view of 'human nature', 156
  'Why ask to know the date – the clime?' (Poem 126), 19, 158–160, 194, 205n16, 218
  'Why ask to know what date what clime' (Poem 127), 158, 159, 160–161
  Wuthering Heights, see *Wuthering Heights*
Brontë, Maria (Maria Branwell), 50, 53, 54
Brontë, Patrick, 33, 46n14, 50, 53, 56, 61, 130–132, 210, 262, 265
  approach to his daughters' education, 55–56
  commitment to education, 53
  The Maid of Killarney, 55
  as teacher and clergyman, 53–54
Brown, Laura
  Homeless Dogs and Melancholy Apes: Humans and Other Animals in the Modern Literary Imagination, 8
Brown, Thomas, 92, 103
  Lectures on the Philosophy of the Human Mind, 89, 104n20
Browning, Elizabeth Barrett, 207, 209
Byron, Lord
  Manfred, 67

Burrows, George Man, 93–94, 103
Buxton, Thomas Fowell, 130, 142n4

Caldwell, Janis McLarren, 12–13, 67
Calvinism, 168, 178–179, 190, 203, 208, 220–221
Carpenter, William, 74–75, 77, 87, 103, 104n12
  Principles of Mental Physiology, 74, 88
Carson, Anne
  Glass, Irony, and God, 23
Carter, Robert Brudenell, 93
cartes de visite photographs, 231–233
cerebral commissurotomy patients, 79
Channing, William Ellery, 18, 176, 187n44
chattel slavery, 125, 130, 134
Chekhov, Anton, 260, 264
  parallels between Brontës and, 261–263
  Three Sisters, 22, 260
child rearing, 12, 27, 33, 39–42
Christian belief in fiction, 167
Christian devotional practices, 168
The Christian Remembrancer, 167
Clapp, Susannah, 260
Clergy Daughters' School, 54, 57, 59
Cobbe, Frances Power, 32, 75–76
Cohen, William, 67
  Embodied: Victorian Literature and the Senses, 10
Coleridge, Samuel Taylor, 72, 75, 168
  The Rime of the Ancient Mariner, 219
Confessions of a Hypochondriac, 96
Conolly, John, 87, 103
Cowper, William, 38, 163
creative imagining, study, 67–69
  neural pathways for, 69
  phrenology, 71
  scientific accounts of unconscious or semi-conscious states, 67, 73–74
  unconscious mind, 74–79

Dallas, E. S., 75–77
  The Gay Science, 77
  'traffic' between conscious and unconscious minds, 75–76
Danahay, Martin A., 3, 4
  Victorian Animal Dreams (with Morse, Deborah Denenholz), 8
Dart, Gregory, 171
Darwin, Charles, 8, 13, 32, 38, 105n39, 127
  The Expression of the Emotion in Man and Animals, 4
Davies, Stevie, 133, 145n38, 191, 205n15, 216, 236
daydreaming, 120

Index

The Declaration of the Rights of Men and of Citizens, by the National Assembly of France, 172
degradation, 6, 15–17, 20, 23, 44, 67, 134, 147, 149–153, 155, 158–160, 163, 179, 202, 238, 246
Dennett, Daniel, 78
De Quincey, Thomas
   Confessions of an English Opium-Eater, 88, 171
Derrida, Jacques, 3
destitution, 15, 21, 234–235, 237, 253
Descartes, René, 10
Dickens, Charles, 108, 169, 251
   Bleak House, 172
   David Copperfield, 172
   Great Expectations, 172
   Hard Times, 175
   Little Dorrit, 258
dogs
   in Brontë novels, 27, 30–31, 41
   in Hogarth's prints, 28–29
   as subjects of drawings and paintings by Brontë sisters, 126–127
   in *The Tenant of Wildfell Hall*, 126–129
Douglas, James, 85, 88–90, 92, 103
   On the Philosophy of the Mind, 88
Dublin Review, 214
Dumas, Paula E.
   Proslavery Britain: Fighting for Slavery in an Era of Abolition, 130

Eagleton, Terry
   Myths of Power, 129
Edelman, Gerard, 78
Edinburgh Review, 110
education, significance of, 48–49
   Brontë family and, 57–64
   educating women, 51–53
Eliot, George, 169, 207, 209, 258
   Daniel Deronda, 88
   Middlemarch, 258
   The Mill on the Floss, 181, 251
   Romola, 169, 178
Ellis, Sarah Stickney, 44
   The Mothers of England, their Influence and Responsibility, 39
Emerson, Ralph Waldo, 109
empathy, 13, 215, 234, 235
Encyclopedia of Feminist Literature, 149
Engels, Friedrich
   The Condition of the Working-Class in England, 49
Evangelical virtues, 168, 190
   self-sacrifice and self-abnegation in, 168–170
Evans, Nicholas, 129
The Examiner, 110–111

female prophecy, 207, 210–216, 222
Fforde, Jasper, 23
fictional autobiography, 9, 13, 85, 91
Fodor, Jerry, 78
folk psychology, 80
Fonblanque, A. W., 110
Forcade, Eugene, 110, 111
Foucault, Michel, 3, 40
Fraser's Magazine, 55, 110, 116, 171
Fuller, Steve
   Humanity 2.0: What it Means to be Human Past, Present and Future, 7
functional Magnetic Resonance Imaging (fMRI), 69

Garrison, William Lloyd, 132
Gaskell, Elizabeth, 42, 56, 74–75, 150, 169, 266
   The Life of Charlotte Brontë, 74–75, 111, 261
   Mary Barton, 175
   Ruth, 111
Gazzaniga, Michael, 79–80
Gezari, Janet, 190
Gigerenzer, Gerd, 77–78
Gilbert, Sandra
   The Madwoman in the Attic (with Gubar, Susan), 191
Gladwell, Malcolm, 78
Glen, Heather
   Charlotte Brontë: The Imagination in History, 5
Godwin, Samuel, 131
Godwin, William
   Caleb Williams, 176
Goldsmith, Revd J.
   A Grammar of General Geography for the Use of Schools and Young Persons, 55
Goldsmith, Oliver
   History of England, 55
Gondal, 12, 56, 159, 163, 218, 219
Goodall, Howard, 266
Graham, Thomas John, 93, 103
   Modern Domestic Medicine, 46n14, 93, 96
Gregory, James R. T. E., 129, 131
   The Leeds Mercury, 131
Grey, Maria Shirreff
   Thoughts on Self-Culture: Addressed to Women (with Shirreff, Emily Anne), 177
Groth, Helen, 14–15, 22, 107
Gubar, Susan
   The Madwoman in the Attic (with Gilbert, Sandra), 191

Hamilton, William, 75, 76–77
Harari, Yuval Noah, 4

Haraway, Donna, 3
Hardy, Barbara, 2, 20, 21, 22, 26*n*33, 251, 258*n*1, 259*n*3, 259*n*5, 259*n*6, 259*n*7, 259*n*8
  Dorothea's Daughter and Other Nineteenth-Century Postscripts, 21
Harman, Claire, 127
Haworth, history of, 50
Hayles, Katherine, 3
H. D.
  Pilate's Wife, 224*n*24
hearing and listening, constructs in Victorian era, 109–122
  acoustic triggers in *Villette*, 120–121
  attentive listening, 118–121
  in Charlotte Brontë's 'Pilate's Wife's Dream', 121–122
  link between mind and sound, 110–111
Heger, Constantin, 54, 59, 150
Henck, K. Cubie, 210
Hogarth, William, 11, 31
  elements of sexual sadism, 28
  The Four Stages of Cruelty, 28–29
  instances of animal cruelty, 28–29
Holland, Sir Henry, 13, 92–93, 103, 105*n*39
  Medical Notes and Reflections, 96
human, idea of, 2–3, 5, 7, 23, 67, 74, 85–86, 103, 105*n*39, 154–156, 172–173, 183–185, 230, 235–237
human fallibility, 1, 10
human rights, 5, 7, 15, 23, 167, 172–175
  language of, 184
  women's rights, 182
Hunt, Lynn, 172
hypochondriasis, 13, 84, 85, 86, 88, 92–99, 103
hysteria, 13, 84, 86, 88, 92–98

individualism, 3, 12, 158, 168, 173

Jane Eyre, 1, 9, 13, 31, 79, 99, 155, 210, 214, 252, 254–255, 257, 264–265
  advocacy of women's rights, 182–185
  analysis of religious self-culture, 18, 167–168, 174–185
  different relations to degradation, 23
  educational experiences, 54–55, 59
  gendered expectations, 9, 54–55, 59
  imaginative visions, 72–74
  Lewes's review of, 110, 115–117, 120
  narrative style of, 107–113, 115–117
  representation of animals in, 127
  slavery in, 129, 132–133
  teaching, using artworks, 20–21, 226–249
  use of literary soundscapes, 14–15, 113
Jones, Gail, 23
The Journal of Psychological Medicine and Mental Pathology, 101

Kant, Immanuel, 10
Kay, James (Sir James Kay-Shuttleworth), 49
Kay-Shuttleworth, Sir James, 12, 49
Keats, John
  'Ode to Autumn', 162
Kierkegaard, Søren, 6
kindness to animals, 12, 35–38, 39, 43
Kingsley, Charles
  Water Babies, 178
Kosslyn, Stephen, 69
Kreilkamp, Ivan, 8, 99, 100, 108–109
Krueger, Christine, 172, 209

Lawrence, D. H., 151
Laycock, Thomas, 77, 92, 98, 103, 105*n*41
Levine, George, 5–6
  Realism, Ethics and Secularism, 6
Lewes, George Henry, 110–111, 113, 115–117, 120
  Jane Eyre, review of, 171
Lewis, Alexandra, 1, 13–14, 26*n*30, 26*n*36, 84, 103*n*1, 103*n*2
Lyell, Charles, 8

Malthus, Thomas
  Essay on Population, 179
Mangnall, Richmal
  Historical and Miscellaneous Questions, 55
Marsden, Simon, 17, 18–19, 189, 205*n*8, 205*n*12, 206*n*32, 216, 224*n*32
Martineau, Harriet, 68
  Household Education, 57, 65*n*20
Marx, Karl, 6
master–slave dialectic, 30
Maurice, F. D., 168, 169
meaning-making systems, 80, 83*n*52
Mellor, Ann, 209
memory, 8, 10, 85–90, 92, 101, 102, 103
Mermin, Dorothy, 211
Methodist Magazine, 210
Meyer, Susan L., 129
Milbank, John, 18, 192, 196
Mill, J. S., 88, 113, 169
  The Subjection of Women, 132
Miller, Andrew, 181
  The Burdens of Perfection, 6
Miller, J. Hillis, 190–191, 198–200, 206*n*30
  The Disappearance of God, 190
  Fiction and Repetition, 198
Millingen, John Gideon, 87, 98, 103, 105*n*41
Mitchell, W. J. T., 69, 78
Molloy, Claire
  Beyond Human: From Animality to Transhumanism (with Blake, Charlie and Shakespeare, Steven), 7

Index

More, Hannah, 57, 220
  on light-mindedness of women, 52
  Moral Sketches of Prevailing Opinions and Manners, Foreign and Domestic: With Reflections on Prayer, 53
  Strictures on the Modern System of Female Education, with a View of the Principles and Conduct Prevalent Among Women of Ranks and Fortune, 52
  on women's education, 52
Morgan, William, 131
Morrison, Blake, 2, 20, 21–22, 26n35, 260
  We Are Three Sisters, 22
Morse, Deborah Denenholz, 3–4, 15–16, 22, 24n11, 125, 142n7, 142n8, 143n13, 143n14, 145n38, 145n42
  Victorian Animal Dreams (with Danahay, Martin A.), 8
Mott, Lucretia, 132
Mousley, Andy
  Literature and the Human: Theory, Criticism, Practice, 7

National Review, 111
Neill, Anna, 68
Nussey, Ellen, 260

Ong, Walter J., 108
Otis, Laura, 78
  Rethinking Thought, 78

pain, 5, 11, 35, 38, 95, 96, 99, 100, 112, 133, 140, 150, 182, 237, 241, 254, 258
Paine, Thomas
  Rights of Man, 173
Parker, Theodore, 221
Pearce, Lynne, 189, 190, 204n2
Penner, Louise, 68
phrenology, 68, 71, 88, 118
Pickstock, Catherine, 18, 192, 199, 200
Plasa, Carl, 129, 132, 144n16, 212
post-humanism, 2, 3
Prichard, James Cowles, 86–87, 103
  A Treatise on Insanity and Other Disorders Affecting the Mind, 86
privation, 95, 196, 206n23, 234
prophecy and women's poetry, 208–210
prophet with Romantic ideals, 208–210
Providence, 32, 52, 55, 121, 178, 254

rabies, 11, 41, 43
racist humanism, 2
Rayfield, Donald, 261
realism, 6, 72, 150, 159, 176
Reid, John, 84, 89–91, 102–103
  Essays on Hypochondriasis, 90
Reign of Terror, 171

repetition, 18, 192, 197–199
Rhys, Jean
  Wide Sargasso Sea, 22, 129, 183
Richardson, Alan, 68, 69
Richardson, David, 129
Richardson, Samuel
  Clarissa, 130
Ritvo, Harriet, 32, 41
  The Animal Estate: The English and Other Creatures in Victorian England, 4
Rollin, Charles
  Ancient History, 55
Romantic art and literature, 170–172
Roper, Derek, 158, 218
Roscoe, William Caldwell, 111, 112
Rossetti, Christina, 207, 209
Rousseau, Jean-Jacques, 160, 170–171
Ryan, Robert, 174
Rylance, Rick, 113
  Victorian Psychology and British Culture, 1850–1880, 3

Sartre, Jean Paul, 2
Scarry, Elaine, 13, 68, 79
  Dreaming by the Book, 69
Schmitt, Cannon
  Darwin and the Memory of the Human: Evolution, Savages, and South America, 8
schooling, 12, 48, 50
Schramm, Jan-Melissa, 17–18, 167, 185n9, 185n12, 187n43, 187n53
Scott, Sir Walter, 136, 159
self-control, 13, 42, 43, 86–88, 90, 93
self-culture, 176–177, 182, 187n50
Senf, Carol, 130–131
Sewell, Anna
  Black Beauty, 4
Shakespeare, Steven
  Beyond Human: From Animality to Transhumanism (with Blake, Charlie and Molloy, Claire), 7
Shakespeare, William
  Othello, 134–135
Shires, Linda, 209
Shirreff, Emily Anne
  Thoughts on Self-Culture: Addressed to Women (with Grey, Maria Shirreff), 177
Shuttleworth, Sally, 11, 25n17, 27, 68, 71, 81n4, 81n14, 81n15, 106n59, 118, 124n45, 143n11
Simon, Bart, 2
Slaughter, Joseph, 173, 181, 182, 183
slavery and abolitionism, 15, 129–135
Small, Helen, 16–17, 147, 164n7
Smith, Charlotte, 207

Index

Society for the Prevention of Cruelty to Animals (SPCA), 29, 32, 33, 34, 126
Soper, Kate, 7, 16, 154
  What is Nature? Culture, Politics and the Non-Human, 7
The Spectator, 110
Spivak, Gayatri, 129
Spolsky, Ellen, 78–79
  Gaps of Nature: Literary Interpretation and the Modular Mind, 78
Stanton, Elizabeth Cady, 132
Sterne, Jonathan, 107, 115–116
Stewart, Garrett, 108
Stowe, Harriet Beecher
  Uncle Tom's Cabin, 134, 138
Styler, Rebecca, 19, 207, 223n18, 225n43
sympathy, 10, 12, 18, 43, 61, 133, 148, 154, 174, 180, 189, 201, 211, 220, 246, 253

Taylor, Charles
  Sources of the Self: The Making of Modern Identity, 3
Taylor, Mary, 56
The Tenant of Wildfell Hall, 15, 22, 42, 61–62, 168, 178–179, 181–182, 184, 221
  abolitionist context of, 129–135
  dogs in, 126–129
  heroine's art of resistance, 135–141
Tennant, Emma, 23
Thackeray, William, 108, 174
Thomas, Sue, 129
Thormählen, Marianne, 25n17, 167–168, 184, 191, 205n13, 210, 223n7
  The Brontës and Religion, 191
traumatic memory, 85–86, 91, 93–94, 98–103
tyrannical memory, 87

unconscious cerebration, 13, 74–77, 85
unconscious mind, 67, 75, 76, 77, 79, 80, 87
Universal Declaration of Human Rights, 172
universal salvation, doctrine of, 178

Veith, Ilza
  Hysteria, 93
Victorian women's poetry, 208–210
Villette, 1, 9, 13, 14, 21, 107, 111, 115, 122, 133, 172, 179, 183–184, 252
  centrality of memory, 85
  characters and destinies of Lucy Snowe and Paulina Bretton, 254–258
  conceptions of ill and healthy mind in calamity, 88–91
  dynamics of listening in, 118–121
  failure of communication and interpretation of human trauma in, 98–102
  forgetfulness in, 90
  manifestations of hypochondriasis and hysteria in, 92–98, 103
  references to pain in, 99
  reviews of, 111
  shaping of Lucy's autobiographical record, 98–102
Vincent, David, 107, 108–109
violence against animals, 43
  as assertion of power, 44
  in Brontë novels, 27, 31–32, 39–41, 125–129
  child cruelty to animals, 35–38
  hanging of dogs, 40–41, 43
  shooting of dog, 41–43
  torturing of birds, 39–41
  vivisection, 32–35, 44
Vrettos, Athena, 94

Wallace, Diana, 213
Wallace, Jeff, 2–3
Western philosophy of mind, 80
Westminster Review, 111
Wilberforce, William, 57, 126, 130, 142n4
Williams, Rowan, 192
Williams, W. S., 48, 109
Wilson, Timothy, 77
Wolfe, Cary, 3
Wollstonecraft, Mary, 12, 36
  arguments for a 'rational' education, 52
  on education of women, 51–52
  Thoughts on the Education of Daughters: With Reflections on Female Conduct, in the More Important Duties of Life, 51
  A Vindication of the Rights of Woman: with Strictures on Political and Moral Subjects, 51, 130
women's rights, 17, 125, 132, 141, 172, 238
Woolf, Virginia, 74, 79, 117
Wordsworth, William
  The Prelude, 171
Worrall, Jessica, 266
Wuthering Heights, 6, 9, 11, 27, 29, 32, 42–43, 60, 62–63, 85, 112, 129, 133, 136, 184
  child's unsupervised play, 157
  distortion of Christian social ethics, 192–197
  dual narrative/lyric modality, 163–164
  Heathcliff–Cathy relationship, 189–191
  Heathcliff's campaign of vengeance, 202–203
  narrative drama of degradation, rejection and revenge, 16, 147–164, 202–203
  relationship between first and second generations in, 200–203
  repetition of names and symbols in, 192, 197–200, 202, 204
  romance plot in readings of, 190
  screen adaptation of, 22

CAMBRIDGE STUDIES IN NINETEENTH-CENTURY
LITERATURE AND CULTURE

General editor
Gillian Beer, *University of Cambridge*

Titles published

1. *The Sickroom in Victorian Fiction: The Art of Being Ill*
Miriam Bailin, Washington University

2. *Muscular Christianity: Embodying the Victorian Age*
edited by Donald E. Hall, California State University, Northridge

3. *Victorian Masculinities: Manhood and Masculine Poetics in Early Victorian Literature and Art*
Herbert Sussman, Northeastern University, Boston

4. *Byron and the Victorians*
Andrew Elfenbein, University of Minnesota

5. *Literature in the Marketplace: Nineteenth-Century British Publishing and the Circulation of Books*
edited by John O. Jordan, University of California, Santa Cruz and Robert L. Patten, Rice University, Houston

6. *Victorian Photography, Painting and Poetry*
Lindsay Smith, University of Sussex

7. *Charlotte Brontë and Victorian Psychology*
Sally Shuttleworth, University of Sheffield

8. *The Gothic Body: Sexuality, Materialism and Degeneration at the Fin de Siècle*
Kelly Hurley, University of Colorado at Boulder

9. *Rereading Walter Pater*
William F. Shuter, Eastern Michigan University

10. *Remaking Queen Victoria*
edited by Margaret Homans, Yale University and Adrienne Munich, State University of New York, Stony Brook

11. *Disease, Desire, and the Body in Victorian Women's Popular Novels*
Pamela K. Gilbert, University of Florida

12. *Realism, Representation, and the Arts in Nineteenth-Century Literature*
Alison Byerly, Middlebury College, Vermont

13. *Literary Culture and the Pacific*
Vanessa Smith, University of Sydney

14. *Professional Domesticity in the Victorian Novel Women, Work and Home*
Monica F. Cohen

15. *Victorian Renovations of the Novel: Narrative Annexes and the Boundaries of Representation*
Suzanne Keen, Washington and Lee University, Virginia

16. *Actresses on the Victorian Stage: Feminine Performance and the Galatea Myth*
Gail Marshall, University of Leeds

17. *Death and the Mother from Dickens to Freud: Victorian Fiction and the Anxiety of Origin*
Carolyn Dever, Vanderbilt University, Tennessee

18. *Ancestry and Narrative in Nineteenth-Century British Literature: Blood Relations from Edgeworth to Hardy*
Sophie Gilmartin, Royal Holloway, University of London

19. *Dickens, Novel Reading, and the Victorian Popular Theatre*
Deborah Vlock

20. *After Dickens: Reading, Adaptation and Performance*
John Glavin, Georgetown University, Washington DC

21. *Victorian Women Writers and the Woman Question*
edited by Nicola Diane Thompson, Kingston University, London

22. *Rhythm and Will in Victorian Poetry*
Matthew Campbell, University of Sheffield

23. *Gender, Race, and the Writing of Empire: Public Discourse and the Boer War*
Paula M. Krebs, Wheaton College, Massachusetts

24. *Ruskin's God*
Michael Wheeler, University of Southampton

25. *Dickens and the Daughter of the House*
Hilary M. Schor, University of Southern California

26. *Detective Fiction and the Rise of Forensic Science*
Ronald R. Thomas, Trinity College, Hartford, Connecticut

27. *Testimony and Advocacy in Victorian Law, Literature, and Theology*
Jan-Melissa Schramm, Trinity Hall, Cambridge

28. *Victorian Writing about Risk: Imagining a Safe England in a Dangerous World*
Elaine Freedgood, University of Pennsylvania

29. *Physiognomy and the Meaning of Expression in Nineteenth-Century Culture*
Lucy Hartley, University of Southampton

30. *The Victorian Parlour: A Cultural Study*
Thad Logan, Rice University, Houston

31. *Aestheticism and Sexual Parody 1840–1940*
Dennis Denisoff, Ryerson University, Toronto

32. *Literature, Technology and Magical Thinking, 1880–1920*
Pamela Thurschwell, University College London

33. *Fairies in Nineteenth-Century Art and Literature*
Nicola Bown, Birkbeck, University of London

34. *George Eliot and the British Empire*
Nancy Henry The State University of New York, Binghamton

35. *Women's Poetry and Religion in Victorian England: Jewish Identity and Christian Culture*
Cynthia Scheinberg, Mills College, California

36. *Victorian Literature and the Anorexic Body*
Anna Krugovoy Silver, Mercer University, Georgia

37. *Eavesdropping in the Novel from Austen to Proust*
Ann Gaylin, Yale University

38. *Missionary Writing and Empire, 1800–1860*
Anna Johnston, University of Tasmania

39. *London and the Culture of Homosexuality, 1885–1914*
Matt Cook, Keele University

40. *Fiction, Famine, and the Rise of Economics in Victorian Britain and Ireland*
Gordon Bigelow, Rhodes College, Tennessee

41. *Gender and the Victorian Periodical*
Hilary Fraser, Birkbeck, University of London
Judith Johnston and Stephanie Green, University of Western Australia

42. *The Victorian Supernatural*
edited by Nicola Bown, Birkbeck College, London Carolyn Burdett, London Metropolitan University and Pamela Thurschwell, University College London

43. *The Indian Mutiny and the British Imagination*
Gautam Chakravarty, University of Delhi

44. *The Revolution in Popular Literature: Print, Politics and the People*
Ian Haywood, Roehampton University of Surrey

45. *Science in the Nineteenth-Century Periodical: Reading the Magazine of Nature*
Geoffrey Cantor, University of Leeds
Gowan Dawson, University of Leicester
Graeme Gooday, University of Leeds
Richard Noakes, University of Cambridge
Sally Shuttleworth, University of Sheffield
and Jonathan R. Topham, University of Leeds

46. *Literature and Medicine in Nineteenth-Century Britain from Mary Shelley to George Eliot*
Janis McLarren Caldwell, Wake Forest University

47. *The Child Writer from Austen to Woolf*
edited by Christine Alexander, University of New South Wales
and Juliet McMaster, University of Alberta

48. *From Dickens to Dracula: Gothic, Economics, and Victorian Fiction*
Gail Turley Houston, University of New Mexico

49. *Voice and the Victorian Storyteller*
Ivan Kreilkamp, University of Indiana

50. *Charles Darwin and Victorian Visual Culture*
Jonathan Smith, University of Michigan-Dearborn

51. *Catholicism, Sexual Deviance, and Victorian Gothic Culture*
Patrick R. O'Malley, Georgetown University

52. *Epic and Empire in Nineteenth-Century Britain*
Simon Dentith, University of Gloucestershire

53. *Victorian Honeymoons: Journeys to the Conjugal*
Helena Michie, Rice University

54. *The Jewess in Nineteenth-Century British Literary Culture*
Nadia Valman, University of Southampton

55. *Ireland, India and Nationalism in Nineteenth-Century Literature*
Julia Wright, Dalhousie University

56. *Dickens and the Popular Radical Imagination*
Sally Ledger, Birkbeck, University of London

57. *Darwin, Literature and Victorian Respectability*
Gowan Dawson, University of Leicester

58. *'Michael Field': Poetry, Aestheticism and the Fin de Siècle*
Marion Thain, University of Birmingham

59. *Colonies, Cults and Evolution: Literature, Science and Culture in Nineteenth-Century Writing*
David Amigoni, Keele University

60. *Realism, Photography and Nineteenth-Century Fiction*
Daniel A. Novak, Lousiana State University

61. *Caribbean Culture and British Fiction in the Atlantic World, 1780–1870*
Tim Watson, University of Miami

62. *The Poetry of Chartism: Aesthetics, Politics, History*
Michael Sanders, University of Manchester

63. *Literature and Dance in Nineteenth-Century Britain: Jane Austen to the New Woman*
Cheryl Wilson, Indiana University

64. *Shakespeare and Victorian Women*
Gail Marshall, Oxford Brookes University

65. *The Tragi-Comedy of Victorian Fatherhood*
Valerie Sanders, University of Hull

66. *Darwin and the Memory of the Human: Evolution, Savages, and South America*
Cannon Schmitt, University of Toronto

67. *From Sketch to Novel: The Development of Victorian Fiction*
Amanpal Garcha, Ohio State University

68. *The Crimean War and the British Imagination*
Stefanie Markovits, Yale University

69. *Shock, Memory and the Unconscious in Victorian Fiction*
Jill L. Matus, University of Toronto

70. *Sensation and Modernity in the 1860s*
Nicholas Daly, University College Dublin

71. *Ghost-Seers, Detectives, and Spiritualists: Theories of Vision in Victorian Literature and Science*
Srdjan Smajić, Furman University

72. *Satire in an Age of Realism*
Aaron Matz, Scripps College, California

73. *Thinking About Other People in Nineteenth-Century British Writing*
Adela Pinch, University of Michigan

74. *Tuberculosis and the Victorian Literary Imagination*
Katherine Byrne, University of Ulster, Coleraine

75. *Urban Realism and the Cosmopolitan Imagination in the Nineteenth Century: Visible City, Invisible World*
Tanya Agathocleous, Hunter College, City University of New York

76. *Women, Literature, and the Domesticated Landscape: England's Disciples of Flora, 1780–1870*
Judith W. Page, University of Florida
Elise L. Smith, Millsaps College, Mississippi

77. *Time and the Moment in Victorian Literature and Society*
Sue Zemka, University of Colorado

78. *Popular Fiction and Brain Science in the Late Nineteenth Century*
Anne Stiles, Washington State University

79. *Picturing Reform in Victorian Britain*
Janice Carlisle, Yale University

80. *Atonement and Self-Sacrifice in Nineteenth-Century Narrative*
Jan-Melissa Schramm, University of Cambridge

81. *The Silver Fork Novel: Fashionable Fiction in the Age of Reform*
Edward Copeland, Pomona College, California

82. *Oscar Wilde and Ancient Greece*
Iain Ross, Colchester Royal Grammar School

83. *The Poetry of Victorian Scientists: Style, Science and Nonsense*
Daniel Brown, University of Southampton

84. *Moral Authority, Men of Science, and the Victorian Novel*
Anne DeWitt, Princeton Writing Program

85. *China and the Victorian Imagination: Empires Entwined*
Ross G. Forman, University of Warwick

86. *Dickens's Style*
Daniel Tyler, University of Oxford

87. *The Formation of the Victorian Literary Profession*
Richard Salmon, University of Leeds

88. *Before George Eliot: Marian Evans and the Periodical Press*
Fionnuala Dillane, University College Dublin

89. *The Victorian Novel and the Space of Art: Fictional Form on Display*
Dehn Gilmore, California Institute of Technology

90. *George Eliot and Money: Economics, Ethics and Literature*
Dermot Coleman, Independent Scholar

91. *Masculinity and the New Imperialism: Rewriting Manhood in British Popular Literature, 1870–1914*
Bradley Deane, University of Minnesota

92. *Evolution and Victorian Culture*
edited by Bernard Lightman, York University, Toronto and Bennett Zon, University of Durham

93. *Victorian Literature, Energy, and the Ecological Imagination*
Allen MacDuffie, University of Texas, Austin

94. *Popular Literature, Authorship and the Occult in Late Victorian Britain*
Andrew McCann, Dartmouth College, New Hampshire

95. *Women Writing Art History in the Nineteenth Century: Looking Like a Woman*
Hilary Fraser Birkbeck, University of London

96. *Relics of Death in Victorian Literature and Culture*
Deborah Lutz, Long Island University, C. W. Post Campus

97. *The Demographic Imagination and the Nineteenth-Century City: Paris, London, New York*
Nicholas Daly, University College Dublin

98. *Dickens and the Business of Death*
Claire Wood, University of York

99. *Translation as Transformation in Victorian Poetry*
Annmarie Drury, Queens College, City University of New York

100. *The Bigamy Plot: Sensation and Convention in the Victorian Novel*
Maia McAleavey, Boston College, Massachusetts

101. *English Fiction and the Evolution of Language, 1850–1914*
Will Abberley, University of Oxford

102. *The Racial Hand in the Victorian Imagination*
Aviva Briefel, Bowdoin College, Maine

103. *Evolution and Imagination in Victorian Children's Literature*
Jessica Straley, University of Utah

104. *Writing Arctic Disaster: Authorship and Exploration*
Adriana Craciun, University of California, Riverside

105. *Science, Fiction, and the Fin-de-Siècle Periodical Press*
Will Tattersdill, University of Birmingham

106. *Democratising Beauty in Nineteenth-Century Britain: Art and the Politics of Public Life*
Lucy Hartley, University of Michigan

107. *Everyday Words and the Character of Prose in Nineteenth-Century Britain*
Jonathan Farina, Seton Hall University, New Jersey

108. *Gerard Manley Hopkins and the Poetry of Religious Experience*
Martin Dubois, University of Newcastle upon Tyne

109. *Blindness and Writing: From Wordsworth to Gissing*
Heather Tilley, Birkbeck College, University of London

110. *An Underground History of Early Victorian Fiction: Chartism, Radical Print Culture, and the Social Problem Novel*
Gregory Vargo, New York University

111. *Automatism and Creative Acts in the Age of New Psychology*
Linda M. Austin, Oklahoma State University

112. *Idleness and Aesthetic Consciousness, 1815–1900*
Richard Adelman, University of Sussex

113. *Poetry, Media, and the Material Body: Autopoetics in Nineteenth-Century Britain*
Ashely Miller, Albion College, Michigan

114. *Malaria and Victorian Fictions of Empire*
Jessica Howell, Texas A&M University